New Media and Public Relations

This book is part of the Peter Lang Education list.
Every volume is peer reviewed and meets
the highest quality standards for content and production.

PETER LANG
New York • Washington, D.C./Baltimore • Bern
Frankfurt • Berlin • Brussels • Vienna • Oxford

New Media and Public Relations

SECOND EDITION

EDITED BY SANDRA DUHÉ

PETER LANG
New York • Washington, D.C./Baltimore • Bern
Frankfurt • Berlin • Brussels • Vienna • Oxford

Library of Congress Cataloging-in-Publication Data

New media and public relations / edited by Sandra C. Duhé. — 2nd ed.
p. cm.
Includes bibliographical references and index.
1. Public relations. 2. Internet in public relations.
I. Duhé, Sandra C.
HM1221.N47   659.20285'4678—dc23   2011051548
ISBN 978-1-4331-1627-8 (paperback)
ISBN 978-1-4539-0556-2 (e-book)

Bibliographic information published by **Die Deutsche Nationalbibliothek.**
Die Deutsche Nationalbibliothek lists this publication in the "Deutsche
Nationalbibliografie"; detailed bibliographic data is available
on the Internet at http://dnb.d-nb.de/.

The paper in this book meets the guidelines for permanence and durability
of the Committee on Production Guidelines for Book Longevity
of the Council of Library Resources.

© 2012 Peter Lang Publishing, Inc., New York
29 Broadway, 18th floor, New York, NY 10006
www.peterlang.com

Printed in the United States of America

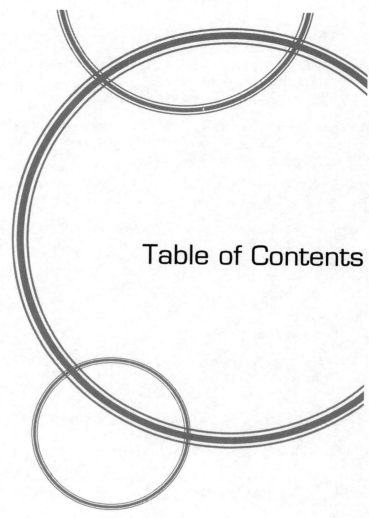

Table of Contents

PART II: Corporate Applications

PART III: Political Applications and Governmental Impacts

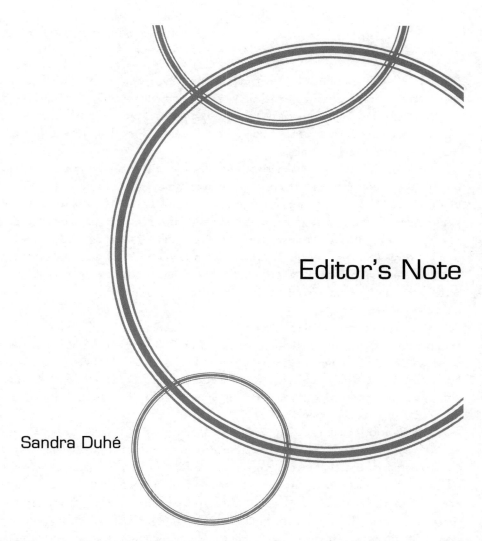

Editor's Note

Sandra Duhé

At first glance, publishing anything other than a blog post, email, or instant message about new media seems counterintuitive. What is "new" from a technical perspective as this book goes to press will not be so new when this book is in the hands of readers. The concepts presented in the following chapters focus not so much on the ever-changing intricacies of communication technologies as they do on the adaptation of time-tested principles that guide a thoughtful, responsible, mutually beneficial approach to building and maintaining public relationships, no matter what the new media vehicle for doing so may be.

Regardless of what technological developments provide as "new media," public relations practitioners will still be relied upon to learn from their operating environments, help their organizations benefit from that flow of information, and, ideally, improve the lives of publics upon whom organizational survival depends. Treating new media as just another one-way information dissemination channel is detrimental to our profession, the organizations we represent, and the publics we aim to serve. Authentic interaction with online publics involves risk, but what part of a worthwhile enterprise is truly risk free? With a thoughtful, strategic approach to new media, public relations scholars and counselors can lead organizations into changing communication environments, but we must likewise be prepared to assess our effectiveness and measure our results. As long as technology evolves, so must our thinking.

The rapid growth and adoption of new communication technologies, particularly in social media, prompted the creation of this second edition. As of this writing, YouTube, Twitter, Facebook, Flickr,

Digg, StumbleUpon, LinkedIn, Delicious, Technorati, Alexa, and Mashable are among the most popular social sites for image sharing, social networking, content sharing, bookmarking, website locating, blog searching, site ranking, and news posting ("Top 15 Most Popular," 2011; "Top Social Media Sites," 2010). These sites consistently appear on a number of ranking lists that are readily available through Google, Bing, and other search engines.

The growth in use of social networking sites since the first edition was published in 2007 is astounding. Between 2007 and 2009, the number of users doubled, with nearly one-third of the U.S. adult population visiting social networks on at least a monthly basis. In 2009, social networking surpassed the popularity of instant messaging among adults (Ostrow, 2009). Between 2007 and 2010, two-thirds of Americans with Internet access reported use of social networking sites, an incredible increase of 230% over the three-year time period (Diana, 2010). In July 2011, Facebook claimed to have more than 750 million active users ("Statistics," 2011). During the same month, Twitter was not as forthcoming about an active user count on its official site but estimated that, on average, 460,000 new accounts were created daily ("#numbers," 2011).

At the time of this writing, social media service Google Plus (Google+) is in its invitation-only field trial period, offering features called circles (selected sharing), hangouts (video chats), and sparks (Web content that matches personal interests) to make sharing on the Web more life-like. Google's new "+1" button is similar to liking something on Facebook and can be used Web-wide (Frenette, Woo, & Shaw, 2011). It will be interesting to see what happens with Google Plus and the extent to which it affects use of popular sites like Facebook.

In light of this rush to social media, it was tempting to change the book's title to *Social Media and Public Relations*, but I prefer "new media" to leave open the possibility of what will come next. The second edition comprises 32 chapters organized in seven parts, each focusing on a particular aspect of public relations. Part I includes discussions of how new and social media impact public relations theory, strategy, and the workplace. The insights offered can be applied to any public relations setting, whether it is corporate, nonprofit, or governmental. Parts II through VII focus on specialty areas, including corporate, political, nonprofit, health, university, and crisis communication applications. Although a number of authors from the first edition have returned to contribute to the second edition, the content of each chapter is entirely new. Authors include long-standing contributors to our body of knowledge as well as newly minted Ph.D.s and graduate students from around the globe, many of whom have worked in the field and/or are actively consulting today.

Chapters have been written for upper-level undergraduate and graduate students to use in both classroom and research settings. Just as public relations students and professionals are growing in number, so, too, are research projects focusing on new media growing in popularity. This makes sense in that most college students are early adopters of new communication technologies and thus have a natural inclination to include these tools as part of their research.

Editing, for me, is an enjoyable process, but it gets a bit more complicated when I am tasked with editing the work of others. Good writers share some part of themselves in each explanation, argument, and conclusion. They trust that the benefits derived from their work will outweigh any risks they take in making their ideas public and open to scrutiny. Each of the authors in this book entrusted me with his or her work, considered my suggestions, and expanded my thinking about new media. For that, and for their contributions, I am grateful. Our months of working together provided an enriching, rewarding, *collective* experience that I alone as a writer could not have enjoyed.

My thanks additionally go to Mary Savigar, Bernadette Shade, Catherine Tung, Maria Abellana, and many others at Peter Lang who helped bring this idea to publication. Their outstanding demeanor, responsiveness, and guidance are much appreciated.

Since the first edition of *New Media and Public Relations* went to press in 2007, I became a mother, earned tenure, and was promoted to associate dean. My heartfelt thanks go to my husband, mother, and mother-in-law for helping and supporting me along the way. As I look back on the crazy hours and moments of panic, frustration, and sheer joy involved in bringing this second edition to press, I realize how very blessed I am.

Sandy Duhé
July 2011

A Thematic Analysis of 30 Years of Public Relations Literature Addressing the Potential and Pitfalls of New Media

Sandra Duhé

An examination of articles published in four public relations journals (Public Relations Review, Journal of Public Relations Research, Public Relations Journal, *and* PRism) *across a 30-year timeframe (1981–2011) revealed the bulk of new media research has focused on applications (47%) and perceptions (27%) much more than proposing theoretical frameworks for relationship building (11%), raising legal/risk/ethics concerns (9%), or addressing usability (3%). Six primary themes found in the literature are discussed, as are suggestions for future research in a public relations environment that is increasingly social, temperamental, and influenced by publics external to organizations.*

In this introduction, I attempt to capture the primary themes found in a three-decade (1981–2011) span of public relations research related to new media. I was curious to investigate how new media have been addressed in the public relations literature over time so that I could better understand not only contemporary thinking about emerging communication technologies, but also our intellectual history and where new media may take us as practitioners, scholars, and educators.

The approach I used is best described as thematic analysis, a qualitative methodology that "focuses on identifiable themes and patterns" (Aronson, 1994, para. 3) and aids in reducing a volume of data into a more manageable, understandable format. The steps are fairly straightforward, though time consuming: collect the data, identify themes, categorize data according to those themes, and then provide justification for choosing those themes. Ultimately, the thematic analysis should result in "a developed story line [that] helps the reader to comprehend" (para. 10) whatever process the researcher is trying to convey. In this case, I applied thematic analysis to describe 30 years of public relations research related to new media. My unit of analysis was a journal article.

I chose four public relations journals for my investigation: *Public Relations Review, Journal of Public Relations Research, Public Relations Journal,* and *PRism.* These were selected because of their specialized focus on public relations (versus other areas of mass communication), the accessibility and depth of their archives, the variety of their formats (print and online), and their frequently observed citations in the public relations literature (rather than any formal ranking system). Searches dated back to 1975 for *Public Relations Review,* 1989 for *Journal of Public Relations Research,*[1] 2003 for *PRism,* and 2007 for *Public Relations Journal.* The oldest article selected was from a 1981 volume of *Public Relations Review,* and the most recent articles were gathered from published issues available in June 2011 (not including in-press articles).

Articles were selected based only on the content of their titles. The titles in each volume and issue of each journal were reviewed using either a university-based database or archives available online. If a title included keywords such as computers, new technology, information age, Web, Internet, virtual, digital, cyberspace, online, new media, social media, Facebook, Twitter, or blog, the article was included in the study. Articles focusing on educational, internal communication, or practitioner-to-practitioner communication (e.g., PRForum) practices related to new media were not included. A total of 150 articles, published over a 30-year time period (1981–2011), was analyzed. I relied primarily on the abstract, discussion/implication, and conclusion sections of articles to record a summary of and general theme for each. Articles were reviewed in chronological order to get a better sense of how theories, methodologies, and findings changed (or did not change) over time.

Once summaries and themes were recorded for all 150 articles, I went through an iterative, reductive process of creating mutually exclusive themes so that each article would fall under one theme that best described its primary intent and contribution. Ultimately, six themes emerged: Early Predictions, Theoretical Frameworks for Relationship Building, Usability, Applications, Perceptions, and Concerns. Literature related to each is discussed in this same order.

Theme 1: Early predictions

The four articles in this theme comprised only 3% of the dataset but were significant in that the authors made bold predictions of how new communication technologies and the increasing use of computers in public relations would affect the practice.

In 1981, BITNET, a cooperative network that allowed email and file transfers, was established at City University of New York and made its first connection to Yale (Zakon, 2010). During this time, Chester Burger (1981) aptly predicted new communication technologies would change corporate America in fundamental ways, including lower communication costs, vanishing paper trails, streamlined communication networks, reduction in the frequency of face-to-face meetings, and a newfound ability to direct messages to specific audiences.

Betsy Plank (1983), then assistant vice president of corporate communications for Illinois Bell, also saw a coming revolution in communication technologies—particularly for public relations practitioners. In a lecture, she spoke of changes in daily work habits, a coming increase in research and measurement capabilities, and new systems for the customization and delivery of messages. She warned practitioners to "brace yourself for the mixed blessing of immediate reaction to our messages" (p. 7) in regard to the instant, sometimes thought-lacking, feedback technology would allow. She expressed concerns about a decrease in human interaction, an information divide between rich and poor, privacy, education, language and writing skills, and information overload. She reassured her audience, however, saying "we are a creative, resourceful breed" (p.7) and stressed that public relations practitioners should be the catalyst, steward, and architect of the coming communication revolution to ensure "technology will improve the quality of American life" (p. 3).

Seventeen years after Burger's (1981) prediction, Robert Heath (1998) and Timothy Coombs (1998) published the now frequently cited articles about the expected impact of new technologies on issues management. Heath acknowledged that technology offered both opportunities for and threats to relationship building. He argued that the interactive nature of the Internet and Web would help democratize discussions, providing participants with a shared "platform of fact and opinion much earlier in issues dialogues" (p. 288) compared to conventional media. Coombs described the Internet not only as a "new weapon" (p. 289) for confronting corporate irresponsibility, but also as a "potential equalizer" (p. 289) that would enhance the power of activists (who were initially regarded in stakeholder theory as being relatively weak) and significantly affect issues management and corporate social performance.

Theme 2: Theoretical frameworks for relationship building

A variety of theories from communication and other disciplines guided research reported in the articles analyzed. The articles that fell under this particular theme, however, focused on providing theoretical frameworks or medium-specific guidance for the study of how Web-based communication contributed to relationship building with external publics. Two (often overlapping) sub-themes surfaced: dialogic communication and interactivity. Seventeen articles were categorized under this theme, comprising 11% of the dataset.

In 1998, e-commerce, e-auctions, portals, e-trade, XML, and intrusion detection were coming into vogue (Zakon, 2010). It was a banner year for public relations scholarship as well. In addition to Heath's (1998) and Coombs' (1998) previously discussed predictions about the effect of new communication technologies on issues management and stakeholder influence, Kent and Taylor (1998) published the first theoretical framework (among the four public relations journals studied) for building dialogic relationships through the World Wide Web. Throughout the 30-year time span studied, their framework of five website design principles (dialogic loop, useful information, generation of return visits, ease of interface/navigation, and conservation of visitors) was prominently referenced and/or tested by researchers. Kent and Taylor distinguished two-way symmetrical communication as a "process" and dialogic communication as a "product" in which a relationship exists (p. 323). They explained that without dialogue, "Internet public relations becomes nothing more than a new monologic communication medium" (p. 325). Five years later, Kent, Taylor, and White (2003) proposed that "the more an organization depends upon its publics for achieving its mission, the more it should employ dialogic features into its Web site design" (p. 75).

Also in 2003, the concept of interactivity, and its connection to relationship building, was woven into the public relations literature examined. Jo and Kim (2003) suggested that the interactive, participatory nature of the Web distinguished it from traditional media, noting its "intrinsic interactivity...can enhance the mutual relationship and collaboration between the message sender (the organization) and the receiver (public)" (p. 202). Using relationship dimensions proposed by Hon and Grunig (1999; i.e., trust, control mutuality, commitment, satisfaction, exchange relationships, communal relationships) and Kim (2001; i.e., community involvement, reputation), Jo and Kim tested the interactivity levels of corporate websites and concluded that enhanced interactivity would lead to improved public relationships. They were among the first to note what became a recurring theme in the literature reviewed—that "public relations practitioners do not fully use the Internet to enhance interactions between organizations and their publics" (p. 217).[2] Gustaven and Tilley (2003) used Dholakia, Zhao, Dholakia, and Fortin's (2000) dimensions of interactivity (i.e., user control, responsiveness, real-time interaction, connectedness, personalization, playfulness) in their attempt to operationalize the concept and likewise found that corporate websites were not as interactive as they could be.

Galloway (2005) approached the concepts of dialogue and interactivity when he proposed that public relations practitioners should focus more on designing virtual *experiences* as part of relationship building with publics. He drew upon psychology and ergonomics to argue that "dynamic communicative touch" (p. 573) was not limited to the physical realm and could be achieved online through *cyber-haptics*, or experiences that stimulate "feelings such as connectedness, involvement, appreciation, and meaningfulness" (p. 573). He suggested that cyber-haptics could occur through venues such as online polling, games, voice and video messaging, email, or live chats. He acknowledged his article as an initial step in what he hoped would be the beginning of continuing research into actual applications of this approach. Cyber-haptics, per se, was not the primary focus of any other articles in the dataset, however.

It was Seltzer and Mitrook (2007) who suggested that weblogs, more commonly known as blogs, were potentially more effective for relationship building than websites because of their inherently responsive design. Kent (2008), too, recognized the potential of blogs, but mainly for research, framing, persuasion, issue monitoring, and environmental scanning. He discussed the dialogic and interactive nature of blogs, but also highlighted "the exaggerated significance" (p. 37) associated with blogs as a public relations tool as well as the risks involved with blogging. Kent concluded: "A blog will only be useful to an organization if it has someone to maintain it, someone trained in effective dialogic communication, and someone who has the trust of individuals and publics" (p. 39).

Yang and Lim (2009) proposed and empirically tested a theoretical model of Blog-Mediated Public Relations (BMPR) that included relational trust as an outcome of effective blogging practices. They found that bloggers who exhibited a "dialogical self" (p. 345) enhanced interactivity, which, in turn, enhanced trust. A dialogical blogger, they explained, focused on mutual understanding rather than persuasion. Yang and Kang (2009) validated a four-dimensional scale to measure blog engagement and found that interactivity positively enhanced one's connection to, attitudes toward, and word-of-mouth intentions about a company.

Interestingly, Xifra and Huertas (2008) found that although most public relations blogs were written by public relations professionals, they lacked development in interactive resources, and research was rarely addressed. Similarly, Xifra and Grau (2010) observed that Twitter discourse related to public relations contributed more to practice than theory. Wakefield (2008) called for a reconstruction of international public relations theory in light of Internet-based effects on communication.

Hickerson and Thompson (2009) suggested that wiki sites were untapped sources for dialogic communication and relationship building but were somewhat risky in that creation of content is a shared process. In their investigation of health wikis, sites perceived by respondents to be more dialogic were also perceived to be more valuable. Respondents additionally expressed "a significantly higher commitment to future usage for wikis than non-wikis" (p. 8).

In 2010, Smith (2010b) used Twitter involvement in Haitian earthquake relief to propose a *social model* of public relations and expand thinking about interactivity. He differentiated a social model from a dialogic model as follows:

> In this social model, public relations-related activities are initiated by an online public, facilitated by communication technology, and based on user interactivity (or the searching, retrieval, and distribution of information online).

Whereas other online models consider the organization as source {i.e., the dialogic web model (Kent et al., 2003)}, social public relations are based on user-initiation and comprise three concepts: viral interaction, public-defined legitimacy, and social stake. (p. 333)

Smith observed that communication power was shifting away from public relations practitioners to social media users whose organizational interests or roles may not be well defined. What results, then, "is a social model of public relations in which traditional public relations responsibilities are distributed to social media users" (p. 329). He stressed, too, that scholars "move beyond efforts to simply translate public relations models into the online sphere…[and] consider this an opportunity to consider new levels of risk, relationship, and interactivity" (p. 334).

Similarly, Grunig (2009) admonished the practice of using new media in the same way traditional media have been used for "dumping messages on the general population" (p. 1). He asserted that the use of social media could make the practice of public relations more dialogic and interactive and argued that any "illusion of control" (p. 4) is misplaced in a paradigm that views public relations as a messaging, rather than a strategic, function in organizations. In his review of interdisciplinary Web 2.0/3.0 research, Macnamara (2010) emphasized the need for organizations to abandon the control paradigm in light of "social and cultural shifts" (p. 1) taking place in communication. He suggested that "relinquishing control is a much greater challenge for practitioners and the management groups in which they operate than adapting to new technologies" (p. 8). Although neither Grunig nor Macnamara was proposing new theoretical models for relationship building in new media environments, each emphasized paradigmatic shifts in thinking that are relevant to and necessary for theory building.

Theme 3: Usability

Five articles were categorized under this theme, accounting for 3% of the dataset. The thematic distinction between interactivity (as discussed in Theme 2) and usability is an important one. Gustavsen and Tilley (2003) explained that usability refers to ease of operation whereas interactivity refers "to the levels of reciprocity provided by a site" (p. 2) while using it. Therefore, a site may be highly interactive but difficult to navigate, or vice versa. Usability and interactivity are not synonymous.

Articles falling under this theme discussed how a "trial and error" approach to website development was more commonly found than one based on research, planning, and evaluation (White & Raman, 1999); how usability research used in product and software development could be applied to examine the effectiveness of websites (Hallahan, 2001b); how XML could be customized for public relations (Gregory, 2004); how an experience-centered methodology could test the usability of websites (Vorvoreanu, 2006); and how usability software could assist in online data gathering (Moayeri, 2010).

Theme 4: Applications

The largest number (71, or 47%) of articles studied fell under the applications theme, with the sub-theme of corporate applications (23 articles) being the most popular, followed by crisis/risk applications (14 articles), activist (10 articles) and government/political (10 articles) applications, nonprofit applications (8 articles), and college/university applications (6 articles). The primary focus of each of these articles was how new media were used in a particular communication environment.

Corporate applications

The popular sub-theme of corporate applications began in 1982 with Glenn, Gruber, and Rabin's finding that the adoption rate of computer-based technology as a "new management skill" (p. 34) was slow among companies. Nelson and Heath (1984) were the first in the dataset to use "new media" in a title and discussed how the use of emerging cable networks allowed corporations to circumvent Fairness Doctrine restrictions on commercial networks for corporate advocacy and advertising messages.

Esrock and Leichty (1998) were the first in the dataset to examine how corporate social responsibility (CSR) was conveyed on websites, finding the predominant corporate model was "top-down/information push communications" (p. 317) rather than interaction between organizations and publics. Nine years later, Capriotti and Moreno (2007) found CSR website communication still to be unidirectional and lacking in interactivity. In their study of corporate websites in Asia, Europe, and North America, Kim, Nam, and Kang (2010) found most to be lacking in dialogic features for communication of environmental responsibility. Pan and Xu (2009) observed that U.S. corporations were more likely than Chinese corporations to highlight social responsibility on their websites. In 2011, Gomez and Chalmeta found that although most U.S. companies had operational CSR websites, interactive features were uncommon.

Esrock and Leichty (2000) were also the first in the dataset to study the relationship between corporate Web pages and publics, finding investor/financial publics to be a primary audience. Kim, Park, and Wertz (2010) likewise found shareholders, compared to other stakeholders such as consumers and activists, to be a priority for corporate websites.

Research investigating corporate use of the Internet for media relations began with Callison's (2003) finding that dedicated press rooms were "a rarity" (p. 40). In the first international analysis of virtual press rooms, Alfonso and Miguel (2006) found shortcomings in reliability and timeliness. Four years later, Pettigrew and Reber (2010) reported that use of dialogic components was improving and suggested a sixth principle, relationship initiation and enhancement, be added to Kent and Taylor's (1998) framework. In 2010, Waters, Tindall, and Morton introduced the idea of "media catching" based on their observation of how reporters sought information for stories on Twitter, thus illustrating a reversal of traditional communication patterns in media relations for a variety of organizations, including corporations.

Cultural influences on corporate Web use were examined in Maynard and Tian's (2004) study of how global companies "glocalized" (p. 285) websites by incorporating cultural, political, and economic nuances into their brand strategies for local audiences. A few articles explained how corporations attempted to manage their image and/or reputation online, including Connolly-Ahern and Broadway (2007) and Gilpin (2010a). DiStaso and Messner (2010) were the first in the dataset to analyze how corporate image was shaped on Wikipedia.

Park and Reber (2008) found that higher ranking companies (in the *Fortune 500*) utilized dialogic features but made only moderate use of feedback or response capabilities. Other corporate application articles focused on how organizations communicated with sports fans on message boards (Woo, An, & Cho, 2008), monitored public opinion through social media (Lariscy, Avery, Sweetser, & Howes, 2009b), used Facebook (McCorkindale, 2010a) and Twitter (Rybalko & Seltzer, 2010), and practiced stewardship with virtual stakeholders (Waters, 2011b). Many of these findings indicated that corporations were using online venues more so for one-way than two-way communication.

Crisis/risk communication applications

The opportunity to leverage just-in-time information technology before, during, and after a crisis was first proposed in the dataset by Calloway (1991). Thereafter, articles focused on use of the Internet as a crisis management tool for Y2K (DiNardo, 2002); how airlines utilized online communications after 9/11 (Greer & Moreland, 2003); best practices for Internet use in crisis response (Taylor & Perry, 2005); how Spanish companies conveyed chemical-related risks through the Internet (Capriotti, 2007); how companies used the Web in the aftermath of Hurricane Katrina (Greer & Moreland, 2007); the role of emotion in online bulletin boards during a toy recall (Choi & Lin, 2009); the interrelationship between crisis type, online media type, and public trust (Oyer, 2010); and the importance of government officials establishing online dialogues with citizens before crises occur (Tirkkonen & Luoma-aho, 2011).

The study of blog use during crises started with Sweetser and Metzgar (2007) and later included how emotional support was sought in blogs during a pet food recall (Stephens & Malone, 2009) and how newspapers compared to blogs in their crisis coverage (Liu, 2010). In 2010, Jin and Liu introduced the first Blog-Mediated Crisis Communication Model (the later development of which they discuss in this book), which was "based on the assumption that engagement is the most effective way to manage crises" (p. 450). The role of Facebook and Twitter in crisis communication was examined by Muralidharan, Rasmussen, Patterson, and Shin (2011).

Activist applications

In the dataset, Taylor, Kent, and White (2001) first examined how activist organizations were using the Internet to build relationships. Although they found most activists met technical and design requirements for dialogic communication, they observed that activists did not fully engage publics in two-way communication. Several studies found activist organizations, including NGOs and advocacy groups, good at providing information but lacking online interactivity with their publics (Bortree & Seltzer, 2009; Naudé, Froneman, & Atwood, 2004) and doing little to activate offline participation in social movements (Yang & Taylor, 2010). Reber and Kim (2006) found that activist groups were more likely to be dialogic with the general public than with journalists. Zoch, Collins, Sisco, and Supa (2008) and Zoch, Collins, and Sisco (2008) argued that activists were underutilizing the potential of framing techniques on their websites.

In contrast, Han and Zhang (2009) highlighted a successful case of activist use of the Internet to close a Starbucks site in Beijing's Forbidden City, thus exposing public relations complexities related to the interplay of "new communication technology and globalization" (p. 400). Seo, Kim, and Yang (2009) found promoting organizational image and funding were the two most important functions of new media for NGOs, and Sommerfeldt (2011) discovered that "activist groups most often take a reactionary or confrontational approach to establishing identification with publics" (p. 87).

Overall, analysis of articles in this sub-theme revealed that, similar to corporations, many activist groups were foregoing online opportunities to improve relationship building by not being interactive. Bortree and Seltzer (2009) succinctly captured this point:

> Most of the advocacy groups in our study seem to adopt the position that the mere creation of an interactive space via a social networking profile is sufficient for facilitating dialogue. However, these organizations are missing a significant opportunity to build mutually beneficial relationships with stakeholders by failing to effectively utilize the full gambit of dialogic strategies that social networking sites offer. (p. 318)

Government/political applications

Harris, Garramone, Pizante, and Komiya (1985) were the first in the dataset to discuss how computers could provide a two-way flow of information between elected officials and their constituents. Articles that followed explained how 1996 U.S. presidential candidates used the Web to reach voters during the general election (McKeown & Plowman, 1999), how blog-based attacks were utilized during the 2004 U.S. election (Trammell, 2006), how the Obama campaign utilized the Internet for grassroots efforts in 2008 (Levenshus, 2010), how Middle East (Curtin & Gaither, 2004) and UAE (Ayish, 2005; Kirat, 2007) governmental organizations used the Internet, as well as the role of culture in country-sponsored tourism websites (Kang & Mastin, 2008), the diffusion of social media in public health communication (Avery et al., 2010), and the impact of transparency laws on Latin American government websites (Searson & Johnson, 2010).

The Obama campaign (Levenshus, 2010) was noted as one that fully leveraged the Internet's potential, but not every government/political application focused on interactivity. Several studies that did, however, found communication to be primarily asymmetrical in nature and lacking in dialogic features (e.g., Curtin & Gaither, 2004; Kirat, 2007; McKeown & Plowman, 1999).

Nonprofit applications

Nonprofit use of the Web was first discussed in the dataset by Kang and Norton (2004), who found that nonprofits were effective in presenting traditional public relations materials online but "were largely unsuccessful in making interactive and relational communications with publics" (p. 279). Subsequent articles examined how nonprofits used websites for fundraising (Ingenhoff & Koelling, 2009), dialogic website design (Auger, 2010), online social good networks (Branston & Bush, 2010), and social media for relationship building (Briones, Kuch, Liu, & Jin, 2011; Curtis et al., 2010; Hovey, 2010; Waters, Burnett, Lamm, & Lucas, 2009).

In their study of nonprofits' use of Facebook, Waters et al. (2009) remarked on organizations' failure "to take advantage of the interactive nature of social networking" (p. 105). Likewise, Ingenhoff and Koelling (2009) observed that although Swiss nonprofit organizations were efficient at serving the information needs of their donors online, their use of dialogic technologies was lacking.

College/university applications

In the dataset, Kang and Norton (2006) investigated how colleges and universities used the Web to accomplish public relations goals. Other studies found prospective donors to be the primary audience targeted on university sites (Will & Callison, 2006) and historically black colleges and universities missing opportunities to present themselves adequately online (Brunner & Boyer, 2008). Several studies found interactivity and two-way communication lacking, including McAllister and Taylor's (2007) analysis of community college websites; Gordon and Berhow's (2009) study of dialogic features on university websites; and Chung, Lee, and Humphrey's (2010) study of Web-based recruiting efforts by universities in the US, UK, and South Korea.

McAllister and Taylor (2007) cautioned that presence alone of dialogic principles does not make a website dialogic "if it does not offer and follow through with two-way communication" (p. 232). They added, "The absence of feedback opportunities essentially makes these sites one-way communication tools. They are not much different than a printed brochure. This sender-to-receiver focus is not helping to build relationships among key publics" (p. 232).

Theme 5: Perceptions

Articles focusing on the perceptions of various participants in the public relations/new media process comprised the second largest theme with 40 articles, or 27% of the dataset. Five sub-themes were identified, with practitioner perception studies being the most popular (24 articles), followed by consumers (6 articles), college students (4 articles), journalists (3 articles), and bloggers (3 articles). The primary aim of these articles was to ascertain how certain users/audiences felt about some aspect of new media use.

Practitioner perceptions

Practitioner perceptions were assessed in 1993 when Ramsey found issues management practitioners highly likely to use "advanced communication technologies" (p. 261) on the job[3]. In recent years, Wright and Hinson (2008b; 2009a; 2009b; 2010a; 2010b) have provided continuous updates to streams of longitudinal data regarding how public relations practitioners around the globe are using new and social media[4].

Several studies reported that practitioners perceived improvements in their managerial (vs. technician) status, involvement in executive decision-making, and/or personal power as a result of using new media, including Thomsen (1995); Johnson (1997); Sallot, Porter, and Acosta-Alzuru (2004); Porter and Sallot (2005); Porter, Sweetser-Trammell, Chung, and Kim (2007); and Diga and Kelleher (2009). Other researchers found Web-based tasks to have a low priority in the workplace (Hill & White, 2000), discovered the importance of the Web and email in science public relations (Duke, 2002), explored the roles of practitioner gender and institutional types (Ryan, 2003), assessed the extent to which practitioners who blog are more accommodative to publics (Kelleher, 2008), reported publicists' perceptions of how online communication positively impacts reputation (Aula, 2011), revealed agreement among executive-level practitioners that Twitter is a useful communication tool (Evans, Twomey, & Talan, 2011), and provided practitioner insight into how technology has impacted the practice of crisis communication (Young, Flowers, & Ren, 2011). Other studies investigated how new and/or social media are used by practitioners in certain geographies, including Singapore (Fitch, 2009), Israel (Avidar, 2009), and Greece (Kitchen & Panopoulos, 2010).

Consumer perceptions

In the dataset, Park and Lee (2007) first examined consumer perceptions of information provided through new media. They found that positive comments about a company in an online news forum led to positive consumer perceptions of the company. Alternatively, Cho and Hong (2009) found online readers were cynical about CSR activities, including monetary donations, undertaken by companies in the aftermath of a crisis.

In her study of expectations of college websites, McAllister-Spooner (2010) found that active use of dialogic features could increase high school student applications to colleges. Hong and Rim (2010) observed a direct, positive link between consumers' use of corporate websites, perceptions of a company's CSR, and trust in the company.

In 2011, Freberg, Graham, McGaughey, and Freberg suggested that public perception of social media personalities could affect organizational responses to social media influencers. Schultz, Utz, and Göritz (2011) tested user perceptions of crisis-related messages in social versus traditional media.

College student perceptions

Kiousis and Dimitrova (2006) observed no significant differences in how college students perceived online stories from public relations versus news sources. Kennan, Hazleton, Janoske, and Short (2008) found college students highly attached to new communication technologies, particularly for maintaining their social contacts.

In her analysis of community college websites, McAllister-Spooner (2008) reported that undergraduates reacted negatively to a lack of dialogic loop features. Lewis (2010) found that public relations and advertising majors had a more positive view of social media compared to other majors.

Journalist perceptions

Hachigian and Hallahan (2003) were the first in the dataset who studied how journalists used sponsored websites for newsgathering. Their survey research discovered computer industry journalists were only "moderately reliant" (p. 59) on websites for information. In 2007, Chen found that websites could enhance the political candidate-journalist relationship but asserted that "websites have a long way to go before being accepted by journalists as newsgathering tools" (p. 105).

As social networking sites were gaining popularity, Lariscy, Avery, Sweetser, and Howes (2009a) stated that "journalists embrace the concept of social media more than they enact the practices" (p. 316). The authors found that business/financial journalists actually preferred "non-interactive online information sources" (p. 316) such as websites for their work.

Blogger perceptions

Sweetser (2007) was the first in the dataset to study bloggers. In her analysis of their coverage of the 2004 U.S. presidential nomination conventions, she discovered party-based bias in credentialed bloggers' reports. Steyn, Salehi-Sangari, Pitt, Parent, and Berthon (2010) surveyed active B2B bloggers, finding 57.5% of their 332 respondents had not yet seen a social media release.

Based on interviews with bloggers from a variety of fields, Smith (2010a) outlined "an evolutionary process" (p. 175) that bloggers experienced (introduction, community membership, and then autonomy), which in turn affected their willingness to work with public relations practitioners. Smith suggested that "practitioners may find optimal reception from bloggers in the community membership stage, because desires for new content make a practitioner-blogger relationship mutually beneficial" (p. 177).

Theme 6: Concerns

Thirteen articles, or 9% of the dataset, were categorized as relaying some type of concern about the use of new communication technologies in public relations. Three sub-themes emerged: legal (5 articles), risk (5 articles), and ethics (3 articles).

Legal

Hallahan (2004) identified "five major culprits" (p. 255) that posed legal problems to organizations via the Internet: "attackers, hackers, lurkers, rogues, and thieves" (p. 255). For each, he described legal ramifications and possible means of protection.

Other articles discussed the role of the Internet in litigation public relations (Reber, Gower, & Robinson, 2006), the dangers of assuming business blogging is fully protected by the First Amendment (Terilli, Driscoll, & Stacks, 2008), commercial speech concerns related to CEO blogging (Terilli & Arnorsdottir, 2008), and cautions in responding to anonymous Internet speech (Terilli, Stacks, & Driscoll, 2010).

Risk

Using a cultural studies approach, Mickey (1998) took a critical look at the relationship between the technology industry and education. He warned: "In the 1930s many argued that television would make for a more educated society but television became a vehicle to sell goods and services. The Internet is moving along the same path" (p. 335). Strobbe and Jacobs (2005) raised concerns about news becoming commodified through online press release services.

In a BledCom keynote address, Hiebert (2005) suggested that new communication technologies could "save democracy" (p. 1) but cautioned that "the pathway ahead for public relations is strewn with landmines" (p. 1). Included in his landmines were privacy invasion, identity theft, information inequities, and cyberterrorism. He also acknowledged that new media could "become tools of tyranny and suppression" (p. 8) but that "what happens will ultimately depend on what we let happen, how vigilant we will be, how much we listen, and how much we participate in the world around us" (p. 8).

Robards (2010) wrote about privacy concerns in his examination of how young Australians used social media. In their content analysis of PRSA's *Public Relations Tactics*, a publication widely distributed to college students, Taylor and Kent (2010) observed a blatant lack of discussion of the risks associated with using social media in public relations practice and called for a reconsideration of how the profession is socializing students.

Ethics

In 1995, Judd wrote about the role of ethics in the information age: "While we may view it as amoral, a means to an end, technological innovation unsettles old values and creates new views of the world. Practitioners face a challenge" (p. 36). He went on to suggest ways to establish credibility for organizations in the midst of technological change.

Related articles revealed that nondisclosure harms relationship building in social media environments (Sweetser, 2010) and the importance of corporations not only discussing ethical parameters on websites, but also building relationships with publics beyond just shareholders (Bowen, 2010).

Conclusion

Figure 1 provides a pictorial representation of how six major themes have been distributed across 30 years of journal articles addressing new media and public relations. Clearly, application and perception studies comprise the bulk of our body of knowledge, with theory building at a distant third.

Admittedly, it is tempting to call for more theory building as it relates to building relationships online, as many authors have done. But, some caveats are in order. Of important note is that the distribution of scholarship shown in Figure 1 represents what has been published through the peer-review process of four public relations journals. This study examined articles only from select journals and not books, additional journals, or other scholarly venues (e.g., theses and dissertations) that are advancing our theoretical understanding of new media. It is also important to note that application and perception studies can and do contribute to theory building, though they were not necessarily classified as such in this particular analysis that focused on the title and primary intent of each article.

There are other limitations to this study. I was the sole reviewer and coder for these articles. Despite a conscientious effort to include every eligible article with a new media focus/title and accurately capture its primary intent, the potential for human error in this process is notable. Any oversight on my part was certainly unintentional, though possible. Others conducting this analysis may have chosen different articles, different themes, different ways of selecting and/or organizing the articles, or different publications to study. I looked only at general themes, leaving open the possibility for others to survey theoretical frameworks or research methodologies employed over time. My hope is that this analysis, which is unique to my knowledge, will be of some benefit in understanding where we have been as a discipline and where we have yet to explore.

With these caveats in mind, this literature review alludes to opportunities for future research. Studies focusing on applications and perceptions of new/social media are popular, and their methodologies and "lessons learned" are undoubtedly instructive for scholars and practitioners alike. There is a rich array of literature upon which to build. Research that offers new theoretical frameworks, addresses various concerns regarding new media, or suggests how to improve usability is less populated, leaving room for

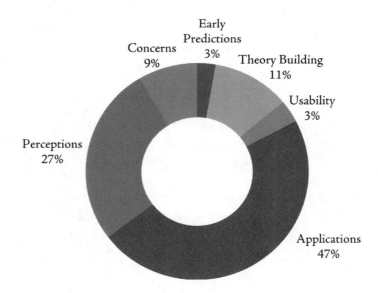

Figure 1. Thematic Distribution of Public Relations Journal Articles on New Media, 1981–2011

Note: Journal articles selected from Public Relations Review, Journal of Public Relations Research, Public Relations Journal, and PRism.

new contributions to be made. If recent trends continue, there will be no shortage of "new" media to analyze in years to come.

Certain calls to action caught my attention during this analysis because of their unique, repetitious, or compelling nature. I conclude by addressing each of these as a research question.

What is interactivity? A recurring theme in the literature is that organizations—both profit-seeking and nonprofit—are not taking advantage of the interactivity new technologies allow. That point is clear. But, as is the case with CSR, there are multiple definitions of interactivity afloat, making operationalization of the concept difficult. Having some consistency in a definition of interactivity *as it relates to public relations* would be helpful, particularly in regard to the next research question.

How do we measure interactivity? A rigorous, valid measurement of interactivity, and the related concept of engagement, eludes a number of disciplines for which online communication is relevant. There is a broad, eager audience waiting to learn how to measure this essential concept beyond hits, likes, followers, and views. How can we determine if and when interactive communication converts to actual *behavior* that benefits both organizations and society? When organizations have a business case for engaging in social media, they will be more likely to commit the resources required to genuinely participate in two-way communication with publics.

What can other disciplines offer to help advance our understanding of new media? Some authors have drawn upon intriguing theoretical concepts found in sociology, psychology, technology, management, linguistics, ergonomics, and other fields to examine new technologies, but many of their studies appear to be "one-hit wonders." Responses to calls to continue their line of thinking seem to be lacking in published literature. If we as a field expect organizations to think broadly and relinquish illusions of control, should not we, as scholars, welcome and seek alternative perspectives?

What effect do emerging technologies have on our public relations theories? The dialogic features of website design introduced by Kent and Taylor (1998) remain prominent in our literature. Subsequent research has examined the dialogic potential of blogs, wikis, and social networking sites using their framework. But, as Pettigrew and Reber (2010) suggested, "operationalized elements of dialogic theory as it applies to the Web should be continually revisited as technology develops" (p. 404). And, as stated by Smith (2010b), "the development of public relations will be defined by scholar willingness to reconsider traditional notions of what public relations are in light of an ever-growing technologically empowered world of communicators" (p. 334). Our field is best served when we continue to test, refine, and expand existing theoretical frameworks and boldly propose new ways of thinking that better fit changes in our communication environment.

How can new media be used to increase quality of life? Betsy Plank's (1983) challenge is a broad-reaching one but certainly falls into the realm of public relations practice. Ultimately, the relationships we establish, maintain, and steward on behalf of organizations should have benefits that overflow into society. The places our organizations reside should be improved by our presence and our assistance in fulfilling various social responsibilities. Public relations practitioners, equipped with courage and resources, are uniquely positioned to lead this mission.

ENDNOTES

1 The 1989–91 archives of *Public Relations Research Annual*, the predecessor of *Journal of Public Relations Research*, were included in the search.

2   In her 10-year literature review McAllister-Spooner (2009) concluded that "the dialogic promise of the Web has not yet been realized" (p. 321).

3   The first article in the database that appeared to study practitioner perceptions was Anderson, Reagan, Hill, and Sumner (1989), but it was not accessible through the author's Science Direct account.

4   Eyrich, Padman, and Sweetser (2008) examined practitioners' use of social media, but the article was not accessible through the author's Science Direct account.

PART I

The Big Picture

Contemporary Theory and Thinking in New Media

Sandra Duhé

Overview

The chapters included in this first section provide broad, varied views of how new and social media impact both public relations theory and practice. The unique insights provided by these forward-thinking authors are informative for any area of practice and serve well as either exciting starting points or supplemental perspectives for research endeavors.

Chiara Valentini and Dean Kruckeberg begin by offering conceptual frameworks to differentiate new media from social media. The distinction is an important one if we are to advance scholarship and better inform practice in the field. A shared, consistent vocabulary is essential, and Chiara and Dean make a compelling case for establishing one.

Jennifer Bartlett and George Bartlett raise some fascinating points about the contradictions social media present for public relations practice. Despite myriad praises of social media's networking potential, Jennifer and George argue that social media's impact on publics is less clear; their effectiveness in getting people involved in social movements beyond clicking "like" is lacking; and outlets, such as blogging, seem to be more about individuals expressing their individuality rather than serving as venues for collective outcomes.

Timothy Coombs and Sherry Holladay continue their discussion of Internet Contagion Theory offered in the first edition to bring us the 2.0 version for the social media landscape. They provide specific, measurable markers and describe written and visual elements both companies and activists can use to assess how publics are attempting to change organizational behavior. Their specificity and guidance are welcomed as we attempt to improve our evaluation and measurement capabilities.

Legal counselors and public relations counselors are frequently assumed to have starkly different views on sharing information with publics. Whereas public relations practitioners want to increase the flow of information, lawyers most typically want to restrict it, yet both groups have an organization's best interests in mind. Donnalyn Pompper and David Crider investigated the legal/public relations relationship as it relates to new media. They found a high degree of agreement between these professionals and a prime example of the contingency model at work.

Betsy Hays and Douglas Swanson investigated another phenomenon in the workplace. Specifically, they looked at how traditional and "reverse" mentoring can be used to counter workplace conflict related to new technology and, in turn, restore social order. They offer a number of tips provided by practitioners who have been involved in similar circumstances.

Eunseong Kim and Terri Johnson also drew upon practitioner interviews to share how social media managers are attempting to measure the effectiveness of social media use. These managers stressed that effectiveness is directly tied to goals set. They additionally underscored the importance of demonstrating to employers and clients that whatever resources are dedicated to social media are well spent. The managers also revealed in which areas of public relations practice social media seem to be most effective.

Donna Davis rounds out this section with an intriguing look at the potential of online gaming for public relations strategy. In particular, she notes its ability to engage publics who, for a number of reasons, would otherwise be disconnected from organizations and goes on to suggest that a hyperpersonal model of two-way symmetrical communication is needed to better understand the possibilities of this communication channel.

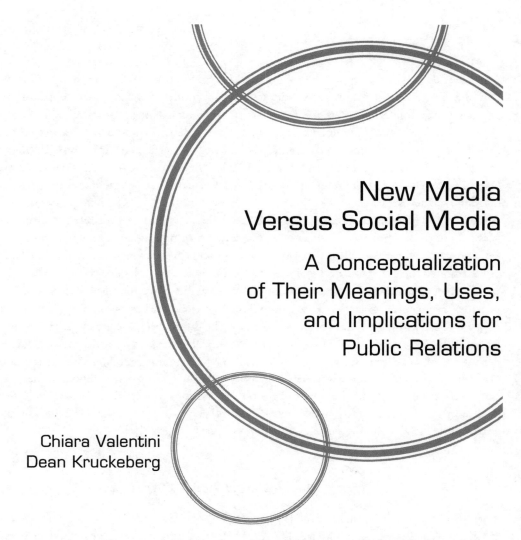

New Media Versus Social Media

A Conceptualization of Their Meanings, Uses, and Implications for Public Relations

Chiara Valentini
Dean Kruckeberg

This chapter provides conceptual frameworks of new and social media. We operationalize the two concepts and explore them in depth to clarify how new and social media should be defined constitutively and what implications they bring to the use and practice of public relations in global society.

Introduction

When thinking about 21st century public relations, one cannot avoid the need to discuss the increasing use and influence of information and communication technologies in the life of organizations, people, and social systems in general. Information and communication technologies have become an essential part of most people's everyday lives, not only in modern societies, but also in the core activities among the dominant social groups throughout most of the world (Castells, 2000a). New and social media are "helping to change people's expectations about sources, availability, and uses of information in all of its forms, both in society at large and in the practice of science" (Lievrouw, 2010, p. 221).

Organizations of all types have increasingly been adopting new communication strategies within their public relations activities that take into consideration the integration of new and social media. For example, the American company Best Buy is using Twitter to enhance customer relations;[1] Procter &

Gamble used new media to launch a campaign in spring 2010 around the theme "The Man Your Man Could Smell Like" for Old Spice products; the film industry is using new communication strategies to engage stakeholders through interactive campaigns such as the 2010 campaign of the animated film "Despicable Me," which provided games, music, and videos with amusing teases.[2] Political leaders are increasingly using new communication technologies to engage in conversations with their constituencies as well as with other potential supporters. Examples are Barack Obama's presidential campaign (Borins, 2009; Harfoush, 2009; Qualman, 2011); the UK labor party's social media campaign in preparation for Gordon Brown's candidacy for election in 2010 (Anderson, 2010); the German parliamentary election campaigns in 2009 (Tumasjan, Sprenger, Sandner, & Welpe, 2009); and the different political communication activities of the members of the European Parliament, commissioners, and political groups in a specifically created Twitter page "Europatweets" as well as on Facebook, Myspace, and YouTube—EUTube.[3]

Although the use of new and social media has increased, considerable conceptual ambiguity still exists within the professional and academic communities about what constitutes these media. This chapter provides a conceptual framework. We intend to operationalize the two concepts and to explore them in depth to clarify how new and social media should be defined constitutively and what implications they bring to the use and practice of public relations in global society. We will first provide an overview of definitions of and discussions about new and social media, taking into consideration a range of theoretical frameworks and perspectives. We will follow with definitions and explanations of the main characteristics of new and social media to help public relations practitioners better exploit the new digital environment in their professional activities. We will conclude with some reflections on the implications of new and social media for contemporary public relations practice.

What are new media?

The term "new media" generally refers to emerging information and communication technologies and applications such as mobile phones, the Internet, streaming technologies, wireless networks, and the high-quality publishing and information-sharing capacities of the World Wide Web (Bennett, 2003; Wardrip-Fruin & Montfort, 2003). Van Dijk (2006) defined new media as those that are simultaneously integrated and interactive and that also use digital code. Therefore, integration, interactivity, and digitalization must all be present to qualify a medium as a part of the new media.

As Silverstone (1999) explained, "new" defines the boundaries by which new media should be understood in relation to other forms of mass mediated communications. Specifically, the prefix "new" refers to those technologies that have been developed in recent years, in particular digital technologies. "New" also defines novel applications as well as the technologies that allow original, innovative ways of performing new tasks. Internet and digital technologies give us new powers—for example, in creating, sharing, and diffusing ideas, opinions, and interests—and they create new consequences for us as human beings.

Definitions of new media remain fluid and are continually evolving, with some definitions focusing exclusively upon computer technologies and digital content production, while others stress the cultural forms and contexts in which these technologies are used (Dewdney & Ride, 2006). In relation to the definitions of these forms of communication, two types of thinking have been thus far postulated: we should cease using the term "new media," or we should use the term "new media" only in relation to recent innovations, i.e., applications in digital environments that provide innovative ways of relating, communicating, and interacting.

The point of departure of these discussions can have both ontological and epistemological dimensions. From an ontological perspective, it is relevant to understand what makes a medium new and which factors affect the extent to which new media lose their novelty. Authors who seek such answers are interested in understanding the relationship between continuity and change and in investigating the complexities of innovations, both as technological and as sociological processes (Livingston, 1999; Silverstone, 1999). Livingston (1999), for example, argued that the novelty in new media is not so much about the technological development per se, but in these new media's relationship to the cultural processes of domestic diffusion and appropriation. Along this line of thought, Peters (2009) argued that "new media" as a technological construction is only a temporary approximation of a medium's modern relevance because technologies change so fast that what is "new" today will shortly become commonplace and then "old." Stöber (2004), in his introduction to the history of new media, also claimed that the innovation—to which the term "new" refers—comes to an end when it is replaced by a more recent innovation or when it is adapted and improved upon to meet new requirements. As a result, these scholars argue for a continuing and constant re-thinking of this term and recommend more attention on what makes certain communication technologies be "new media."

From an epistemological dimension, the focus is on why specific technologies are defined as "new media" and how they affect processes of communication. Discussing the reasons why specific technologies are called "new media," scholars highlight the innovation of use and application to contexts and situations that were not possible before. They further stress that what is currently known as "new media" does not offer particularly innovative uses, since digital communications have become commonplace daily practices. Unless information and communication technologies bear some novelty in their applications and uses, it is not accurate to define "new" if something is not so new anymore. Those who favor this perspective tend to drop the term "new media" for more semantically precise words. However, an acceptable alternative term has not been provided because a widely accepted conceptualization has not yet been agreed upon. Despite lack of agreement on a common conceptualization, scholars who are interested in this subject in an epistemological sense seek an understanding of the consequences and implications of using new media within communication processes. Scholars such as Kennedy (2008), Marvin (1988), Stöber (2004), and Tomasello, Lee, and Baer (2010) have pinpointed that what matters is the innovation that new media bring to different opportunities for communication, specifically "the new media's capacity to be adapted to meet the needs and desires of individual users" (Kennedy, 2008, p. 308).

"New," thus, is a relative and problematic concept (Scolari, 2009) because it conveys different perspectives and research interests. Despite these differences, it appears clear that the term "new media" should be used only when a new technological development and/or use in the sphere of information and communication technologies are actually occurring. Scolari (2009) proposed the use of the term "digital communication" to define the sub-field of mass communication that specifically uses digital technologies. Considering that a better definition of 21st century information and communication technologies and applications is not yet available, we propose to provisionally use "digital media" when referring to those information and communication technologies that can integrate different applications, functions, and content productions and, at the same time, will allow great interactivity among users. Readers should be aware that, as soon as the digital environment becomes outdated and is replaced by a new invention, the term "digital media" will no longer be accurate in defining this new technological development. However, in today's media landscape, we can see digital media as channels of communication comprising Web 2.0, 3.0, mobile communications, computer-enabled user devices, and the social media. In this media landscape, "digital media" is an umbrella term for the digital technology-based environment that allows "networking, multimedia, and collaborative and interactive communications" (Scolari, 2009, p. 946).

Defining the social media environment

Social media are a group of Internet-based applications that builds on the ideological and technological foundations of Web 2.0 and that allows for the creation and exchange of user-generated contents (Kaplan & Haenlein, 2010). Furthermore, social media refer to:

> activities, practices, and behaviors among communities of people who gather online to share information, knowledge, and opinions using conversational media. Conversational media are web-based applications that make it possible to create and easily transmit content in the form of words, pictures, videos, and audios (Safko & Brake, 2009, p. 6).

Both definitions focus upon the behaviors and interactions that are established among individuals, that is, for what individuals use social media. Because Safko and Brake (2009) emphasized behavior as the central characteristic of social media, we could argue that, accordingly, social media do not exist without users. This argument was corroborated by Beckett (2008), who said that social media interactivity also provides a sense of community "that transcends anything offered by mainstream media" (p. 22). Social media can thus provide opportunities to bring a variety of people together from different backgrounds to find common ground in their beliefs and interests (Lee & Lee, 2010). However, interaction is not enough to qualify social media as such. As Wright and Hinson (2009a) explained, the contents posted must also be generated by people themselves. Therefore, social media are referred to as "consumer-generated media" or as "user-generated content" (p. 3).

Terms that are often used interchangeably with social media are "social network site" and "social network." boyd and Ellison (2007) defined social network sites as Web-based services that allow individuals to (1) construct a public or semi-public profile within a bounded system, (2) articulate a list of other users with whom they share a connection, and (3) view and traverse their list of connections as well as those made by others within the system. The nature and nomenclature of these connections may vary from site to site (boyd & Ellison, 2007). According to this definition, Myspace, Facebook, Cyworld, and Bebo, for example, are social network sites because they provide public profiles, make users' connections visible, and pass them through the system.

Social media and social network sites describe similar online environments and often refer to the same digital technologies and applications. However, they define precise boundaries of research interests. The focus on social network sites is on connecting users, that is, attention on the network, whereas the focus of social media is on how users interact, that is, attention on users' behaviors. Although both social network sites and social media connect users, the latter take this connection a step further and use this connection to create channels of communication and information for establishing relationships among individuals and organizations. Another important point of differentiation is between social networks and social network sites. Social networks are not necessarily digital, nor do casual and imprecise uses of these terms help scholars to explain human groupings (in-group, out-group). Social network sites stress the place, that is, the environment where such relationships among people occur, for example, the Web 2.0 environment. The term social network site, therefore, can be considered synonymous with social media, whereas media refer to storage and/or transmission tools used to store and deliver information or data. Social networks, instead, refer to any type of social connection among individuals that define in-group situations.

A wide variety of social media exist, ranging from social sharing sites such as YouTube and Flickr to social network sites such as LinkedIn and Facebook. Social media use forums such as blogs, message boards, podcasts, wikis, and vlogs to allow users to interact. These applications may or may not be integrated into a social network site and, in general, they are used as one of the possible frameworks for classifying social media on the basis of their functionality. For example, Wikipedia is considered to be a reference social media site, Myspace and Gather.com social networking sites, YouTube a video sharing site, Second Life a virtual reality site, Digg a news sharing site, Flickr a photo sharing site, and Miniclip a game sharing site. Social media can also be classified according to their scope. Traffikd[4] organizes them into 38 categories, which can be assembled into five major groups:

- *Informational social media,* such as *Tripadviser, DoctorConnected, BlogBuzz,* and *RateItAll,* in which the scope of joining such communities is to discover answers to problems, issues, or concerns. Typically, people seek information on products and services and the opinions of those who have used or experienced these products and services.
- *Professional social media,* such as *LinkedIn, Academic.edu, Xing,* and *Zigg,* which are intended for those who want to advance their careers, both by establishing professional links with colleagues and potential employers and by providing or asking advice from professional experts in these communities.
- *Educational social media,* such as *Booking, Good Reads, MyDish,* and *MiGente,* which are social network sites that have learning scopes, such as to improve or practice a foreign language, to learn new cooking recipes, and to discuss and share opinions on books.
- *Entertainment social media,* such as *Game Diggity, Filmcrave, 10Tune, Flickr,* and *Photography Network,* which are those social media that are centered on a passion, for example, music, movies, games, or any other hobby or interest. People who join entertainment social media are looking for advice, but also like to communicate with people who have similar interests.
- *Personal social media,* such as *Facebook, Myspace, Bebo, NetFriendships Family 2.0,* and *MyChurch,* which are social media that focus on family, social, and religious relations. People who join these networks are interested in knowing other people, developing "virtual" relationships, or remaining in touch with longtime friends, family, and religious communities.

Functionality and scope are important variables for online public relations, which need to be taken into consideration, both when planning online activities and when evaluating the outcomes of these activities. Trying to push contents about a product into a personal social media, for example, may not be as effective as would be publishing the same content in specific entertainment or informational social media. Also, the format of the content, that is, video, audio, text, or visual content, that public relations practitioners intend to post in social media play an important role in the choice of a social network site. Recent developments in social media environments have made different tools not only more integrated into social platforms, but also within these media environments themselves. Today, most social media offer features in addition to blogs and forums, for example, videos, music, and other artifacts that can be added to an individual's profile or provide games. In practice, many social media are offering similar possibilities for interaction, self-expression, and content creation.

Beside functionality and scope, another important variable is the popularity or prestige of a specific social network site. The more members a social network site has, the greater are the possibilities that online public relations activities address a larger number of targeted publics. Each year, the Social Networking Website Review classifies the top 10 most important social network sites on the basis of six criteria: profile, security, networking features, search, help/support, and legitimate friend focus.[5] For 2011, the most popular ones were Facebook, Myspace, Bebo, Friendster, Hi5, Orkut, PerfSpot, Zorpia, Netlog, and Habbo. However, public relations practitioners should not blunder into the dangerous trap of thinking that choosing the most popular social network site will automatically result in the best online public relations choice. Rather, this choice should be based on a combination of the three above-mentioned variables—functionality, scope, and popularity of a social network site—with an accurate assessment of the underlying goals of online public relations activities for particular stakeholder groups that should drive public relations practitioners' choice of the best digital medium.

Furthermore, while controlling conversations and content creation is quite difficult in the social media, social network sites exhibit dual processes that enable both the creation of new public spaces and the control and monitoring of these spaces through mechanisms facilitated by the architecture of the network itself (Mejias, 2010). Interactivity, as Kaplan and Haenlein (2010) argued, is one of the most important dimensions of social media. Interactivity is hereby related to the "content and applications [that] are no longer created and published by individuals, but instead are continuously modified by all users in a participatory and collaborative fashion" (pp. 60–61). Participation and collaboration should thus be the mantra of online public relations in social media environments. Moving away from a simplistic mass mediated communication perspective, Skoler (2009) postulated that the role of relationships and interactions in social network sites is something beyond the physical mediatization of digital technologies. Accordingly, Skoler (2009) saw social networks as a force for community relations, for enhancing sociability—at least virtually—among people, and for empowering and reinforcing individuals' voices. Social media and social network scholarships share a common social functionality that is not necessarily implied in the new/digital media definition. Social media and social network sites should thus be conceptualized as *online social environments* that enable people to engage in relationships of a different nature, for example, professional, personal, and spiritual ones.

Solving the ambiguity...a clarification of the terms

To clarify the boundaries between digital and social media, we will now discuss what distinguishes true (adequate) knowledge on the digital environment from false (inadequate) knowledge. We thus need to operationalize the two terms and to examine their main characteristics, similarities, and differences as well as the purposes for which they can be used in different public relations activities.

In the previous sections, we identified the main differences between digital and social media, the former intended as a *digital technology-based environment* that enables people to do different activities across time and space, the latter as an *online social environment* that enables people to engage in relations of a different nature, for example, professional, personal, and spiritual relations. Taking this as a point of departure, several substantial differences exist between digital media and social media, for example, on the level of interactivity, the type of communication provided, the direction of communication, the position/power that publics have vis-à-vis to organizations, and the role of publics in these media. Table 1 summarizes the main characteristics of digital and social media.

Digital media allow publics to customize their search for information and delivery methods. Digital media can simultaneously provide audio, visual, and text publication methods as Internet users prefer. Digital media publics can also post comments, for example, in blogs as well as in digital news articles. However, contents published in the digital media have their own existence independent of social interactions. A blog can still exist and present ideas, opinions, positions on issues, persons, institutions, or other content, even if none of the blog followers posts comments and discusses the topic that was shared by the blogger. Organizations can thus blog, podcast, create, and post videos on YouTube and host Blog Talk Radio shows. No one else needs to be involved for them to create useful content. This means that digital media can create different forms and types of online communications, but not necessarily to enhance dialogic ones. When corporations decide to create blogs, they may or may not allow followers

Table 1. Main characteristics of (new) digital and social media

CHARACTERISTICS	(NEW) DIGITAL MEDIA	SOCIAL MEDIA
Content production	Anyone	Anyone
Content control	High for the source, lower for publics	Shared control among community members and organizations, content is continuously modified by all users in a participatory and collaborative manner
Type of communication	Organizational/ Conversational communication (among organizations and publics)	Conversational /Interpersonal communication (among community members)
Direction of communication	One-way when no comment/post or two-way asymmetrical when comments and posts by publics	Two-way asymmetrical or two-way symmetrical
Interactions between people	Medium	High
Public participation	Medium (it is not required)	High (it is necessary)
Power relations	Higher for organizations and individuals creating contents for digital environment	Higher for community members
Capacity for building relationships	Low	High
Capacity for transmitting contents	High, contents can directly reach specific publics	Medium, contents can directly reach specific publics, but they may not be processed and shared if not considered relevant for community interactions
Role of publics	Mostly receivers, publics can however express opinions, agreement or disagreement with contents	Active role of co-creation of contents and meanings. Publics do not only receive contents, but also engage in creating their own contents, participate in content collaborations, share contents with community members
Entertaining function	High	High
Possibility for viral effects—publicity tactics	High	High
Capacity for endorsements	Medium	High, if the content is perceived relevant by community members. Community members trust each other more than organizations.
Market analyses	Medium, studies on consumers or other stakeholder groups' behaviors are dependent on the level of participation of these members	High, organizations can obtain information by simply monitoring community members' activities, posts, comments, and interactions
Organizational purposes	Marketing public relations, marketing communications, reputation management, corporate/political branding, corporate/political image management, product/service related communications, public diplomacy (e.g. nation branding, promotion), information provision, political campaigning	Organization-public relationships, community relations, employee relations, relationship marketing, stakeholder engagement, reputation management, political participation (i.e. e-democracy, e-governance tools)

to post opinions and questions on these corporate blogs. When allowing for comments, organizations can also decide to filter incoming content according to their communicative goals, thereby limiting the power of public participation and freedom of expression. Corporate blogs can provide information that is less official in nature and can give more insights into organizations' behaviors and thinking, but this does not necessarily reconcile with the idea of co-creation of meaning around organizations' products, services, brands, and values.

Differently, social media by nature are interactive and require the participation of others to exist. In social media, the focus is on the community, and communities are established around a common interest, a passion, an idea, or around the human need to be around like-minded people. Social media were, in fact, originally developed to help people connect with friends, professional colleagues, and others with similar interests with the intent to share ideas, opinions, hobbies, and interests within those communities (Stassen, 2010). Members' communications in social media are, therefore, the lymph necessary to make social network sites grow and expand. Without conversations, interactions, and collaborations, social media will lose their function of being social.

Avery, Lariscy, and Sweetser (2010) claimed that from a strategic communication perspective, "social media create an instantly available avenue through which to disseminate messages to target publics" (p. 191). However, this representation of social media is quite old-fashioned and very much anchored on a mass media perspective. The scope of using social media is social in nature. Social media have a social functionality. They exist because individuals decide and have the opportunity and the ability to create social relations with others by building and interacting in virtual communities, not primarily because they want to receive organizations' messages and product information.

Social media need a certain level of interactivity, participation, and engagement by different parties, and organizations must understand that social media environments are multi-vocal and are primarily dependent on community members' willingness to enter into a dialogue with organizations. Organizations can provide the means and the tools to enhance members' conversations, for example, by providing interesting feeds that engage members' interests and enable them to reinforce the community, even creating social capital.

Along with this line of thought, Lee and Lee (2010) claimed that social media are interesting means for creating online communities and facilitating social capital-building activities. Social media members participate in online communities to garner mutual benefits among group members, for example, by strengthening social ties, circulating information, archiving experiences, and exchanging opinions. These elements, Lee and Lee (2010) asserted, distinguish social media—as a means of a social function—from digital media. Relationships in social network sites, thereby, have similarities with real-life relationships when it comes to norms of behaviors and interactions. However, they also occur in a unique context in which time and space are not relevant; in which identity of community members can be real, partial, or completely invented; and in which conversations are opinioned and can be filtered. This has direct consequences on the level of communicative efficacy of an organization and on relationships among an organization's online publics. If communication within an organization aims to contribute to the achievement of organizational goals, that is, with the activities of management and with the coordination of systems of relationships that are active between an organization and its publics, and if trust is fundamental in managing and coordinating these relationships, then the social media environment does not necessarily facilitate the disintermediation that results from a more direct approach between organizations and their multiple publics (Valentini, 2010).

In sum, organizations' effective use of social media to achieve public relations goals and objectives is not as simple as it might appear. Proliferation of these media provides ample and inexpensive opportunities for communication, but not all social media lend themselves to such use. Furthermore, public relations outcomes can be difficult to perceive, let alone measure. Organizations' attempts at participation and community-building can be hampered by the ambiguity of these virtual communities at several levels, particularly because community members construct and may assume different identities in these virtual common arenas, and trust of other community members is not to be assumed. Social media provide a relatively inexpensive means to communicate with, and, more importantly, to enter into a dialogue with, strategic publics. As Vujnovic and Kruckeberg (2010b) have noted, the question remains whether these communities that form in the social media will be used to create a truly virtual public space, that is, a so-called Habermasian public sphere, or whether social media are used in public relations as just another means, for example, for paid corporate bloggers to disseminate information and obtain feedback on an organization's behavior by creating pockets of controlled "private spheres" rather than to utilize these resources to create awareness and build communities around issues within the global context.

Conclusion

Information and communication technologies have cemented their place as an avenue of communication by using a variety of methods, such as email, listserv, instant messenger, online chatting, electronic bulletin board, and weblogs. These interpersonal and interactive functions enable people to actively communicate with others online, at high speed, and with relatively low cost, regardless of time and distance (Lee & Lee, 2010). Furthermore, they have empowered people, and by doing so, the overall role, involvement, and power of the consumer are rising too (Bhagat, Klein, & Sharma, 2009).

Within this digital environment, it is extremely important to have a clear understanding of the meaning, use, and implication of new/digital and social media. As we have tried to demonstrate, a lot of ambiguity remains today about what constitutes new/digital or social media. By failing to acknowledge that these terms—new/digital media and social media—do not accurately describe the same phenomenon, it becomes impossible to retain the same considerations, for example, in public participation, distribution of power among communication participants, level of transparency and truthiness, and control over contents.

Furthermore, even in the *online social environment*, a distinction should be made between social media and social network sites. Although the two terms are used interchangeably when referring to specific digital technologies with social functionality, it is important to bear this distinction in mind when, for example, we need to evaluate the outcomes of online public relations activities. If we want to benchmark our outcomes with other activities in relation to organizations' stakeholder networks, we should use the social network concept, whereas if we are interested in evaluating the response, attitude, or behavior of users, we should label it social media.

It is critically important that scholars studying these media differentiate among the terms new/digital media, social network, social network site, and social media. Each of them pertains to specific characteristics, many are overlapping, but they are also dissimilar. Each can contribute, support, and reinforce particular public relations activities. Each requires different considerations and leads to different communication results and, when possible, measurements.

Social media, more so than new/digital media, must be at the heart of public relations activities because social media can enhance organization-public relationships by increasing and improving community relations. In a virtual community, they provide opportunities for community-building as espoused by Kruckeberg and Starck (1988), who argued that "[A] fundamental reason public relations practice exists today is because of a loss of community resulting from new means of communication and transportation" (p. 21). Ironically, their contention in 1988 that new means of communication have resulted in the loss of community today has the potential of being addressed, albeit "virtually," through new means of communication, that is, through digital and social media.

Kruckeberg and Starck (1988) said that the breakdown of a sense of community that had existed historically was fundamentally destroyed about a century ago, that is, within a compressed timeframe from about 1890 to 1917, thus creating the need for public relations professional practice. This breakdown of communities that were geographic in nature was because of a spike in the evolution of communication technology during that timeframe that fundamentally changed society, resulting in nationalism through mass media, re-segmentization from geographic communities to professional/occupational/avocational communities, and an inversion of what was public and what was private. One hundred years later, globalism has superseded nationalism; re-segmentation has evolved into a chaotic, seemingly infinite, fragmentation of society; and privacy and people's expectations about any privacy have largely disappeared into an all-knowing ether world. Ironically, Kruckeberg and Starck's (1988) call for the restoration and maintenance of a sense of community is more possible technologically than ever before. However, as Kruckeberg and Tsetsura (2009) have argued, an infinite number of volatile "publics" worldwide can form immediately and unpredictably with unforeseen power, and they can act seemingly chaotically. Vujnovic and Kruckeberg (2010a, 2010b) observed that social media have become the point of intersection between global and local, and, because these media are social, they will hopefully push organizations toward bridging participatory gaps, that is, building communities while maintaining good communication strategies. Kruckeberg and Tsetsura (2011) concluded that public relations must no longer be about persuading individuals or being aligned with a corporation's goals, but about supporting communities.

These new forms of media must be used wisely, that is, with understanding and discernment, and they require continuing scrutiny by scholars and communication practitioners alike. To begin this process, however, requires agreed-upon conceptualizations that we hope this chapter addresses.

ENDNOTES

1 See http://twitter.com/twelpforce
2 See http://www.despicableme.com/
3 See http://europa.eu/take-part/social-media/index_en.htm
4 See http://traffikd.com/social-media-websites/#news
5 See http://social-networking-websites-review.toptenreviews.com/

Kaleidoscopes and Contradictions

The Legitimacy of Social Media for Public Relations

Jennifer L. Bartlett
George Bartlett

If the current discourses of progress are to be believed, the new or social media promise a kaleidoscope of opportunity for connecting and informing citizens. This is by allegedly revitalizing the fading legitimacy and practice of institutions and providing an agent for social interaction. However, as social media adoption has increased, it has revealed a wealth of contradictions both of its own making and reproduction of past action. This has created a crisis for traditional media as well as for public relations. For example, social media such as WikiLeaks have bypassed official channels about government information. In other cases, social media such as Facebook and Twitter informed BBC coverage of the Rio Olympics. Although old media are unlikely to go away, social media have had an impact with several large family based media companies collapsing or being reintegrated into the new paradigm. To use Walter Lippmann's analogy of the phantom public, the social media contradictorily serve to both disparate the phantom in part and reinforce it.

Yet despite the influx of social media and the changes they promise, it is important for public relations professionals to consider the role of social media in their communications programs and the multiple contradictions they present. On one hand, a huge amount of information is now available with newspapers alone, each presenting more information in an issue than a person used to learn in a lifetime (Bauman, 2007). Research, however, shows that people who use social media know the same about current affairs as 20 years ago (Pew Research Center, 2007). A Ketchum-USC study ("Media Myths," 2010) also showed that although social media are widely used, traditional media are considered the most credible sources of information. Whereas social media have built legitimacy alongside traditional media, the ways they impact publics are less clear. This chapter explores these issues and considers implications for public relations practice.

Introduction

New or social media have created a kaleidoscope of possibilities for how media and information are used. Opening mass communication channels to multiple users and, in turn, changing the way audiences use media has introduced new challenges, and opportunities, for how public relations is practiced. One of the enormous opportunities is that audiences have access to a great amount and diverse range of information accessible 24 hours a day. New media have also facilitated the proliferation and reproduction of social capital by allowing exchange between and amongst individuals across physical and social boundaries (Hazelton, Harrison-Rexrode, & Kennan, 2007).

The expansion of the social media has seen the reworking and creation of networks that have allowed information to be created and disseminated while avoiding traditional media outlets and organizations. Most prominently, social media such as WikiLeaks have bypassed official channels about government and organizational information. The changes have also meant that the way media gather information has changed. For example, BBC's coverage of the clearing of the Rio slums by police prior to the Olympics was informed by members of the slums tweeting and Facebooking rather than by traditional representatives such as sociologists, anthropologists, or the state. There is also evidence that social media have had an impact on the very viability of traditional media with several large media companies collapsing, including *Newsweek*, America's second bestselling weekly magazine, after revenue dropped 38% from 2007 to 2009 (Lauria & Grove, 2010).

The kaleidoscope of options offered by social media also brings with it a series of contradictions about their impact, which are important considerations for practicing public relations. In this chapter we present an overview of social media phenomena and what it means for public relations practice. We first define what is meant by social media and reveal it as a relatively new phenomenon. Although the diffusion rates of social media are extraordinary, time is needed for sense making, structuration of meanings, and negotiation of legitimate uses. We then discuss what appears to be a kaleidoscope of possibilities that new technology offers to the individual to take part in public discourse. Following this, we present insights from the increasing number of studies that suggest these new possibilities have been tempered by contradictions. This lays the foundation for discussing the implications of these insights for public relations practice, which concludes the chapter.

What are social media?

The term social media is generally used to refer to "the group of Internet-based applications that build on the ideological and technological foundations of Web 2.0 and allow the creation and exchange of User Generated Content" (Kaplan & Haenlein, 2010, p.61). They have grown exponentially as tools for social interaction since 2004 with the emergence of such sites as Myspace and Facebook as interfaces. The term suggests that media are used for social interaction rather than merely a source of information.

The relatively short history of social media has emerged alongside the development of, and adoption of, the Internet as a channel of communication. The origins of the Web go back to around 1979 when the Internet facilitated an online information board where members could post messages, and information could be retrieved by other users (Kaplan & Haenlein, 2010). However, online social networking sites go back much further to the late 1950s when "Open Diary" allowed a community of writers to collaborate. It was here that the term blog emerged from weblog, and the colloquial "we blog" term became popular amongst users (Kaplan & Haenlein, 2010). It was during the 1990s that blogging took off among the broader population, giving everyday people the taste for communicating with one another and the opportunity to create their own mediated material.

The widespread use of the Internet has grown from a source of information with an inherent assumption that the information was provided by an authority, to a place where everyday users can make a contribution to the enormous body of knowledge available via the World Wide Web. The Web gave individuals as well as organizations the ability to develop a Web presence and post information about themselves. Institutional acceptance for the Internet grew particularly between 1995 and 2005 with the spate of corporate and organizational websites that arose using the Internet as a communication channel. However, much of the content was largely static and considered up-loaded versions of traditional print material. Over time this is changing, but the use of websites to provide information about individuals and organizations is both accepted and expected. It was estimated that there were almost 19 billion pages of information on the Web in June 2011 ("The Size," 2011).

The era of Web 2.0 and User Generated Content that emerged around 2003 opened up an entirely new dimension to the Internet. Kaplan and Haenlein (2010) suggested that Web 2.0 might be considered an ideological term to describe how the Internet is used. According to their reasoning, Web 1.0 was about publishing of information, whereas Web 2.0 refers to the interactive use of the Internet. User Generated Content (UGC) was coined as a term around 2004 to describe the vast array of ways that this idea of Web 2.0 was enacted. The emergence of Myspace and Facebook around 2004 are examples of early venues for User Generated Content. This opened a new era where all manner of Internet users could have their say and share information, and it underpinned the explosion of applications such as Twitter, YouTube, and others.

At the time of this writing, just over 30% of the world's population were Internet users ("Internet World Stats," 2011). The world regions with the greatest population penetration of Internet usage were North America, Europe, and Australia/Oceania, as one would expect due to accessibility in these first world economies. Asia had the greatest number of users with more than 920,000,000 people using the Internet. Of the major applications, Facebook has reached the 500 million active users mark, with more than 50% using Facebook daily ("Statistics," 2011), and there are about 200 million Twitter users (Chiang, 2011). These figures reflect the enormity of the adoption of social media by citizens around the world in such a short period of time.

There have been numerous predictions made about the impact of the Internet and inevitable comparisons made to the advent of the printing press, and then later, radio and television (Paine, 2007). Print and electronic media certainly enhanced the availability and accessibility of information to the masses. However, the locus of control of traditional media lies with a comparative few who produce and disseminate information to usually large audience groups who subscribe to those channels. The locus of control of social media rests not with a relatively small group of producers, but instead resides in and amongst the users of a channel. It is here that the enormity of the effects of social media is situated.

Social media—a kaleidoscope of possibility

The term kaleidoscope is symbolic of the social media and brings forth imagery in which the users are involved in a seemingly endless array of options and possibilities that the Web 2.0 ideology facilitates. As its rapid growth suggests, this is a dynamic, colorful, and ever changing space that has been touted as having endless possibilities. Some of these claims have focused on the organization and the role of social media in making communicating more effective (Weber, 2007). Others have claimed that social media can facilitate improved reputation, image, and perceptions of corporate social responsibility (CSR) (Cakim, 2007; Worley, 2007). One study suggested that engaging in social media is beneficial for organizations because 38% of active Internet users think more positively about companies that maintain a corporate blog (Universal McCann, 2008).Such perspectives focus on the organization and its outcomes and align with functionalist perspectives of public relations.

Other scholars have highlighted increased stakeholder and citizen involvement via social media. For example, stakeholder power and involvement could be increased through the use of Web applications, as a way to influence and to shame corporations (Coombs & Holladay, 2007a). Social media applications for democracy and the role of governments at a societal level have also been highlighted. Eid (2007) for example, suggested that the Internet has potential for promoting democracy and the Arabic cause but faces challenges from the very type of democracy operating as an institution in the Middle East. However, it has been social media that have been attributed with providing the opportunity for the Arab Spring events of early 2011.

A common theme of facilitating transparency in organizations, government, and society has run throughout the promises that social media offer. For the public relations practitioner, tapping into these options as both a channel and form of transmission has been imbued with possibility. Using the Internet to promote transparency (Capriotti & Moreno, 2007; Wang & Chaudhri, 2009) is one of the central concepts within the CSR discussion. It does this by allowing organizations to publish a wide range of information about their economic, social, and environmental practices that can be retrieved by stakeholders as needed.

An online survey of public relations practitioners (Eyrich, Padman, & Sweetser, 2008) identified 18 different types of online tools that public relations practitioners used. These included blogs, intranets, podcasts, video sharing (e.g., YouTube), photo sharing (e.g., Shutterbug, Flickr), social networks, wikis (e.g., Wikipedia), gaming, virtual worlds (e.g., Second Life), micro-blogging/presence applications (e.g., Twitter, Pownce, Plurk), text messaging, videoconferencing, PDAs, instant message chat, social event/calendar systems (e.g., Upcoming, Eventful), social bookmarking (e.g., Delicious), news aggregation/RSS, and email. We suggest this wide variety of tools can be categorized into four types of applications: social networking (e.g., Facebook), knowledge collaboration (e.g., wikis, podcasts), content sharing (e.g YouTube, Flickr), and blogging (e.g. blogs, Twitter).

These four types of applications—knowledge collaboration, content sharing, blogging, and social networking—offer a range of tools for engaging in social media. Knowledge collaboration through wikis and podcasts facilitate sharing of information. Groups can be formal, such as in a work situation, or form around a special interest. Content sharing allows the sharing of information in a public forum. Those sharing the information do not need their own site but can instead provide information where others might access it. Well known sites, such as YouTube and Flickr, largely deal in images, which act as powerful forms of communication.

Blogging, or the use of weblogs, is one of the oldest forms of interactive Internet use. It allows participants to share information and ideas. Twitter is, as of this writing, the dominant microblogging site, allowing users to follow and share short message tweets. Social networking adoption has extended from the personal to the organizational. For example, Facebook has been adopted by organizations as a way to create social linkages between the organization and amongst stakeholders. With such a range of ways for individuals and organizations to engage with one another, it would seem that social media provide a variety of means for public relations practitioners to inform and engage with their publics. Three of the key practice areas in which these possibilities have been investigated are relationship management, media relations, and crisis management.

Numerous studies attest to the potential for public relations to harness social media. Wright and Hinson (2008a) have called on public relations practitioners to incorporate social media in their communication and relationship building. Wright and Hinson have noted that "the potential impact of blogs on public relations and corporate communications is phenomenal" (p. 4). Direct communication with stakeholders and publics is expected be more dialogic as Web 2.0 platforms provide the means for informing and responding. This theme of relationship building possibility is predominant in the public relations literature. The claims for greater interactivity have also been noted in relation to CSR as one of the prominent management issues of the 20th century. The two-way capacity of the Internet allows for greater stakeholder engagement around other organizational issues as well (Capriotti, 2011).

Another area where social media have transformed practice is in media relations. Trammell (2006) suggested, "With the rise of personal publishing, practitioners need no longer rely on media for transmitting those messages and reaching their public" (p. 402). Indeed, even the Commission on Public Relations Education report (Turk, 2006) noted, "Often, new technological forms and channels, such as electronic pitching, podcasting, and blogging, prevail over traditional news releases and media kits" (p. 31). For broadcast messages, gatekeepers within traditional media can be bypassed, and stories can be provided to large audiences. Whereas previously just a few journalists and editors may have seen a media release when it was distributed by the public relations practitioner, the presence of social media means that the same information can reach an enormous audience without reliance on traditional media outlets (Wright & Hinson, 2008a). With social media, the power to communicate rests in the capabilities of the organization and its technology and the expertise of the public relations practitioner. In addition, the phenomenon of "media catching" (Waters, Tindall, & Morton, 2010) has emerged where public relations practitioners are contacted by journalists as a result of the journalists following the organizations using social media.

Some of the significant public relations research has focused on crisis response as a central task of public relations practitioners in managing impressions of organizations. Wang and Chaudhri (2009) identified the Internet as one of the top five tools Chinese organizations use to communicate in a CSR crisis. Likewise, Branco and Rodrigues (2006) used legitimacy theory to explain the use of reactive Web-based communication around CSR crises. Schoenberger-Orgad and McKie (2005) highlighted that it is not only corporations that are using the Internet, but also activist groups that use the net as a way to highlight CSR transgressions. These studies highlight the way that social media provide platforms for stakeholders and citizens as well as opportunities for organizations to communicate via the Internet to deal with issues. Social media have also been used in high profile crises such as an epidemic in Finland (Tirkkonen & Luoma-aha, 2011) and the use of Twitter and Facebook during the natural disaster in Haiti (Muralidharan, Rasmussen, Patterson, & Shin, 2011). However, these large scale crises have also revealed some of the issues, such as trust, emerging from the use of social media and the need for a prior online relationship to be in place in order for communications to be effective (Tirkkonen & Luoma-aha, 2011). These are some of the contradictions that the opportunities of social media have revealed.

Contradictions

In general, the social media promote the modernist discourse of enlightenment through access to knowledge, individualization, and freedom. This discourse can be traced as far back as the 1960s and the beginning of the Internet (Castells, 2001). The rapid uptake of new and social media has also been blamed for the predicted demise of traditional media. Here we explore these major areas of contradiction in the social media discourse.

The advent of the social media has created a new expanded public sphere that promises an expansion of both information and interactivity. However, like any new paradigm, it creates a host of contradictions between its promise and practice. Individuals in the current age experience more knowledge in a single newspaper than was once encountered in one lifetime (Bauman, 2007). In practice, this discourse is challenged by research that indicates that individuals know roughly the same amount of information as they did in 1989 (Pew Research Center, 2007). In addition, this deluge of information creates legitimacy problems for those with the authority to possess certain information. For example, certain professions, such as medical doctors, are facing legitimacy problems due to both the widespread access to once specialized knowledge, and the blurring of what Goffman (1959) called the front sphere and the back sphere.[1] The collapse of the front sphere into the back sphere, to which the social media appear particularly well suited, causes the legitimacy of the actor to be challenged by the inability of the audience to believe in his or her role.

While traditional sources of authority and legitimization are challenged, new forms of communication take their place, often failing to fill in the gap they vacate. This can be seen in research that shows that the most popular and grassroots source of information across the Web, the blog, has low levels of access to a mass audience. For example, 52% of bloggers suggested that they blogged for themselves only, while only 32% suggested that they blogged for an audience (Castells, 2009). As such, the new social media enlarge the public sphere, albeit a sphere still largely centralized on media that have created a monopoly for themselves. Contradictory to the idealized atmosphere of the public sphere of discussion, social media in many ways reaffirm the individualization process and serve to confirm that "I am truly I" rather than facilitate truly collaborative discussions and debates that surpass, through synergy, the contribution of any one individual.

Further to the point of actual involvement, social media practitioners cite the 1% rule of social media usage (Arthur, 2006). According to this principle, 1% of users are creators, or those developing original material used on the Internet. This would include activities such as development of podcasts, tweets, wikis, etc. Another 9% are editors. These people will forward original material or respond to material developed by others. The greatest proportion of Internet users, that is the remaining 90%, are viewers only, meaning they read others' online material but do not engage with creating or editing material themselves.

The subject of a great deal of the debate over the relevance of social media lies on the door mat of Facebook. A plethora of reporting suggests it played a key role in assembling social movements in the Arab Spring, Democracy Now! in Spain, the student protests in Britain, and the Tea Party movement in America. On the more institutional side, Barack Obama's presidential campaign was largely driven by Internet campaigning. Although Facebook serves as an excellent source of networking, research indicates that it is only good for accessing people who already know each other (Castells, 2001). As such, social media network use is often limited by access to networks that, in turn, are limited by different strata's access to social capital. In other words, participation in social media networks tends to maintain, rather than expand, one's social reach and social capital, thereby strengthening weak links between individuals but often not connecting individuals to people they did not previously know.

One of the most powerful factors in social media has been Facebook's ability to broker information. The main source of information brokerage lies in users' ability to network with other likeminded people by forming groups. Although this has been used as a powerful lobbying tool with some success in changing commercial behavior, there is a comparatively smaller recorded success in reinvigorating movements outside of cyberspace (Bauman, 2007). In general, the social media have failed to reinvigorate broad scale social movements and long term participation in institutions. One can easily join a Facebook group protesting greenhouse gases, but few people go beyond clicking "Like" and become active in related offline activities (e.g., letter writing campaigns, protesting, lobbying, donating) that require a more significant commitment of personal time and resources. Perhaps it is somewhat unfair to suggest that the Internet is the root technological cause of the declining role of social movements and traditional institutions. This decline can be traced back at least as far as the 1970s. The emergence of identity politics and the collapse of the public sphere into the private sphere (Sennett, 1986) have long been noted as contributors, and many of these issues predate the social media. As such, the social media discourse of the renewal of the public sphere and of the solution to the legitimacy crisis of the state via citizens' involvement remains a myth.

So although there is limited hard evidence of the political influence of social media, the greatest strength (and weakness) of the new social media lie in their ability to quickly reorganize a network, release data, and ignore some of the regulations of the traditional media field. The classic example of this is the 1998 release of President Bill Clinton's Monica Lewinsky scandal by the then relatively new Internet news site—*The Drudge Report*—when *Newsweek* editors refused to publish it (Press & Williams, 2010). Since that time, the media have seen many changes, chief among them being the rise of the 24-hour news cycle. Alongside this has been the mainstreaming of certain bloggers such as Arianna Huffington, the widespread use of news websites often without charge, and the merger of old print media with new media such as *Newsweek's* merger with *The Daily Beast*. It is also interesting to note that organizations such as WikiLeaks, which assembles a network that bypasses traditional state boundaries and jurisdictions, are intimately linked with old newspaper organizations such the *Le Pais*, *Le Monde*, *Der Spiegel*, *The Guardian*, and *The New York Times*. As such, to ensure legitimacy, online organizations are actually using the already established cultural capital of the journalistic field and traditional media.

This leads to the second major contradiction of the future of traditional media given the prevalence, immediacy, and reach of social media. The recent narrative of the collapse of the newspaper due to the advent of social media can be seen as a myth. The newspaper has been in decline since at very least the widespread use of television in the 1950s (Press & Williams, 2010). This decline has been accelerated by the proliferation of personally customized news (and other) products based on "pull" technology. This can be seen by the advent of a variety of 24-hour news channels often owned by the same parent corporations seeking out new customers by catering to niche markets. Paradoxically, this has not resulted in the growth of a variety of different producers offering a variety of products. Instead, a limited number of companies are monopolizing the market and expanding by creating networks catering to different tastes. This media monopoly is not limited to traditional media such as the established press. Instead, we see such phenomena as the vertical integration of media networks such as Time Warner, which owns several cable and satellite stations including Time Warner Cable and CW Network, Newline Cinema, 80 magazines in the UK and Australia, Bebo, AOL, Netscape, and a variety of other products (Castells, 2009). Another example of the integration of different media outlets can be seen in big budget movie releases, which, in order to remain profitable, must not constrain themselves to the traditional medium of film but also expand via merchandise, books, magazines, and often interactive advertisements. This suggests that traditional forms of media are shifting structure and reach to cater to new audience demands. It also suggests that these examples of media focusing on entertainment are reflective of predictions that social media and the Internet would have a strong entertainment focus (Putnam, 1995; 2000).

Interestingly, research into the credibility of social and traditional media suggests that although there is high usage of social media, there is greater credibility attributed to traditional media by audiences ("Media Myths," 2010). This annual USC study of media usage revealed findings that follow in the flavor of Putnam's comments from the late 20th century about the overall demise of interest in the public sphere and political life with shopping sites being some of the most referred to even in traditional media. In addition, the study revealed that despite the hype about the possibilities of social media, public relations practitioners' use remains emergent.

Conclusion

Taylor and Kent (2010) reinforced the Ketchum-USC ("Media Myths," 2010) findings, showing that within industry there is little evidence of either effectiveness or of widespread use of social media in public relations practice despite the high expectations of their contribution. Whereas social media have built legitimacy alongside traditional media, the ways they impact publics are less clear. As noted earlier, social media are being used during crisis, but there are limitations to their effectiveness in such situations (Muralidharan et al., 2011; Tirkkonen & Luoma-aha, 2011). This is an area for future research, not just for the sake of new findings, but to also track the way that social media are used by practitioners and by publics in a way that meets their emergent expectations.

An area in which there appears to be significant change is in the way that the news media and public relations people work together to share information. The phenomenon of declining media budgets is not expected to go away. For journalists, the Internet provides a wealth of information about events and issues. Social media, and journalists linking into particular networks, potentially offer not only fruitful avenues for understanding how information is gained, but also for how agendas and issues are formed.

As such, perhaps the initial optimistic promises of a functionalist tool for achieving organizational goals may not be the best lens through which to consider social media and public relations. No doubt, over time there will be rich material that public relations scholars can bring to inform effective practice. Perhaps now the focus of public relations research and insight lies in the ways that public relations practitioners act as agents within the emergent space of making sense of and legitimating practices and relationships facilitated by new technology.

ENDNOTE

1   e.g., Guests would be received in the front sphere (living room) vs. the back sphere (bedroom).

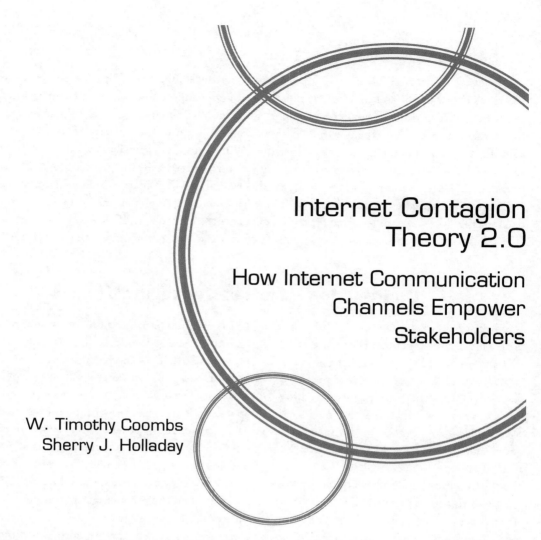

Internet Contagion Theory 2.0

How Internet Communication Channels Empower Stakeholders

W. Timothy Coombs
Sherry J. Holladay

Internet Contagion Theory (ICT) provides a framework for assessing how Internet-based communication can influence the organization-stakeholder power dynamic. The burgeoning social media landscape prompted an update to ICT 2.0 that includes both written and visual elements of these social media tools. ICT uses stakeholder theory and network theory to explain why and how activist stakeholders can harness Internet-based communication resources to influence organizations to modify behaviors. Reputational threats pose exigencies for organizations that are difficult to ignore, and effective use of social media tools can intensify these threats. Social media channels and content can be examined for specific markers of stakeholder salience, relationship building, communication skill, and issue contagion success. ICT can be used by both stakeholder activists and organizations to evaluate the potential for influence in their relationships.

Corporate managers have become increasingly concerned with what "average" stakeholders say about them online. What makes this interesting is that although organizations have the more powerful position in the organization-stakeholder relationship (Coombs & Holladay, 2010b), the proliferation of social media capabilities has minimized organizational control over messages regarding the organization. The concern over social media is simultaneously warranted and absurd. Clearly not every stakeholder is a social media message master. Of the billions of Internet messages posted, few are viewed by a significant number of other stakeholders. However, there is always the potential that a social media message may resonate with stakeholders and become a contagion that spreads to others (Watts, 2007).

Internet Contagion Theory (ICT) originally was developed to explain the dynamics behind the Internet's effects on the stakeholder-organization power dynamic and move beyond simplistic descriptions of the influence of word-of-mouth communication (Coombs, 2002; Coombs & Holladay, 2007a). ICT's framework can be used to explain *why* the Internet can empower stakeholders and *how* that dynamic operates. This revised chapter elaborates on the stakeholder empowerment dynamic and demonstrates how social media, which experienced rapid growth since 2007, can be easily integrated into ICT's explanatory framework. We term this newest version of the theory ICT 2.0. The chapter is divided into three sections: (1) context and theories that influence ICT's development, (2) explanations of how communication is used to empower stakeholders, and (3) ethical concerns associated with ICT. It is important to note that ICT can be used by both stakeholders and corporations to understand and manage their relationships.

Context for Internet Contagion Theory

Organization-stakeholder relationships are complex, as organizations simultaneously have relationships with a myriad of stakeholders. Relationships exist because the stakeholders and organizations are interdependent. Interdependence sets the stage for potential conflicts of interests. Interests can be values, beliefs, expectations, etc. that lead people or corporations to act. When some stakeholders perceive conflicts of interest, they may be motivated to act to influence the organization in some way.

Organizations depend on stakeholders for support and to not interfere with their operations (Pfeffer & Salancik, 1978). A stakeholder may perceive that an organization poses a threat to his/her interest(s) and decide to challenge the organization. Stakeholder challenges—or the potential for challenges—may necessitate a response. An issue, or point of contention, has now developed (Coombs, 2002). Ideally, the organization recognizes and addresses the problems. However, organizations lack the time and resources to address all stakeholder concerns, especially when those concerns are not perceived as central to organizational operations. When stakeholders decide to push the point of contention or issue, they can be defined as activist stakeholders. Activists are people who stand up for a cause and work for change (Raymond, 2003).

An isolated activist stakeholder may not be perceived to pose much of a threat. However, if several stakeholders share their concerns and mobilize, they can gain power (Hallahan, 2001a). Stakeholders can create awareness of a problem through Internet messages and move issues into the organization's consciousness. To do so requires stakeholders to alter their salience (importance) for the organization. Stakeholder theory explains how stakeholders come to be viewed as powerful and thereby important to management.

Mitchell, Agle, and Wood (1997) proposed three dimensions for evaluating stakeholder salience: (1) power (the ability to get an actor to do something s/he would not do otherwise), (2) legitimacy (the perception that actions are appropriate or proper within the context of some belief system), and (3) urgency (the call to action due to the importance of the concern to stakeholders). Stakeholder salience is a function of the ability to demonstrate these attributes. The more attributes the stakeholders are perceived to possess, the greater their salience and the greater the pressure on management to deal with them and their issues. Salience is complicated by contradictory demands of different stakeholder groups. The importance of various stakeholders can change depending on the situation. The issues advocated by competing groups must be prioritized according to their threat level as determined by their ability to damage the organization and their probability of developing momentum. Managers must assess the potential impact (amount of damage) and likelihood (odds of the negative effect occurring) of the threat (Coombs, 2012).

ICT posits that stakeholders can increase their salience by managing the three attributes (i.e., power, legitimacy, and urgency) through their Internet activities. Savvy stakeholders can engage in coalition-building and meaning management to enhance their perceived salience (Mitchell et al., 1997). For example, by using Internet communication, activist stakeholders can demonstrate that their claims are legitimate and can damage the organization's reputation. When management perceives activist stakeholders as salient, they will be motivated to address the issue. At this point, we have identified the three factors that constitute stakeholder salience but have little insight into what actually allows stakeholders to alter those factors. Network theory and reputation management provide insights into the dynamics of altering stakeholder salience.

Rowley (1997) argued that a network perspective should be used to understand the power and potential influence of stakeholders. Network analysis reflects a systems perspective and focuses on the structural relationships, or interconnections, among various stakeholders and the organizations that comprise the organization's environment. Because organizations have relationships with many different stakeholders, they must consider the entire web of relationships. However, because relationships can evolve and change over time, relationships with some stakeholders gain salience and are difficult for an organization to ignore (Rowley, 1997).

Network structure represents opportunities and constraints for actors within the network (Wasserman & Galaskiewicz, 1994). The organization is an actor, as are the other stakeholders, and they operate within a larger environment consisting of similar corporations, suppliers, other relevant stakeholders (Rowley, 1997). For our purposes, the network boundaries are defined as those who share an interest in the activist stakeholder's issue (Knoke, 1994).

In network terms, power is a function of the activist stakeholder's position within the larger network (system) of stakeholders. *Centrality* is a critical variable because it enables communication and is an indicator of an actor's power and importance within the network (Coombs, 2002; Rowley, 1997). The Internet offers a wide range of communication opportunities for activist stakeholders to develop relationships with others and to other organizations that may support the issue.

Centrality is a function of an activist stakeholder's closeness and degree (Rowley, 1997). The *closeness* aspect of centrality refers to the ability of the stakeholder to independently access others across the network. An activist stakeholder who demonstrates closeness can spread information quickly and directly without intermediaries. *Degree* centrality indicates the extent to which the activist stakeholder is well-connected. Degree refers to the number of ways a stakeholder can be linked to others in the network. It signals the potential for access to many different sources of information (Rowley, 1997). A stakeholder who demonstrates the degree dimension of centrality can use a variety of Internet-based communication tools to facilitate access to others, including issue websites, social networking sites, content sharing sites, discussion groups, email alerts, blogs, content sharing sites, and social bookmarking sites (Coombs, 2002).

Corporations that monitor the Web activity of activist stakeholders will be able to observe this centrality and make inferences about power. Indicators of centrality include website traffic, links to the website search engine results placement (it should be close to the organization being challenged and within the top 10 placements), number of subscribers to email alert systems, discussion group traffic, followers on Twitter, friends on Facebook, and views on YouTube. The indicators signal the ability of activist stakeholders to connect with others through a variety of methods. Reputations are one way centrality can be translated into power.

Developing and protecting reputations are critical functions in modern organizations because of the value favorable reputations bestow upon organizations. Favorable reputations are associated with positive outcomes such as financial success and community support (Fombrun & van Riel, 2004; Watson, 2006).

Because reputations are valuable, reputational threats are serious concerns. If activist stakeholders can convert their issue into a reputational threat or a potential threat, an organization will be motivated to take some form of action on the stakeholder's concern. Activist stakeholders can find leverage in reputations. It is now time to address the specifics about how communication and the Internet provide the forum for altering stakeholder salience and leveraging an organization into addressing stakeholder concerns that were previously ignored.

Internet Contagion Theory in action

Illustrating the application of ICT involves explaining how Internet communication empowers activist stakeholders. It is by amassing and utilizing communication resources that the Internet becomes a tool of empowerment (Coombs, 1998; Heath, 1998). The application of ICT depends upon utilizing the Internet communication resources of channels and message content.

Channels

Prior to the Internet, it was difficult for activist stakeholders to build their networks and reach additional stakeholders. Media advocacy was the best way to reach a wide array of stakeholders (Ryan, 1991). However, media advocacy is a form of uncontrolled publicity that cannot guarantee story placement or how the story will be framed in the media (Treadwell & Treadwell, 2000). Activist stakeholders did not control if the message was sent to others or the content of that message. There were limited communication channels for connecting with other stakeholders and altering the network dynamic.

In contrast to traditional media advocacy, the Internet affords activist stakeholders greater control over their communication by eliminating gatekeepers. Activist stakeholders communicate directly to a potentially large audience using a wider range of Internet communication channels. The Internet enables a greater degree of control and presents more options for stakeholder activists to create awareness of their concern or issue. It can become a contagion when it spreads rapidly among stakeholders. Activists can use the Internet to demonstrate that their concerns pose real threats to organizational reputations thereby increasing the pressure on managers to take their demands seriously.

Channels build power by increasing the centrality of the activist stakeholders. Internet channels increase closeness by providing direct access to other stakeholders in the network. Internet channels also can increase degree centrality by providing multiple links to other stakeholders in the organization's network. In the language of social networking theory, channels are about nodes and ties. Nodes are the people in the network, in our case organizational stakeholders. Ties are the connections between the stakeholders.

Activist stakeholders use Internet communication channels to increase their *radiality*, the degree to which their network reaches out into the larger organizational network. Each Internet communication channel seeks to add more nodes and ties to the activist stakeholders' network. Various Internet channels have the potential to reach new stakeholders and increase centrality. Ideally, these new stakeholders then pass the information (and/or contribute additional information to the original message) to people in their networks and create a contagion. Message redundancy results from the increased use of multiple Internet channels. Repetition is an advantage when communicating about an issue, and multiple channels that reach the same stakeholders increase degree centrality. Because power is in part a function of centrality, efforts to increase centrality should magnify activist stakeholder power.

Markers of Issue Contagion Success

- Traffic to the websites
- Links to the issue websites
- Search engine results placement of the issue websites
- Number of subscribers to the email alert systems
- Traffic to discussion groups
- Traffic to complaint portals
- Popularity of blog site
- Followers on Twitter
- Popularity of Twitter followers
- Number of trackbacks to a blog
- Number of retweets
- Number of times appearing at social bookmarking sites
- Number of people who "like" a social networking site
- Number of appearances in social bookmarking sites
- Popularity of people who "like" the social networking site
- Search engine results placement (it should be close to
- the organization being challenged and on the first page)
- Number of subscribers to email alert systems
- Number of views on YouTube and Flickr
- The variety of Internet-based channels used (e.g., websites, blogs, discussion groups, YouTube, email lists, etc.)

Figure 1. Evidence of Issue Contagion Success (Power through Connections)

Using various Internet channels can alter a network dynamic, thereby increasing activist stakeholders' power and making them more salient to the organization. Figure 1 reviews some basic markers of success in using Internet channels to spread a contagion.

Urgency can be addressed through channels too. Urgency includes showing a commitment to the issue. Commitment indicates that the activist stakeholders will stay with the issue and not quickly lose interest and move on to other things. Activist stakeholders can use their communication skill to create a sophisticated network of Internet channels to demonstrate their commitment to the issue. They are building a communication structure designed to build pressure over time. It is not just an angry message that is sent once and forgotten. A complex communication infrastructure is a reflection of urgency.

Social media messages operate in short bursts. For instance, Twitter is limited to 140 characters. Blogs are longer but still are relatively short messages. Websites can contain copious amounts of information organized clearly for easy navigation. The website is the primary repository for information about the concern or issue. Interested people can visit the website and access whatever material they want. The other Internet channels are spokes that can lead people to the website hub. When Greenpeace pressured Nestlè to change its sourcing of palm oil, the Greenpeace website contained details about the problem while the YouTube video and Facebook posts created attention and potential links to the website hub (Owyang, 2010b). The website-as-hub is a marker of communication sophistication because it demonstrates various social media channels are being employed to not only create awareness, but also to drive potential supporters to the website for additional information. The total number of unique Internet channels utilized by activist stakeholders and structure of those channels can be taken as an indicator of urgency.

Legitimacy: Content concerns

Discussions of contagions focus almost exclusively on the channel. An example is the current preoccupation with social media in public relations. The talk centers on the need to *use* the new channel, not the actual *content* of the messages. But what a message says is also important. Because legitimacy is critical to stakeholder salience, the perceived *legitimacy* of the issue must be a preeminent concern for activist stakeholders. For the activist stakeholders, the legitimacy of the issue may seem self-evident. However, creating a contagion that is difficult for an organization to ignore requires more than the involvement of a few committed "true believers." Creating legitimacy for an issue and the activist stakeholders themselves requires effective *framing*. Activist stakeholders must be concerned with how the issue is described, how the involvement of relevant organizations is depicted, and how recommendations for managing the issue are presented. The framing process requires skill in managing perceptions and meanings.

Overall, issue framing, including the types of support provided and the language used, should be viewed as a *social construction process* (van de Donk, Loader, Nixon, & Rucht, 2004). Communicators contribute to the definition and development of the issue. This social construction process holds significant implications for the perceived legitimacy of the issue. Legitimacy resources should be used to demonstrate that an issue is worthy of public concern. The incorporation of legitimacy resources aids the cause by demonstrating that the issue is not merely an obsession of the lunatic fringe. For this reason, careful attention should be devoted to the process through which legitimacy is socially constructed in cyberspace. Figure 2 describes the two common strategies for developing legitimacy: *endorsement* and *self-evidence*.

Attempts to build issue legitimacy via the Internet will rely on a variety of channels that use a Web-as-hub design. Legitimacy efforts do not fit well into 140 characters or other short message formats. The Web-as-hub design can incorporate a variety of legitimacy resources to support the effort. The need to create legitimacy has an obvious, concomitant persuasive dimension. The hope is to attract and win over additional stakeholders who can benefit the cause and contribute to the contagion (Coombs, 2002).

Legitimacy Strategies

ENDORSEMENT: a person who is perceived to possess legitimacy supports the issue. Endorsers' legitimacy may derive from their position (e.g., a member of a regulatory board), credibility (perceived trustworthiness and expertise; e.g., a noted biologist offers testimony), and/or charisma (they possess extraordinary characteristics that attract others; e.g., popular politicians, celebrities).

SELF-EVIDENCE draws upon features of the issue itself.

  Tradition: uses precedents that demonstrate that things have been done this way in the past and there is no reason to change.

  Rationality: provides logical reasoning and evidence to support the legitimacy of an issue.

  Emotionality: creates strong feelings and reactions to effectuate persuasion.

Figure 2. Legitimacy Strategies

Based on Coombs, 1992.

For example, the website may include: links to news media stories related to the issue; links to scientific reports appearing in established research journals; maps of the history of the development of the issue (e.g., a time line demonstrating destruction of rain forests, a time line of regulatory actions or litigation); video testimony from scientists who are working on research related to the issue; comments from government officials in countries where the issue has gained prominence; and stories from individuals who have personally experienced adverse effects from the failure of corporations to address the issue (e.g., survivor and victim stories).

High quality websites are more likely to be seen as legitimate. Witmer (2000) contended that a quality website includes qualified creators, objective information, offers evidence to support claims, and is up-to-date. Although stakeholders who are already strongly committed to the issue may enjoy reading diatribes by other like-minded individuals, websites that are composed only of highly biased information or opinions may fail to attract and retain new stakeholders. Corporations can more easily ignore the rants of zealots than the well-reasoned, fact-based arguments of scientists, respected humanitarians, and other trustworthy sources. Witmer's (2000) criteria for a quality website can be extended to evaluate nearly any type of online message, including most social media.

Markers for the success of legitimacy reflect that others are accepting an issue and suggest an issue is building momentum. Markers of legitimacy success include: a) the number of different people posting or commenting on a blog, micro-blog message, or other social media message; b) the valence of the posts or comments (supports or refutes the issue); c) the length and valence of threads (exchanges of messages between people); d) the valence of crossover stories; and e) repeating or connections to the message via trackbacks to blogs, tags in social bookmarking sites, or retweets (any form of social media echoes). The number of different markers indicates growing interest rather than a limited debate between a few people. Negative valence indicates people are not seeing the issue as legitimate while positive valence indicates acceptance of the issue's legitimacy. Crossover stories are when the news media report on the website. Media stories that are positively valenced signal that the media are endorsing the issue's legitimacy (Coombs, 2002). Moreover, traditional news stories can spur interest in the online content and widen the audience beyond Internet users. Trackbacks indicate other people believe a blog is worthy of attention.

Content provides an opportunity to build urgency as well. Part of urgency is the need to take action. Consistent messaging about the need to take action and for the organization to reform its behavior reflects urgency. Repeated calls for action indicates the issue remains a priority and has continued urgency for the activist stakeholders. Ultimately, activists should take some action beyond reading or posting messages. A final marker is action. Sending online messages is a weak marker but does count as action. Showing up physically at rallies or hosting viewing parties for educational DVDs are strong markers because they show greater commitment to the social concern (Gladwell, 2000).

Legitimacy Markers

- Issue websites (quality of website, utilization of legitimacy resources)
- Discussion groups (quality of the post, utilization of legitimacy resources)
- Blogs (quality of blog writing, utilization of legitimacy resources)
- Other written social media (quality of the writing, utilization of legitimacy resources)
- Other visual social media (quality of the images, emotions evoked, and utilization of legitimacy resources)

Figure 3. Indicators of Attempts to Build Legitimacy

Applying ICT to analyze activist stakeholder actions

ICT can be used to gauge potential stakeholder salience shifts. Either stakeholders or the organizations they oppose can use ICT to track network changes indicating salience shifts. We can use the information from Figures 1 through 3 to assess activist stakeholder power and to identify changes in that power.

The first step is to build or to assess legitimacy. Activist stakeholders need to build legitimacy by utilizing legitimacy resources and demonstrating quality of online messages. Organizations need to assess if the activist stakeholders have legitimacy and the likelihood of other stakeholders accepting those legitimacy claims. Again, the legitimacy resources and quality of the online messages would be used to make the assessment. Activist stakeholders with quality messaging that uses a variety of legitimacy resources would constitute a greater threat than those with crude messaging that do not make any attempt to build legitimacy.

The second step is to build or to assess power. Activist stakeholders can build power with a sophisticated communication approach that extends the potential reach of its message. Markers of a sophisticated communication approach include: a Web-as-hub communication system, utilization of multiple online communication channels, the potential reach of message, and the amount of social media echo. Web-as-hub and using multiple online communication channels are efforts to increase power through network centrality. The idea is to reach more people through more channels. Moreover, the Web-as-hub strategies uses short messages to drive potential supporters to a central repository of knowledge. Organizations should rate activist stakeholders as more powerful when they utilize the Web-as-hub system along with a diversity of online communication channels.

The potential reach of the message would be predicated on such factors as the number of Twitter followers and the followership of those following the activist stakeholders, the number of people who "like" an activist stakeholder's Facebook page and the number of friends of those liking the page, amount of traffic to a website, number of positive links to a website, and number of email alert subscribers. Activist stakeholders seek to extend the potential reach and organizations evaluate power in terms of potential reach. Reach can be amplified through social media echoes. Social media echoes are repeats of the message that would include retweets, trackbacks, comments, and social bookmarking posts. Activist stakeholders want to increase the positive echo while organizations must assess the amount of positive echo. However, echo can be negative if others are critical of the activist stakeholder messages. Negative echo decreases the activist stakeholder power by reinforcing the organization's position on the issue. Crossover to traditional news media would be a variant of echo as well. We can include in this section whether or not the activist stakeholder's website appears on the first page of a search related to the target organization. If a search of the organization's name or products includes the activist stakeholder's website on the first page of the search, potentially more people are exposed to the message.

Urgency is a combination of channels and content. A sophisticated use of interlocking online communication channels demonstrates commitment to an issue. Content signs of urgency include frequent updating of material (active on the issue) and repeated calls for action. Activist stakeholders need to maintain an active presence in a number of online communication channels and repeat calls for action. Actions taken by other stakeholders serve as urgency markers. Physical activities such as protests are considered stronger urgency markers than just cyber activities. Organizations can assess urgency by the presence or absence of these urgency markers.

Activist stakeholders can alter their perceived salience for organizations through their use of online communication channels. Organizations may need to track activist stakeholders over time to determine if they are taking actions that are increasing their salience. For instance, have the activist stakeholders increased the number of communication channels and quality of those channels? Have the activist stakeholders strengthened their legitimacy claims or increased urgency? ICT identifies factors that can be assessed over time to identify shifts in stakeholder salience.

We need to add one final layer to the organizational assessment—strategic impact. If the activist stakeholders want organizations to engage in behaviors that are too expensive or run contrary to organizational strategy, the organization must fight back rather than make changes. Organizations cannot risk implementing changes that can severely damage their operations. Moreover, if the activist stakeholders are targeting critical stakeholders, such as customers, who may support the issue, the threat is increased. The concern is over the strategic impact of the changes. If critical stakeholders eventually support the activist stakeholder issue, the organization will suffer more reputational damage than if marginal stakeholders are drawn to support the issue. Again, stakeholder salience does influence organizational decisions (Mitchell et al., 1997).

ICT and ethical concerns

Any time you broach the subject of creating power there is a need to consider the ethical ramifications of doing so. No communication tool can be inherently ethical. There are always ways to abuse communication and twist it an unethical fashion (Palenchar & Heath, 2006). This section recognizes three ethical concerns related to ICT: (1) exaggeration, (2) suppression, and (3) astroturfing.

Activist stakeholders have been accused of exaggerating dangers to win support for their issue. They may evoke emotion through words and images to scare people into taking action on an issue. For instance, some critics claim the 1989 ban of Alar, a chemical used to help fruit ripen, was based on exaggeration and fear rather than the scientific evidence ("How Chemical Industry Rewrote History," 1998). If you search the term Alar even today you can find information on this ongoing debate. Exaggeration and fear can circumvent the legitimacy process, even though emotion can be a form of legitimacy. People are asked to act on emotion not the critical thinking associated with legitimacy assessments. Activist stakeholders must resist the temptation of a quick emotion-based communication effort built on exaggeration. The ethical approach is to have a full airing of the various viewpoints and to let people make informed choices in the marketplace of ideas (Heath, 1992).

Ever since activist stakeholders first used the Internet to challenge organizations, organizational agents have tried to suppress its use. It began with legal intimidation to remove unfavorable websites and extends to efforts to restrict social media content. Nestlè tried to have Greenpeace's YouTube video about irresponsible palm oil sourcing harming orangutans removed from view and tried to limit how people responded on the Nestlè Facebook page (Owyang, 2010b). Both actions were met with cries of censorship and negative social media posts about Nestlè. Free speech means respecting even those who disagree with you. All parties must respect the rights of other parties to post messages online as long as no laws are violated.

Astroturfing is a term Senator Lloyd Benson used to describe grassroots efforts that were really driven by some group with a vested interest in the issue ("Astroturfing," 2011). The idea is to create the appearance of public support for an issue when there really is little support for it. Astroturfing can include paying people to post messages and posting messages under aliases. Some critics extend astroturfing to any Internet-based campaigns that make it easy for people to express concern about an issue by simply clicking an icon and sending a message. By making it easy for people to be active, the true support for the issue is compromised. Making it easier for people to support an issue is not artificial. These are real people who do care. True, they may not be as motivated as the activist stakeholders promoting the issue, but the concern is authentic. Astroturfing in its classical sense is unethical and should be avoided.

Conclusion

When ICT was initially developed, social media were not a part of the Internet communication mix. The growth and influence of social media required an update to our 2007 chapter that appeared in the first edition of this book (Coombs & Holladay, 2007a). The leverage activist stakeholders could generate from social media had to be integrated into the theory.

ICT is a tool for mapping specific changes in the stakeholder-organization relationship because it helps both practitioners and stakeholders to understand and to evaluate the relationship. By increasing their salience to management, activist stakeholders are altering the stakeholders-organization relationship. In turn, managers must be able to recognize these shifts and consider how the shifts affect their communication with stakeholders. The revised ICT 2.0 helps both parties to better understand and evaluate aspects of their relationships.

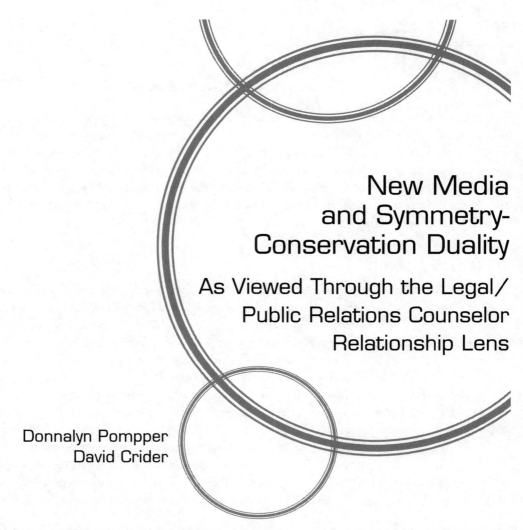

New Media and Symmetry-Conservation Duality

As Viewed Through the Legal/ Public Relations Counselor Relationship Lens

Donnalyn Pompper
David Crider

Based on interviews with 77 public relations and legal counselors, assessment of the oft-categorized adversarial "oil and water team" relationship revealed high degrees of cooperation and like-mindedness with regard to new media. Findings suggest that relationship outcomes in connection with new media default to conservation—as characterized by systems theory's emphasis on homeostasis. Yet, among voices also emerged evidence of symmetry-conservation given new media's omnipresence and utility. Overall, findings support using new media as a portal through which to view the contingency model at work in organizations—where public relations counselors conservatively err on the side of caution nearer to the pure asymmetry end of the continuum. Thematically, contributed are three inter-related elements of conservation in new media contexts: a) reputation preservation, b) offensive/defensive monitoring, and c) litigation avoidance.

The new media environment and its evolving online atmosphere have impacted conventional public relations thinking and may further aggravate tenuous relationships within organizational management ranks. Fitzpatrick (1996) recommended that researchers further scrutinize "particular areas of agreement and accord, as well as reasons for and frequency of conflicts" (p. 7) among management colleagues—as internal dynamics have implications for organizational outputs. Historically, public relations' most serious interpersonal challenges have been with lawyers. The infamous "court of law versus court of public opinion" barrier was placed at this study's center to scrutinize ways legal and public relations counselors negotiate their functions to affect organizational decisions involving new media.

* *Funding for this study was provided by the Arthur W. Page Center at the Penn State College of Communications.*

Review of literature

Overall, findings have been mixed about legal/public relations counselor relationship quality. Both counselors communicate uniquely and inherently disagree on strategy because of their different orientations (Reber, Cropp, & Cameron, 2001). In practical terms, defining new media and understanding its role for organizations may help practitioners of both stripes to navigate relationship conflict. Focusing on new media as it plays out in the legal/public relations counselor relationship by grounding it in systems theory, the contingency model, and symmetry-conservation duality makes a useful contribution to public relations theory building.

Oil & water team

Called "the oil and water team" (Simon, 1969, p. 7), public relations counselors lack fundamental understanding of laws/regulations—negatively impacting their ability to effectively manage the function (Fitzpatrick, 1996) and inviting encroachment (Lauzen, 1992). Comparatively, organizational responses undergirded by *traditional public relations strategy* focus on the court of public opinion to gain forgiveness and rebuild credibility by responding candidly during crises, admitting problems, stating organizational policy, launching an investigation, and implementing corrective measures—whereas, organizational responses using a *traditional legal strategy* focus on courts of law in order to avoid adverse consequences of admitting guilt by saying nothing, quietly saying as little as possible, and shifting or sharing blame with a plaintiff (Fitzpatrick & Rubin, 1995).

Despite legal/public relations counselor differences, some studies have suggested that their relationship quality is "excellent or good" due to "relatively harmonious and collaborative" (Lee, Jares, & Heath, 1999, pp. 1, 7) exchanges with high degrees of cooperation for shared or equal influence on decision making — often with public relations counselors playing a subordinate role to legal counterparts (Fitzpatrick, 1996).

Defining new media's role in public relations

Relationship building—public relations' raison d'être—has been likened to new media's intrinsic communicative, interactive, and social qualities (Avery et al., 2010). Indeed, both definitions of *public relations* and *new/social media* emphasize engagement, participation, and two-way communication (Kelleher, 2010).

No doubt, public relations is undergoing a revolution with new media complementing face-to-face interaction. Perhaps it is the Internet's "conduit of transparency" (Duhé, 2007, p. 57) and two-way communication potential that makes it so suitable for public relations work. Therein lies the rub—for transparency is risky for organizations that prefer to limit interaction with publics and minimize information disclosure. This reality challenges boundary-spanning public relations counselors who recognize that publics demand both legal *and* communication transparency from organizations (Allen, 2006)—because they must convince organizational management of advantages inherent in openness when building and maintaining relationships.

Public relations' new media research tends to focus on three areas: 1) media relations, 2) personal and public relationship building, and 3) ethical standards.

Media relations. Public relations practitioners cannot hope to control media now any more than they ever did, but can use the Internet to circulate current, instant, expert information. Some critics have declared traditional *media pitching* outdated, succeeded by *media catching* online services that connect journalists and sources (Waters, Tindall, & Morton, 2010). Also, the *social media news release* enables readers to interact and build on organizations' content—with multimedia attachments and embedded photos, audio, video, and blog links. Deadline pressures, smaller staffs, and research capacity challenges make these appealing to journalists (Jewett & Dahlberg, 2009). Large corporations have come far in maximizing websites (Alfonso & Miguel, 2006), yet some not-for-profit organizations (Reber & Kim, 2006) and government agencies (Avery et al., 2010) have been slow to adopt new media.

Personal and public relationship building. New media supplement the interpersonal element of relationship building. Organizations' websites foster dialogue, a key to developing mutually beneficial relationships (Park & Reber, 2008). Indeed, Levenshus (2010) used relationship-management theory to examine the Obama campaign's strategic use of the Internet to build public relationships. Similarly, activist groups use public relations and websites to link diverse groups and promote social change (Oliveira, 2007). Also, organizations may defensively react to Web attacks by consumers/resistors and offensively use new media during crises (Coombs & Holladay, 2007a). Still, a gap exists between practitioners' beliefs in the Internet's relationship building potential and their actual use of it (Kent, Taylor, & White, 2003).

Ethical standards. Disclosure is a cornerstone of public relations ethics (Grunig, 1993)—yet, some organizations capitalize on online anonymity and unethically mask their true identity to represent themselves as an individual or grassroots group. Such moves prompted updated new media communication guidelines by the Federal Trade Commission (2009)—and by trade organizations like Public Relations Society of America (2000), Word of Mouth Marketing Association (2009), and Social Media Business Council (2009). BMW and Wal-Mart learned that unethical online behavior negatively affects organization-public relationships (Sweetser, 2010).

Systems theory, symmetry-conservation duality, and the contingency model

Systems theory suggests that organizations—as systems—seek to establish and maintain symmetrical, mutually beneficial public relationships. Organizations adjust to environmental changes to achieve goal states (Broom, 2009) and in the process, may attempt to maintain a stable equilibrium (homeostasis) without too much change, which is a conservative standpoint. *Conservation* means that an organization holds fast to its principles, is unlikely to alter its mission in response to external pressures, and exists primarily to accomplish its own goals (Sha, 2007). Alternately, Sha (2004) advanced *symmetry-conservation duality* given that public relations may be symmetrical even though organizations and publics conserve/maintain their fundamental values.

The contingency model illustrates degrees to which organizations give and take in public relationships. At one end of its continuum is pure *asymmetry* (where public relations benefits only the organization) and at the other end is pure *cooperation* (where public relations benefits only its publics). The contingency model's continuum center is a win-win zone where organizations and publics stand to benefit equitably (Grunig, Grunig, & Dozier, 1996)—even though laws of conservation imply that some organizational aspects remain unchanged. So, knowing to which side of the contingency model continuum's win-win center that legal and public relations counselors feel most comfortable with regard to new media—closer to pure asymmetry or closer to pure cooperation—is uncertain.

Research questions

This literature review suggests two important research questions:

RQ1: In what ways might the legal/public relations counselor relationship be impacted by the new media environment?

RQ2: To what degree do legal and public relations counselors hold similar/ dissimilar views on benefit-risk associated with organizational use of new media?

Method

To collect legal and public relations counselors' perceptions, the qualitative research method of in-depth telephone interviews was used. This method is well suited to: (1) gaining a simultaneously wide and sharp view of phenomena in context, (2) capturing data on perceptions "from the inside" (Miles & Huberman, 1994, p. 6), (3) collecting rich, textured data of in-depth responses, (4) facilitating deep probes of participants' comments, and (5) serving as an initial phase in developing hypotheses for later testing.

Non-random snowball and stratified sampling techniques were used. Snowball sampling involved industry colleagues recommending names and contact information for recruiting legal and public relations counselors. Some research participants then recommended others' names. Stratified sampling involved creating four organization categories and then using directories {*PR Week's Press & Public Relations Handbook* (Haymarket Media, 2010), *Corporate Yellow Book* (Leadership Directories, 2010), and *Hoover's Handbook of Industry Profiles* (Hoover's Business Press, 2010)} to cold-call recruit participants for an equitable number of participants in each category: 1) business-to-business (B2B), 2) business-to-consumer (B2C), 3) government/regulated (G/R), and 4) not-for-profit (NFP).

Although literature searches revealed that organization typologies rely mainly on size/profits (large vs. small), ownership (private vs. public), environment structure (mechanistic vs. organic), and industry sector (e.g., banking, airline), some have posited that *government* is a distinct organization type due to its punitive and legal powers (Smigel, 1956) and that government-regulated organizations respond to incentives or penalties according to government-chosen goals (Tomer & Sadler, 2007)—and that *not-for-profit* (NFP) groups are unique service providers rather than pursuers of profit (Simon, 1997). Moreover, customer types and media used to appeal to them contrasted B2B environments (selling to other businesses) by primarily using trade media, with B2C environments (selling to general consumers) by using mass media.

Selection criteria for participants were: (1) at least five years of experience as a legal or public relations counselor; (2) working relationship with a legal or public relations counterpart; and (3) diverse organizational contexts (B2B, B2C, G/R, NFP). Separate in-depth telephone interviews were hosted with legal and public relations counselors, lasting from 20 minutes to 1.5 hours during September 2010–May 2011. Participants were not compensated for their time, but were promised a copy of study findings. Participants agreed to telephone conversation recording and were assured confidentiality inasmuch as the research report would not identify them by name or employer. Participants are identified by generic descriptors they provided. Protocol and research design were approved by the researchers' university Institutional Review Board. Both researchers are Caucasian/White (female Ph.D./APR, 51, and male doctoral student, 32).

Interview procedures

An interview protocol was tested, with minor phrasing changes. The first 9 questions probed participants' work experience (number of years, work settings) and demographic features (age, education, ethnicity). A subsequent set of 22 questions probed participants' perceptions of working with their legal or public relations counterpart with regard to new media.

Data analysis

Six hundred seventy-nine single-spaced pages were transcribed verbatim from taped recordings. The study's research questions served to navigate transcript readings (but not to the degree that larger themes became invisible), and a hermeneutic phenomenological theme analysis was performed on data. Inductively, comments from transcripts were arranged according to patterns and themes, with anomalies noted (Glaser & Strauss, 1967). Deductively, researchers worked from the larger body of scholarship to contextualize participants' voices and experiences.

First, researchers read transcripts independently to get a sense of the data and then collaboratively discussed preliminary themes. Inspired by Glaser and Strauss' (1967) grounded theory approach to qualitative data, one researcher used an Excel spreadsheet to record observations and the other used index cards, colored markers, and a Word document to cut/paste emergent patterns/themes. Throughout both steps, researchers used a selective technique of pondering statements and phrases in transcripts that seemed particularly revealing, essential, or remarkable, thereby reducing and consolidating the vast amount of data down to an essence, with proposed theme labels rejected, resurrected, and modified along the way (Van Manen, 1990). In all, multiple data readings took place, until both researchers agreed 100% that data were adequately organized for responding to research questions.

Verification is a valuable, quality standard for qualitative research. Two forms of verification were used (as recommended by Creswell, 2007): (1) a constant comparative technique (Lindlof, 1995) of developing alternative themes to arrive at those most parsimonious, and (2) close attention to anomalies (rather than ignoring them) and integrating them into findings for enhanced understanding and nuance (Miles & Huberman, 1994).

Findings

Among research participants ($N = 77$), 32 were legal and 45 were public relations counselors. About half ($N = 38$) were members of 19 dyads—wherein both the legal and public relations counselor in the same organization were interviewed, separately. The other half ($N = 39$) were not members of a dyad—so that either an organization's legal or public relations counselor was interviewed, but her/his counterpart did not accept our invitation. For legal counselors, the mean age was 49.7, the mean number of years practicing law was 23 years, and the mean number of years working in her/his respective industry was 16.1 years. Twenty-eight were Caucasian/White, 2 were African American/Black, 1 declared an ethnic identity of "Italian," and I declined to answer the question. For public relations counselors, the mean age was 50, the mean number of years practicing public relations was 21.7 years, and the mean number of years working in her/his respective industry was 19.7. Thirty-eight were Caucasian/White, 5 were African American/Black, 1 was "Japanese American," and 1 declared an ethnic identity of "American."

Sixty different organizations were represented among research participants: B2B (N = 15; 3 dyads), B2C (N = 17; 1 dyad), G/R (N = 17; 12 dyads), and NFP (N = 11; 3 dyads).

For the first research question about ways the legal/public relations counselor relationship might be impacted by new media, legal and public relations counselor dyads shared high degrees of anxiety that was somewhat offset by support in the form of cooperation, trust, and respect for their counterpart's expertise.

For example, a G/R global offshore drilling company's public relations counselor told how legal/public relations teamwork plays out in new media messaging: "[W]e're able to pick and choose among messages, the component parts of which have been previously approved [by Legal], to select the right ordering of messages to respond in real time." This organization's legal counselor likewise expressed appreciation for her/his counterpart's media relations know-how, characterizing new media as "a very important tool... [P]eople don't read op-ed pieces in the *Washington Post*, they read Twitter...[T]hey're different arrows to pull out of the same quiver."

A national health-focused NFP legal counselor also described a collegial legal/public relations bond, critical when trying to navigate unchartered new media territory: "[I]t's collaborative...a very easy relationship...I'll go to whatever meetings they let me come to...just try to provide good service without being pushy about it.... One of the things for us as a non-profit that's a little bit different than the for-profit world...relates to cause-marketing deals...we don't want to pay tax. So there's that tension there about what we can and cannot do." The organization's public relations counterpart spoke of a necessity for symbiosis: "I feel like at any point I can pick up the phone or shoot an email to my legal partner and get extremely responsive feedback.... Their need to counsel us on social media has increased ten-fold since two years ago. It seems that everything that I'm reaching out to them about is about Facebook or something we want to do on Twitter."

Amidst positive outcomes described, G/R industry dynamics seem to strain the bond. A G/R city law enforcement public relations counselor explained how s/he has acquiesced to legal perspectives: "[O]utbound messaging only is the proper strategy for us, and I came to that decision based on the advice of counsel...[O]nce it's clear that we're not engaging and there's no two-way communication...I'm in much better stead both from a public records perspective and [for avoiding] a potential problem." A G/R global pharmaceutical corporation's legal counselor described the legal/public relations connection as: "Very close, very close, so first of all, almost anything that they put out, especially if it mentions a product, will be reviewed by Legal." Many regulated environments have not kept pace with the new media growth—as described by her/his public relations counselor counterpart:

> [T]ensions are coming to the fore because there are no defined regulations for the pharmaceutical industry from the federal government...[T]he lawyers are very concerned about risk in those scenarios—particularly with pharmaceuticals because someone cites what is considered a side effect from a drug [and] that triggers a whole reporting process that's under FDA regulations.

From non-dyad counselor interviews, fewer anecdotes of full-on collegiality emerged. Instead, research participants described new media as a potential backdrop to public relations' enhanced organizational status, as well as improving legal/public counselor relationships. A B2B independent public relations consultant suggested that organizations need policies unrestricting public relations: "Social media requires such speed and interaction and relationship building that your policy has to say 'These are the people that we trust to act on our behalf at all times without further approval' so they can tweet to their heart's content and blog." Also, a G/R global healthcare organization's public relations counselor said: "[Y]ou need to build that trust with your Legal team so you as a communicator can move without constantly asking for approval...[T]he Legal team trusts me to act appropriately...[I]f speed is part of the equation...you need to completely transform Legal approvals...[and inspire] a mindset change that our Legal teams need to have."

Alternately, a B2C retail grocery store chain's legal counselor criticized public relations firms: "[They] are still relying on old methods of doing business to respond to crises sparked on social media. If you don't start thinking out of the box and understanding how to handle social media from a real-life perspective, your issues...become company-wide very quickly through social media."

Regarding this study's second research question, about degrees to which legal and public relations counselors hold similar/dissimilar views on benefit-risk associated with organizational use of new media, participants shared comparable outlooks on proceeding with caution since new media are unfettered and ubiquitous.

Among the advantages, both legal and public relations counselors value new media as vehicles for monitoring organizational reputation, message testing, and error correction. A B2B specialty steel company's legal consultant described uses for surveillance: "You don't want employees badmouthing one another, things like that, so we do monitor what's going on out there." Whereas, a G/R Fortune 500 healthcare company public relations counselor shared other benefits of new media tracking: "[W]e are monitoring it constantly, and that does factor into the way we decide to present things to the public. We've got two people who are dedicated to it, and then another couple who work on it on a part-time basis." A leading B2C electronics retailer's public relations counselor also said: "Sometimes new media becomes the way we learn about things that are bubbling up...we use it as a communications tool for straightening out the facts.... We're a very active company when it comes to using Facebook, community forums, Twitter; our CEO has a blog." Similarly, a Christian-based NFP's public relations consultant said: "We have a full-time media monitoring person. We send out a daily media analysis on who's saying what about [organization]...[I]f it's something very negative or critical, we don't typically include it in the daily report. We may take it offline with an email, and a heads-up to a key department."

In addition to benefits, both sets of counselors also identified many new media risks—so much so that some participants admitted that their organization intentionally works slowly and cautiously toward incorporating those vehicles into campaigns. Among the most-noted new media risks were lack of control and negotiating a 24/7 news cycle. A B2C apparel company's public relations counselor said: "[T]oo often, it's personal opinion not based on facts and reality, but as a company, if you are not prepared to deal with it quickly, it can rapidly escalate beyond your control...[B]eing swift and being responsive to all of that is becoming increasingly imperative." Likewise, a B2B professional services firm's legal counselor said: "[T]here's no control, there's no restriction...there's a lot of side noise kicking around in the social media world...[O]ur message can get blurred and perhaps even diluted by some of those social media outlets....We're no longer dealing in a world where the only thing that's going to come out for public consumption is what we put out there." A large B2C manufacturer's public relations counselor viewed a strong legal/public relations counselor relationship as a means to minimize risk:

> The fact that the Library of Congress archives every tweet means that this isn't just something that's passing. It isn't something that can just be erased...I think it's good for us to have a good understanding of regulations as communications people, and not to see the lawyers as an enemy because they will sensitize us to the kinds of risk that are out there.

Among anomalies, a few counselors dismissed new media offhand as irrelevant to their organization—or attributed their lack of enthusiasm for it to a generation gap. For example, a G/R pharmaceutical company public relations consultant revealed: "I think it's overrated...[W]e continue to operate under the existing guidelines that have been established for more traditional media, in terms of promotional materials like brochures and press releases." A G/R major U.S. cable company's legal counselor opined: "I don't think that it's had much of an impact on us day-to-day. I mean to me it seems like there's a relatively small portion of the populace that is constantly Twittering and flittering...I kind of view it as largely as like a bit of a lunatic fringe...[M]aybe I'm just a little too old to take that too seriously." A B2B independent public relations consultant said that public relations' role is to educate organizations so that they can make informed new media benefit-risk analyses:

> [T]here needs to be some focus on the generational aspect...[T]raditionalists are moving out of business now; they're retiring. You have the baby boomers who [are] running the businesses; very anti-social media . . . . Then you have Gen X who hates email and loves social media...the millennials who only want to communicate via mobile devices and social media . . . . So, it's a generational issue.

Conclusion

In sum, this study's findings about—one, ways the legal/public relations counselor relationship is impacted by the new media environment—and two, the degree to which legal and public relations counselors hold similar/dissimilar views on benefit-risk associated with organizational use of new media—make significant inroads toward understanding inter-management dynamics that have fascinated public relations scholars for nearly 45 years.

Assessment of the so-called "oil and water" legal/public relations counselor relationship revealed many similarities and high degrees of cooperation on new media issues. Fitzpatrick (1996) hypothesized that a need for collaboration in organizational decision-making was evolving. Indeed, we posit that uncertainties and fears attached to new media draw together legal and public relations counselors—perhaps like never before. Findings here suggest that outcomes of the relationship pertaining to new media issues default to conservation, as characterized by systems theory's emphasis on homeostasis. Among voices there also resonated acknowledgment of new media's omnipresence and utility, so we present this as further evidence of symmetry-conservation duality. Thematically, three interrelated conservation elements guide decision making among legal/public relations counselors in conjunction with new media—and seem to be strong ties that bind counselors today: a) *reputation preservation*, b) *offensive/defensive monitoring*, and c) *litigation avoidance*.

First, concerns about *reputation preservation* routinely surfaced throughout interviews with both sets of counselors as a common factor when deciding what should/not be done with new media—from agreeing to have organizations opt out of new media two-way communication completely, to using new media to set agendas or to set the record straight, to CEOs directly blogging with publics, to strategically selecting messages for new media delivery in defense of attacks on an organization's good name.

Second, *offensive/defensive monitoring* of new media provides organizations a perceived modicum of control over a 24/7 news cycle that knows no geographic bounds—from surveillance of employees' online activities, to environmental scanning for issues management, to alerts for internal problem solving, to message testing strategies to gauge reception among key publics. Perhaps this theme best exemplifies solace that counselors of both stripes take in an arena where little control is to be had and is consistent with excellence theory's advice that public relations counselors proactively monitor organizations' surroundings to avoid being considered outdated or irrelevant (Dozier, Grunig, & Grunig, 1995).

Third, *litigation avoidance* exercised by both legal and public relations counselors thematically guides all new media decisions—as violations of laws and regulatory agency policy compliance (including tax rules) are strict thresholds and representative of conservation in a *traditional legal strategy* (Fitzpatrick & Rubin, 1995) sense.

Findings also position new media as a lens through which to view the contingency model at work in organizations—with its continuum of pure *asymmetry* at one end and pure *cooperation* at the other—flanking a desired win-win zone at the center where organizations and publics stand to benefit equitably (Grunig, Grunig, & Dozier, 1996). We suggest that public relations counselors conservatively err on the side of caution nearer to the pure *asymmetry* end of the continuum, where public relations primarily benefits the organization. When it comes to new media, public relations counselors are challenged to find the sweet spot between too much risk and not enough benefit. Future studies are needed to further examine this dynamic.

Finally, implications of such high degrees of anxiety/caution and conservatism's potentially stifling effects on public relations creativity and autonomy require greater scrutiny to ensure that the function does not become entirely encroached upon by lawyers. Moreover, public relations consultants must seize opportunities via new media to enhance credibility of the function, resulting in greater trust and latitude in the highest levels of organizational decision making.

Despite this study's valuable findings, method-inherent limitations must be noted—including recognition that comments expressed by participants are not generalizable and instead are a point of departure for future study. In addition, when findings are published, the act of generically identifying participants by industry risks assertions among readers that voices are complete and universal.

Using Reverse and Traditional Mentoring to Develop New Media Skills and Maintain Social Order in the Public Relations Workplace

Betsy A. Hays
Douglas J. Swanson

New media understanding and technical skill are critical to the success of any public relations professional. Many seasoned professionals have expressed anxiety over their perceived lack of technical ability. Younger, entry-level professionals are typically proficient with new media but lack the knowledge of how to interact with older professionals who represent earlier generational groups. As a result, two distinct types of conflict occur within the social order of the public relations workplace. The appropriate use of traditional and reverse mentoring allows many of the conflicts to be avoided. An active mentorship program between entry-level workers and seasoned professionals allows the entry-level workers to understand and successfully negotiate a workplace where leadership has contrasting value and reward systems. Likewise, a reverse mentoring program with younger workers tutoring seasoned professionals allows a collegial transfer of technological skills that might not otherwise occur. Our chapter offers recommendations based upon the tenets of social order: division of labor, construction of trust and solidarity, regulation of power, and legitimization of social activity among humans (Eisenstadt, 1992). Throughout, we include real world practice examples in the form of "Pro Tips" that illustrate important concepts. These "Pro Tips" were developed via a series of in-depth interviews with public relations professionals who have engaged in traditional and reverse mentoring.

Times of uncertainty and change in public relations

In the media workplace, rapid technological change has resulted in a faster, broader shift to new communication technologies than has been previously witnessed in human history. This shift is especially pronounced in the public relations profession, where workers at all levels of the hierarchy must quickly acquire and learn to use new knowledge and skills. Public relations professionals of the 21st century work in a fast-paced, competitive environment. Their strategic planning, writing, and analytical skills must constantly be updated because social and economic pressures force constant adaptation to meet the changing expectations of clients and publics. Public relations professionals who do not quickly adapt to the new workplace expectations can quickly find themselves unemployed with no way to re-enter the profession. In the most recent economic downturn, it was not uncommon for established senior public relations executives who thought they had stable jobs to find themselves suddenly unemployed (Sweeney, 2010; Woloshin, 2009). Many of those put out of work learned too late that one of the most important contributors to job security in public relations is possessing the social media skills that allow someone to be perceived as competitive in the marketplace. Public relations professionals must have the ability to "actively engage in social media in a pragmatic, controlled and open manner" (Cahill, 2009, p. 26). This necessitates that public relations practitioners understand not only the use of new and emerging media, but also have the ability to use media tools in ways that were formerly more associated with advertising and marketing.

Pro Tip #1: *Always treat others with respect, regardless of age differences*

"Being respectful is so important," said Tori Randolph, public relations/social media specialist for Everloop.com, recent college graduate, and former social media specialist for 5 Rockets, Inc. "It's a delicate situation, teaching something this important to people who have been practicing PR longer than you've been alive," she said. To avoid this, Randolph makes sure that she is sensitive to the fact that it might be uncomfortable for more seasoned professionals to accept knowledge from those younger than they are. She uses a "soft approach" and makes sure she doesn't come across as abrupt.

Social media skews traditional expectations

The emergence of social media with its inherent links to advertising, marketing, and promotion has erased the traditional perception that public relations doesn't deal with consumer-related issues. In fact, just the opposite is the new reality. A 2009 survey by Korn/Ferry and the Public Relations Society of America found that public relations professionals "are taking the lead over marketing and other departments in managing an organization's use of social media channels" (Digital Readiness Report, 2009, para. 4). The survey concluded that social networking, blogging, and microblogging may be even more important in public relations than traditional media relations skills.

A public relations professional's lack of understanding of the appropriate social media "action" or "response" to take on behalf of a client can be devastating. There have been numerous examples of this in recent years.

One of the most prominent is the 2009 case involving Domino's Pizza, in which employees in a North Carolina store performed unsanitary acts with food and then uploaded a video of their kitchen hijinks to YouTube (Jacques, 2009). Domino's corporate public relations team took several days to react to the incident and was later criticized in the news media and in social media communities for that slow reaction. Company executives later acknowledged that they had a less "aggressive stance" (Jacques, 2009, para. 14) than would have been ideal.

Pro Tip #2: *Speak management's language*

When working to "sell" the value of social media to others more senior, a proven success strategy is to speak the language of your supervisor. Lisa Alvey, social media manager for Fresno Pacific

University, attributes her continued success to her ability to show the value and relevance of her work to the mission of her organization. She enables her bosses to "get it" by relating "followers" and "fans" to the achievement of the department's communication goals. As a result, the higher-ups have developed tremendous trust and faith in her ability to handle the function, and they advocate on her behalf regarding its value.

Impact of staff restructuring/downsizing

Expectations of public relations professionals have shifted as a result of organizational and personnel changes within the workplace. Although there have been significant number of high-profile acquisitions and mergers in the public relations industry (Gofton, 2000), there has also been some dramatic restructuring within organizations ("A Dynamic Industry in a Turbulent Economy," 2001; Jacobson, 2000; Paulden, 2006). The public relations labor force grew significantly during the 1960s and 70s. As the 21st century began, many of those workers were nearing retirement age. One estimate showed that one in four U.S. workers was nearing retirement in 2009 (Miller, 2009). As these experienced workers leave the workplace, a great deal of experience and knowledge is lost in their organizations. As a result of emerging new media, technological change, organizational restructuring, and retirement of experienced workers—many public relations professionals today feel a high level of workplace-related anxiety. Even seasoned professionals are not always certain about what they are expected to do, and how to do it. Some of this anxiety can be lessened by thoughtful consideration of how social order manifests itself in the public relations workplace.

Pro Tip #3: *Don't hide from change—seek it out*

Public relations and marketing entrepreneur Gregg Champion epitomizes the new public relations mindset of bravely seeking out change, and encouraging experimentation among employees. "All of my employees are 20-somethings," said Champion, who owns Champion Media & Entertainment in Santa Monica, California. "The relationships are perfect," he explains. "I provide them with an opportunity to work with cool brands via clients I obtain through my strategic experience and expertise, and they make sure I am up to speed on everything cool in pop culture and everything new in social media. We learn from each other each and every day." Champion even has a scheduled time each week for the "Champion Media Think Tank" where he and his employees/interns brainstorm ideas about clients and projects. "The best idea wins," he said. "It doesn't matter what your title is."

Social order and its relevance

Social order refers to a theoretical framework that seeks to identify predictable, coordinated actions in a place where humans live and work. Over the years, theorists and researchers have concluded that the socially ordered environment is one of predictability and coordinated action. A socially ordered workplace manifests itself through a division of labor, an establishment of trust among people, a regulation of power for decision-making, and a set of systems though which social activity is made legitimate (Cowan, 1997; Eisenstadt, 1992). Written and unwritten rules establish and maintain social order (Edgerton, 1985). Social order is demonstrated directly and indirectly through culture, which can be defined as "an organized set of meaningfully understood symbolic patterns" (Alexander, 1992, p. 295).

Pro Tip #4: *In reverse mentoring, culture is king*

Public relations managers should establish a culture of learning that is not just top-down, but bottom-up and side-to-side. That's the strategy used by Katie Johnson, former director of social media relations for California State University, Fresno. When she hired a new intern or student assistant, she explains early on that she expected to learn from them as well as to teach them. Johnson was careful, however, to let the students who worked for her know that she wasn't necessarily going to use *all* of their ideas, just the ones that are the best strategically. "This sets up the relationship early-on to be collaborative and not hierarchical," said Johnson. In addition, Johnson and her staff played games on Twitter every Monday from the book "Caffeine for the Creative Mind" to keep everyone fresh and energized. The book authors then played with them as well.

At the most basic level, an acceptance of social order as a guiding construct in the public relations workplace tells us that nothing happens by accident. Every choice that public relations professionals make about their conduct on the job communicates something about the social order of their workplace. In public relations, the ideal socially ordered environment is one in which people agree to communicate well and work productively together, no matter what economic or technological uncertainties may exist. Establishment of mentoring relationships can support social order while easing the anxiety felt by professionals Not all public relations professionals are prepared for new organizational and technological demands at work. As a result, many public relations professionals have turned to mentoring as a way to get "up to speed" on expectations of the 21st century workplace.

Generally defined, workplace mentoring is any situation in which "a mentor helps a protégé or mentee become more professionally competent" (Cotugna & Vickery, 1998, p. 1166). Mentoring can involve general business or organizational understandings, as well as specific skills needed for the employee to complete tasks or be more valuable within the organization. In addition to the potential benefit for the individual employee participants, mentoring can have value for the organization itself. Mentoring can result in a situation where the organization is more in tune with the needs of its customers (Parekh, 2007).

Corporate executives in particular need to seek interactions—through mentoring and other interpersonal communication—so that executives can obtain different perspectives that "offer support for managing change" (Carter, 2004, p. 85).

Pro Tip #5: *When mentoring, start with clear goals*

In order for any type of mentoring to be successful, both parties should establish concrete goals at the beginning of the relationships. Whether the situation is formal or informal, each person should identify and articulate what they want to achieve. This keeps the relationship focused, according to Janelle Guthrie, communications director for the Washington State Attorney General's Office, who has participated in a formal mentoring program between the Public Relations Society of America and the Public Relations Student Society of America chapters. Guthrie has formally mentored four PRSSA students and informally mentored many others. She believes this type of goal setting is one of the most important things to discuss at your first meeting with a mentee. "This way everyone knows where the relationship is headed and what it is trying to accomplish," said Guthrie. "Understanding each other's goals keeps everyone on task and focused on what is most important."

Reverse mentoring is a concept initially introduced by former General Electric CEO Jack Welch (Greengard, 2002). Reverse mentoring involves a structured workplace relationship between senior staff members and younger/less experienced workers. Typically, the younger workers have less expertise within the organization but more technological familiarity and skills. The pairing of these workers brings about the education of "older folks who can't figure out technology" (Pyle, 2005, p. 40). Writer Rupal Parekh (2007), in an article profiling Unilever's reverse mentoring program, referred to mentees as "trendslators" (p. 3) because they translate trends and related technology for senior staff.

Technological skill development is not always the focus of reverse mentoring, however. At Procter & Gamble, CIO Steve David participated in a reverse mentoring program in which a dozen P&G scientists taught top managers about the ethical implications of biotechnology. David (as cited in Solomon, 2001) said the experience made him "much more knowledgeable, [and] much more able to address with our customers and our suppliers the issues associated with this bio tech revolution" (p. 41).

Pro Tip #6: *Practice good time management*

Time management is a challenge in any area of business, but when working with social media it can be especially perilous. Lisa Alvey, social media manager for Fresno Pacific University, recommends that all those newly in charge of social media incorporate strict strategies into their workday so they escape the temptation to get lost on the Internet. "It's easy to lose an hour clicking if you are not careful," said Alvey. She employs a loose formula and timeline each day that enables her to spend time on all of her priorities.

Mentoring can build a bridge between younger workers and more seasoned professionals

Everyone's workplace perceptions are influenced to some degree by their generational experiences. There are at least four different generational groups represented in today's workplace, even though there is some disagreement about the exact date range for each group. These recognized workplace groups are: The Silent Generation (born between 1920 and the end of World War II), the Baby Boom Generation (born during the prosperous post-World War II years), Generation X (born between the mid-1960s and the early 80s), and the Millennial Generation (born between the mid-1980s and late 90s) (see Evans, Schmalz, Gainer, & Snider, 2010).

Each of these generational groups has different perceptions of the workplace, different expectations of reward from work, and different understandings of how technology plays a role in task completion.

Though experts don't always agree on all of the differences between generational groups, there are some widely accepted generalities. For example, members of the Millennial Generation tend to have moderate expectations at the outset of their careers, but seek rapid advancement and a strong career/life balance (Ng, Schweitzer, & Lyons, 2010). Millennials tend to be strong team players, but often work best if information is "cut into bits of what they need to know" (Abaffy, 2011, para. 17) and delivered on a schedule of when they need to know it.

Baby Boomers, on the other hand, tend to view their work and careers with a more long-term perspective (Oblinger, 2003). A Boomer wants to see the big picture, but can also easily grow disenchanted about career progress—and that feeling of disenchantment can quickly grow to include personal life issues (Wright, 2005).

Pro Tip #7: *Slow the pace and teach the vocabulary*

When you're working with people of different generational groups, don't move too fast into new concepts. This advice comes from Kent H. Moore, owner of CYC Public Relations and www.

oneeyedblog.net. Moore first became a reverse mentor when he was asked by the Entertainment Publicists Professional Society (Los Angeles Chapter) to help the organization get involved in social media. "I think they first asked me because I was the youngest one in the group," he said, "but it worked because I also had the knowledge." Moore quickly found that moving slower than his normal pace was critical to his success. "I discovered that I couldn't move through things as fast as I normally do," he said. "I needed to take each step one at a time, bit by bit, and walk them through it. I also assumed that 'everybody' knew the necessary vocabulary but that wasn't the case as well. Once I made it a point to explain all of the concepts as we came to them, things went very smoothly."

Clearly, in a socially ordered public relations workplace, the astute executive would seek to balance the breadth of knowledge of senior practitioners with the technological expertise and change-friendly perspective of the younger generation. A reverse mentoring program that does this will help develop opportunities for workers of different generations to contribute strongly to organizational success.

Establishing mentoring relationships to enhance predictability and control in a socially ordered workplace

Although researchers in the business and health professions have shown a lot of interest in mentoring, the subject has not received too much attention among public relations scholars. Even less attention has been given to reverse mentoring, which is troubling for the public relations profession—given the increasing importance of knowledge and skill transfer relating to new technology. Recent research has found many public relations professionals either do not know what reverse mentoring is, or have not been involved with it.

A 2011 survey of public relations practitioners in the U.S. and Canada found that almost 60% of respondents had never heard of reverse mentoring. Fewer than 25% of respondents said reverse mentoring was practiced at their place of employment. Most of the responding professionals who worked where reverse mentoring was practiced said there was no formal structure to measure outcomes or assess the success of the effort (Hays & Swanson, 2011).

Even though reverse mentoring is an emerging concept in the public relations workplace, we can still look to existing literature and make some sound generalizations about the best practices.

Mentorship results from a structured plan that is adaptable. Any organization that wants to advance its goals and improve the knowledge and skills of its people needs to do so with a structured, systematic plan for mentorship. At the same time, because the root of reverse mentorship in particular is personal relationships, everything contained within a mentorship plan needs to be adaptable to fit the personalities and learning styles of the participants. It is important to realize that attitudinal similarity and even the gender of participants have an impact on mentorship outcomes (Avery, Tonidandel, & Phillips, 2008).

Mentorship is about relationships more so than tasks. Mentorship allows for the development of "personal influence" that helps workers at all levels feel as if they are contributing to the success of the workplace. Employees who feel "in the know" (White, Vane, & Stafford, 2010, p. 80) are more likely to feel respected and appreciated at work. Reverse mentoring in particular allows for this knowledge and skill transfer between different generations of workers, and also serves as a way for less experienced workers to learn what is needed to make wise career decisions (Peroune, 2007). Development of successful mentoring relationships involves sensitivity and empathy, two key components of effective leadership (Jin, 2010).

Mentorship requires a complete organizational commitment. How mentoring is viewed in the organization has a direct impact on how it is practiced and experienced (Lai, 2010). Everyone involved either directly or indirectly in mentoring needs to have shared expectations for the experience and its individual and collective goals. Leadership consultant Margaret Wheatley (2008) reminds us that uncertain situations "require people to access their maximum intelligence, to be able to think well in the moment and alongside their colleagues" (p. 45). Knowledge gained from these experiences needs to be widely disseminated throughout the organization.

Mentorship must take cultural considerations into account. As workplace expectations change and organizations become more internationally and interculturally focused, it is important to remember that mentorship should focus outward as well as inward. Mentorship is not so much about "how we work" as much as it should be about "how we interact with others to get work done" (see Kim, Kwak, & Yun, 2010).

Mentorship works only when relationships are honest and open. Mentorship partners allow for the development of accountability and perspective. "An honest partner keeps you from getting in your own way. This is a favor even the brightest of us can't do for ourselves," Kepcher notes (2011, para. 4). At the same time, the honesty that comes about through mentorship must address issues of "how" as well as issues of "why" we do certain tasks. Workers need to know what actions to take and why to take them. Theoretical and practical understandings must work together. Rhodes, Liang, and Spencer (2009) argued that mentorship programs that do not encompass the ethical dimensions of work could potentially do more harm than good.

Mentorship has been proven to work individually and collectively. A network of relationships can be formed through mentoring at the individual and group level. This can bring about enhanced leadership ability that goes beyond the initial performance objectives of the mentorship (Hall & Jaugietis, 2011; Leh, 2005). Remember, too, that mentorship doesn't just happen at work. Individuals can gain workplace-applicable knowledge and skills from mentors in a variety of settings in life (Corney & du Plessis, 2010).

Mentorship's successes and failures must be accounted for. No public relations professional would conduct a communications effort that didn't measure "success." The quantification of success (as well as the determination of what did not work well) leads to the educated evaluation of communication impact. The same is true for mentorship relationships, which are essentially communication efforts showing employees how to create order and productivity in uncertain times. While measurement of success is critical, we must be wary of instituting too much structure in the process. In a powerful article offering guidance to nonprofits in difficult economic times, Wheatley (2008) reminds us that "every time we attempt to control chaos with controls and oversight, we create only more chaos" (p. 45). In other words, assessing our successes is important, but excessive focus on yesterday's accomplishments can limit our ability to see the work that needs to be done today.

Conclusion

Public relations agencies and other organizations in the profession need to make sure that we update our knowledge and skills fast enough—that our customers are not "out front" with more knowledge than we have (see Dozier, 2002). Mentoring and reverse mentoring allow for this to happen.

Maximizing success takes both commitment and sensitivity. It requires sensitivity to social order in the workplace and the inherent complexities of non-traditional mentor/mentee relationships. Savvy public relations professionals—those who are seasoned as well as those who are new to the industry— recognize that learning comes from not only top-down but bottom-up and side-to-side. Utilizing different forms of mentoring, both reverse and traditional, will allow practitioners to arm themselves with the knowledge, skills, and abilities needed to thrive in today's dynamic environment.

The Pro Tips and best practices in this chapter will enable you to do just that.

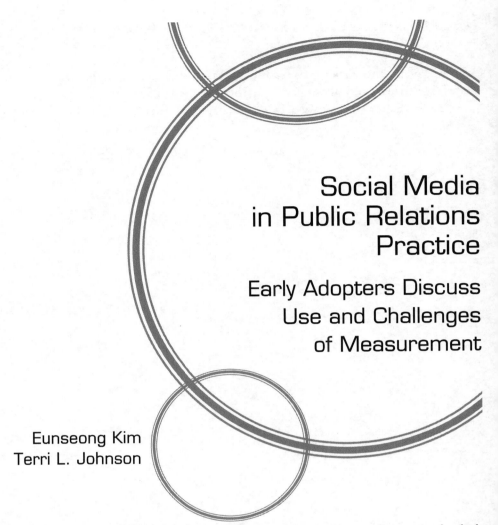

Social Media in Public Relations Practice

Early Adopters Discuss Use and Challenges of Measurement

Eunseong Kim
Terri L. Johnson

Social media, the newest tool in the public relations toolkit, have many proponents across the practice, but little is known about how well social media campaigns work. Interviews with some of the active proponents of social media use reveal that as with any tool, the effectiveness depends on the goals of the campaign. Measuring the effectiveness may be a challenge, but all agree that social media use is effective and that social media will grow in importance. Measurement techniques are evolving as social media use develops, and they are important in demonstrating that time and money spent on social media efforts are, indeed, well spent.

Introduction

The popularity of social media in the Web 2.0 era is strikingly apparent among Internet users. Based on a survey conducted in 2009, the Pew Research Center reported that 73% of online American teens and 72% of online young adults aged 18–29 years old use social networking websites (Lenhart, Purcell, Smith, & Zickuhr, 2010). In a more recent survey, the Pew Research Center found that 47% of Internet users ages 50–64 and 26% of users age 65 and older use social networking sites (Madden, 2010). In other words, the popularity of social media exists across all age groups and is apparently a phenomenon that is here to stay.

Responding to the changing communication environment of Web 2.0 is essential to all communications professionals, and public relations practitioners have been including new media technologies in their tool kits. From early social media, such as email and websites, to social networking sites including blogs and wikis, to the currently most popular of social media sites—Facebook and Twitter—public relations practitioners are adopting and using these tools in a big hurry. A 2010 FedEx/Ketchum social media benchmarking study of social media use in leading companies such as PepsiCo, GE, and P&G found that 100% of participants were using social media to communicate with their publics regardless of their industry and that they plan to redesign their intranets in the next one to two years to include greater social media capabilities (Dworkin, 2011).

Although it is clear that public relations practitioners recognize the popularity and significance of social media as an additional way to reach their publics, it is not clear how they use social media and how they measure its effectiveness. This study reports the findings of in-depth interviews with practitioners who specialize in social media implementation within their organizations or with their clients.

Public relations and Web 2.0

The Pew Research Center explained "Web 2.0" as "an umbrella term that is used to refer to a new era of Web-enabled applications that are built around user-generated or user-manipulated content, such as wikis, blogs, podcasts, and social networking sites" ("Research on Web 2.0," 2011). Tim O'Reilly and John Battelle first articulated the term in 2004 by emphasizing the significance of user participation in the Web 2.0 era, and since then concepts such as user-generated content, openness, and social networking have become key characteristics of the World Wide Web's new era (O'Reilly & Battelle, 2009).

Advances in media technology during the past few years have certainly encouraged this trend. The ease of use of various social network sites to create, disseminate, and share information with others attracted Internet users to these sites, and as of 2009, 46% of online American adults age 18 and older used a social networking site like Myspace, Facebook, or LinkedIn (Lenhart, 2009). The study reported Bebo, Last.FM, Digg, Blackplanet, Orkut, Hi5, Match.com, YouTube, Flickr, and Tagged as additional social networking sites that Internet users use for online social interaction.

To public relations practitioners, this trend has massive significance at several levels. First, Web 2.0 and social media have expanded the parameters of communication and provided public relations practitioners with a wide range of options for reaching their publics. Many practitioners welcome the potential of social media, as these applications enable them to directly communicate with their audiences without going through traditional media gatekeepers (Gillin, 2008). At the same time, Web 2.0 technologies enable audiences to become active participants in the communication process, and this challenges public relations practitioners to re-think their relationships with their audiences. Public relations practitioners have to give up control over their messages and allow audience feedback and participation while determining the most effective way to deliver the core messages (Solis & Breakenridge, 2009).

Second, as user participation is one of the most important characteristics in the Web 2.0 era, public relations practitioners have to work with new influencers and the information created by them. As Solis and Breakenridge (2009) explained, "social media and Web 2.0 are altering the entire media landscape, placing the power of influence in the hands of regular people with expertise, opinions, and the drive and passion to share those opinions" (p. 1). Indeed, audiences in the Web 2.0 era are no longer passive recipients of practitioners' messages, but instead are active collaborators who create and share their own content in their communities. In this environment, Solis and Breakenridge explained, the new goal of public relations (PR 2.0) should be "to understand the communities of people we want to reach and how to engage them in conversations without marketing at them" (p. 38).

Third, public relations practitioners need to reconsider traditional approaches to relationship building. In the social media environment, relationship building is made both easier and more difficult. Social media make it easier to build relationships through the ease of communication and accessibility, but it is more difficult because of the huge numbers of audiences that practitioners may be interacting with, and because of the constant demand for attention. Additionally, the relationship between the organization and its stakeholders in the Web 2.0 era should be based on two-way symmetrical communication—the concept known as an ideal mode of communication (Grunig, 1992). When the communication between the organization and its stakeholders is two-way symmetrical, the organization listens as much as it speaks, and the communication is a dialogue, not just a monologue. Social media make this dialogue possible and ensure more two-way communication than was allowed by traditional media. When stakeholders participate in communication, they are engaged in dialogues, and a sense of loyalty and trust can be developed (Solis & Breakenridge, 2009). Engagement, trust, and loyalty are key concepts that strengthen the relationship between the organization and its key publics in the Web 2.0 era.

Use of social media in public relations

A wide variety of social media applications are available to public relations practitioners today. Some of the communication venues available to a typical public relations practitioner include: discussion forums, message boards, blogs, podcasts, vodcasts, Real Simple Syndication (RSS), photo sharing, audio and video sharing, search engine marketing, wikis, social networks (e.g., Facebook), professional networks (e.g., LinkedIn), and micro-blogging (e.g., Twitter), to name only a few.

Although social media applications are abundantly available to practitioners, previous studies have shown that practitioners' adoption and testing of social media tools and their switchovers to other social media happen quite quickly—at least among some practitioners. For example, Eyrich, Padman, and Sweetser reported in 2008 that practitioners had widely adopted such media as email and intranet, but they were slower to adopt such tools as text messaging, social networks, and virtual worlds.

Only one year later, however, McCorkindale (2009) reported that, based on a content analysis conducted in February and March of 2009, "most companies in the Fortune 50 are taking advantage of the opportunities of Facebook" (p. 352) although these companies did not always utilize the medium to its fullest extent. Wright and Hinson's (2010b) two-year longitudinal analysis of public relations practitioners also found that practitioners considered the social networking site Facebook to be the most important new communications medium for their practice (as of 2010), followed by Twitter, LinkedIn, and YouTube. On the other hand, Myspace, once a massively popular social network site, is no longer considered to be a major competitor of Facebook, as Facebook surpassed Myspace in popularity in April 2008 (Wright & Hinson, 2009b). In 2010, Facebook reached 500 million users worldwide (Mui & Whoriskey, 2010), claiming more than twice as many active users than Myspace (Wright & Hinson, 2010b).

Other studies have shown that public relations practitioners: recognize the significance of social media in their practice (Briones, Kuch, Liu, & Jin, 2011; Wright & Hinson, 2009b, 2010b), try to use social media to engage their stakeholders in conversation and to build relationships with them (Briones et al., 2011; Fitch, 2009), and are in the process of determining the best way to implement social media for their organization's best interest (Fitch, 2009; Stelzner, 2011; Taylor & Kent, 2010).

Effectiveness of social media in public relations practice

The measurement of public relations campaign effectiveness has often been discussed both in academia and industry, but it has always remained a difficult task throughout public relations history. With advocacy as the focus of public relations campaigns, many practitioners agree that effects or outcomes of public relations campaigns are hard to measure or quantify (Solis & Breakenridge, 2009). As Return on Investment (ROI) has become a popular term in discussing the effectiveness of public relations campaigns, research has found that publicity measurement is often the norm (e.g., media clipping, printed or online media coverage, the reach of publications via circulation numbers). The Advertising Value Equivalence (AVE) is still a popular measure when evaluating the effectiveness of public relations, in spite of efforts to highlight problems with it (Solis & Breakenridge, 2009; Watson, 2011).

Recent adoption of social media in public relations practice seems to have made the measurement of public relations effectiveness even more difficult. It seems reasonable to argue that public relations practice in the social media environment differs significantly from the practice in the traditional media or Web 1.0 environment, and thus the effectiveness of public relations should be defined and measured differently in the social media environment. In their book, *Putting the Public Back in Public Relations*, Brian Solis and Deirdre Breakenridge (2009) explained the changed definition of and measurement methods for public relations effectiveness in the social media environment:

> "Everyday people" with access to Social Media tools, in addition to traditional experts, contribute to the public definition and perception of a brand. Perception equals the sum of all conversations in the Social Web, and those who participate steer its definition… PR 2.0 favors engagement more than hits, referrals more than eyeballs, activity more than ad value, sales more than mentions, and market and behavioral influences more than the weight and girth of clip books. (p. 249)

Industry experts have also argued that old metrics of online ROI such as the number of unique visitors, page views, and cost per click, may not work for measuring public relations effectiveness because engagement and participation are vastly important in social media (Fisher, 2009). Rather, many suggest that concepts such as engagement, tone, authority, attention, interaction, relationship, and participation should be measured (Fisher, 2009; Solis & Breakenridge, 2009).

Although industry experts are making recommendations as to what *should* be measured to gauge the effectiveness of public relations campaigns in social media, recent studies indicate that public relations practitioners are not actively pursuing the measurement of social media ROI. Briones et al. (2011) found that the American Red Cross used social media such as Twitter and Facebook to communicate with key publics, but practitioners at the organization failed to create two-way, dialogic relationships with their publics because of a lack of resources such as time and staff and because of difficulties in convincing chapter or board members of the need for social media. After examining one year of issues of *Public Relations Tactics*, PRSA's monthly newsletter, Taylor and Kent (2010) concluded that the articles that specifically addressed social media topics often included claims about the power of social media, but they did not provide specific evidence about the outcomes of social media. The difficulty of measurement of social media ROI seems to be a universal issue. Interviews with Singaporean public relations practitioners revealed:

> They [practitioners] do little to evaluate the perceptions and emotional dimensions in terms of engagement, support or opposition for an issue or organisation. At this stage, it is the ability or capacity to "start conversations," engage users and participate in social media, ensuring an online presence for an organisation or client, that is valued. (Fitch, 2009, p. 11)

Research questions and methods

As social media built its popularity in such a short time, public relations practitioners, at the moment, appear too busy establishing their presence in social media platforms to think much about how best to use social media in their practices and how to evaluate its effectiveness. Through in-depth interviews with 10 social media experts, we sought to gauge three aspects of social media effectiveness: How practitioners view their goals and objectives when using social media, where in their practices practitioners find social media effective, and what measures are taken to evaluate social media's effectiveness.

As public relations agencies and in-house public relations departments often have digital teams who take a large portion of responsibility for using social media, a purposive sampling was a logical choice for this study (Patton, 1990). The sample included a variety of industry experts representing a wide range of geographical areas. They also represented a range of practice types—independent practices, agencies, and in-house public relations departments.

Findings

Several themes run through responses to the inquiries about social media use. Overriding those is the idea that public relations practitioners must become savvy with social media to compete in today's communications environment. Social media have become a critical tool in practitioners' toolkits, and various interviews with practitioners demonstrated the importance of social media. Every practitioner who responded to questions about the importance of social media noted how it has grown in importance and predicted that it will continue to do so. All respondents predicted that social media will continue to change and evolve, and that practitioners will always be scurrying to keep up, while stressing, however, that social media are a supplement or complement to traditional communications methods rather than a replacement.

Social media have become a tool of authenticity. As often as not, the stakeholders start the conversation. Responses to questions and concerns are only effective when they are seen as transparent and forthright. The responses can be instantaneous, as in direct messaging, or constant, as in a series of tweets, but they must be seen as responsive and credible. When asked how long they have used social media, most practitioners reported using it for two to four years through several social media interfaces.

Three areas seemed most important to the discussion of social media use. First, practitioners were goal-oriented in their use of social media, as they looked for varied but specific effects for themselves and their organizations. Second, they each believed that social media are very effective as a tool in reaching their goals for social media campaigns. Third, each practitioner believed that the results need to be measured, but they agreed that measurements can be elusive and often depend on what goals they are seeking at the time.

Social media objectives: To supplement, not replace, traditional media

When respondents talked about goals and objectives, all emphasized that social media provide public relations practitioners with additional channels to reach key publics rather than replace traditional venues they have used in the past. The comment by Rob Logert, Senior Digital Media Strategist at M Booth, a public relations agency based in New York, reflected this point: "We are trying to use it as another communication channel for our clients. It is a tool, not the 'be all, end all.' It is a means to connect with customers and shareholders in an authentic way."

Respondents were well aware that the strength of social media is user engagement and participation, and their comments indicated they were working to promote these characteristics. Deborah Fleischer at the Centre for Sustainability Leadership in Melbourne, Australia said that social media are all about "two-way communication. Interactivity is key to engagement. Effective social media use not only provides information but invites response." The response by Shonali Burke, an independent public relations consultant based in Washington, D.C., reflected Fleischer's point much the same way: "To broaden one's reach, to build community, to build relationships ...to establish thought-leadership."

While focusing on audience engagement and participation through social media, some practitioners pursued further ways to meet their organizations' goals. Jon Greer, managing director at a California-based consulting company, noted that various social media work together in creating social media synergy that supports the organization with a wide variety of audiences. He defined his use of social media interaction as "a method of content marketing, or getting important, relevant and useful content out to as many people as possible, in as many ways as possible." Stephanie Wonderlin, who is best known for her YouTube channel, Tweetheart TV, the first interactive social media show, said that the company she works for, 44Doors, uses social media to acquire new customers. The company encourages employees to build "personal brands" and use their networks to find new customers.

Social media effectiveness: Versatile in many public relations practices

Though all respondents agreed that social media are effective tools, they noted that social media are more effective in some areas—such as internal, community, and media relations—than others.

Internally, public relations practitioners used social media to communicate news and ideas efficiently. Alison Hamer, senior account executive at L.C. Williams & Associates in Chicago said, "We use Twitter Handle, LinkedIn, and Facebook for internal communication. We use Twitter Handle for talking about clients and industry related news, LinkedIn for straightforward and basic information, and Facebook for one-way communication, information, or fun-aspect of working here." Matt Kelly, public affairs social media specialist at State Farm Insurance, explained the effectiveness of social media in both internal and external relations in his company:

> Internally, we have a daily communication vehicle called the "News Hub." It works like a traditional newsroom, with a daily budget meeting, beats and the like. That meeting and communication channel have transformed from a traditional content meeting into a blend of social and traditional. Everything we share internally has a social media angle to it—whether that means allowing employees to comment on stories, or re-purposing internal stories via external channels. State Farm also enables all employees to use an internal social media platform. This platform offers users the opportunity to connect based on shared interest, departments, existing projects and other commonalities. It's tied directly into current enterprise tools and is supported at all levels of the organization.

Respondents' comments made clear that public relations practitioners use social media heavily to communicate with their current and potential publics. Wonderlin stated, "Social media gives companies such a brilliant way to stay in touch with their current clients as well as finding new ones. It isn't anything new, but the technology allows relationships to be taken further than they could before—and faster." Longert added that he finds social media most effective in customer service, innovation, and community building. Burke also touted the effectiveness of social media in solving problems and reaching out to people.

Media relations is another area where public relations practitioners routinely use social media. Respondents noted that they use Twitter for pitching to reporters, monitoring what reporters and partners say and responding as necessary, posting stories, and for research. They also use Facebook pages to search for and research relevant organizations.

However, none of the respondents participating in this study said that they use social media in financial relations.

Social media measurement: Not yet established, but on the way

In measuring the effectiveness of social media campaigns, some practitioners noted that they used Web metrics such as numbers of followers, retweets, page views, likes, comments, active users, monthly views, and other numbers available from social media pages. However, respondents acknowledged that there isn't any industry standard to measure social media effectiveness. Just as Taylor and Kent (2010), we found that practitioners were excited about the potential of social media, but they did not necessarily have or use ways to determine where and why social media prove effective. This sentiment was reflected in Hamer's comment: "Current methods of measurement might be inaccurate. Interaction should be measured to track users' engagement and figure out they're better brand advocates. Measuring commenting, sentiment, tone, loyalty will give us a lot of information." Amy Hodges, media relations coordinator at Millikin University, also struggled with this dilemma: "When you engage community, how do you put a dollar value on it? Instead of ROI, we need to think about measuring ROE, Return on Engagement."

Although the public relations industry has not established the standards to measure social media effectiveness, it is clear that it won't be too long in coming. Longert said, "There isn't a [social media measurement] standard because there is a lot to measure and… the measurement of ROI must map back to specific goals of the campaign or the ongoing strategy, so it is different for everyone." Kelly agreed that there is no universal way to measure social media effectiveness because of the ad hoc nature of media. However, he also argued that evaluating social media campaigns does not have to be completely different from evaluating traditional public relations campaigns. He explained:

> Practitioners would go through the same steps of research, strategy, implementation, and evaluation process. But in each step, make sure to think about social media strategies. Sometimes, social media might not even be the right choice for the public you wish to reach. But when it is, there are many tools and free services you can use to strategize your campaign. If you want to increase social media impressions 50 percent by 1st quarter 2011, for example, you need to establish where the program is currently. For existing programs, it's easy to use an online measurement tool like "Compete" to gather approximate unique visitor figures for any given Web site mentioning your program. The free version of the service gives you figures back one year. If a current search for the site does not exist, you'll have to create one. It takes a month to aggregate new data. If this is the case, begin anyway. Relationship building cannot start too soon.

Fleischer suggested that Google Analytics can be an important measurement tool: "SM [social media] is highly measurable. Just as all good objectives are measurable, plan your social media objectives so they can be evaluated. You can use Google Analytics as a key way to monitor your website hits following both launch and completion of a SM campaign."

Social media practitioners agreed with Katie Delahaye Paine (2011), a leader in public relations measurement: "The future of public relations lies in the development of relationships, and the future of measurement lies in the accurate analysis of those relationships" (p. 219). Smart communicators are already pushing beyond measuring outputs and outtakes and learning to measure the feelings, perceptions, and relationships that they generate.

Conclusion

Social media have become an important public relations tool, joining traditional media as a way to establish and maintain symmetrical two-way communication with important stakeholders. With advantages and disadvantages, social media can be a critical part of a campaign. As with many public relations activities, evaluative measurement of social media engagement has proven to be elusive. Exploratory Web metrics and software are available, but users have mixed opinions about how useful they are and how well they work.

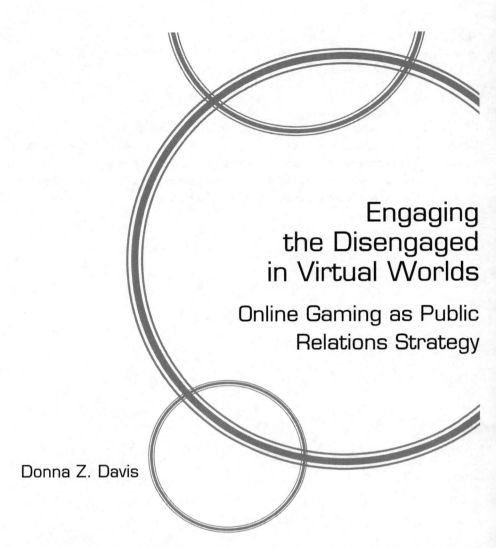

Engaging the Disengaged in Virtual Worlds

Online Gaming as Public Relations Strategy

Donna Z. Davis

Corporate, non-profit, educational, and political organizations are each facing unprecedented challenges in ways to strategically engage both internal and external audiences in the online global environment. One social medium that continues to be overlooked in this quest is online gaming. Yet, analysts estimated that in 2010, 3 billion hours were spent weekly in playing online games. Although 3-D online immersive environments such as Second Life were originally met with unrealistic media exuberance, the adoption of these environments continues to build. Organizations and educational institutions are finding these spaces can engage individuals in unexpected ways, including harnessing untapped perspectives in an environment that steps beyond traditional boundaries. Online virtual worlds provide opportunities for artistic, altruistic, and intellectual collaboration as well as development of community, enhanced by the power of telepresence. Additionally, research continues to show, as with other online interactivity, individuals report a sense of trust that is often stronger than what is experienced in real world interaction. This chapter will reveal the opportunities and challenges of gaming as a public relations strategy.

Networked Virtual Environments (NVEs)

When we think of virtual environments, many people define them as any digital destination, be it company intranets, social networking sites, blogs, microblogs, or even email. These are tools that exist on the Internet, also known as cyberspace. Cyberspace is not just the virtual environment in which people operate, rather it is a complex system in which individuals worldwide increasingly work, play, and interact. Novak (as cited in Packer & Jordan, 2001) wrote, "Cyberspace is a habitat for the imagination—the place where conscious dreaming meets subconscious dreaming, a landscape for rational magic, of mystical reason, the locus and triumph of poetry over poverty, of 'it-can-be-so' over 'it-should-be so'" (p. 254). These environments are not only a place of information exchange, social gathering, and online fantasy, but also have created a new culture that is changing the way we identify with ourselves and engage in our professional and social communities.

3-D virtual worlds, including Massively Multi-player Online Role Playing Games (MMORPGs) such as Second Life, Sims, and World of Warcraft, are online communities, or Networked Virtual Environments (NVEs), where multiple users can interact as avatars simultaneously. Castronova (2001) identified the defining features of virtual worlds as *interactivity*, which provides simultaneous remote access to one shared environment by a large number of people; *physicality*, which allows people to access a program that "simulates a first-person physical environment on their computer screen" (p. 6), central to presence; and *persistence*, or the ability for the program to run whether or not in use while storing data regarding individual users and their online objects. Bell (2008) further defined virtual worlds as a synchronous, persistent network of people (represented as avatars) facilitated by networked computers. These environments offer common-time communication in a sense of real space that continues to exist and function whether or not the participant is online. Virtual worlds are connected by networks of people, via networks of computers, and digital animated representations of individuals, or their avatars. Some may argue there is little distinction between the virtual or real world. As anthropologist Tom Boellstorff explained, "I talk about virtual worlds and the physical world because it's all real" (ChicagoHumanities.org).

Much media and research focus has been given to the explosive growth of social media including blogs, Facebook, Twitter, and YouTube, yet not as much attention has been given to the growth of online gaming. However, an estimated 183 million gamers in the U.S. reported spending, on average, thirteen hours a week playing computer or video games (McGonigal, 2011). Statistics gathered globally between 2008 and 2010 reflected

> more than 4 million gamers in the Middle East, 10 million in Russia, 105 million in India, 10 million in Vietnam, 10 million in Mexico, 13 million in Central and South America, 15 million in Australia, 17 million in South Korea, 100 million in Europe, and 200 million in China. (McGonigal, 2011, p. 3)

What is it about online video games that has drawn the rapt attention of more than a half a billion people worldwide, and why is it important to the public relations profession?

Media effects in 3-D environments: Presence, immersion, and interactivity

3-D virtual environments are rich in visual and contextual interactivity. Individuals create a virtual "self" that is able to interact with others in a 3-D world, where they are able to immerse themselves in a realistic environment and animate their avatar in ways that create what many participants describe as similar to an "awakened dream." Imagine how many times individuals have fantasized about being able to jump into the television, into the book, or on to the big screen and become a part of the story. In essence, for many who enter online virtual environments, that is exactly what they do. Analysts are predicting that the emerging tracking and rendering technologies offered by Nintendo's Wii, Microsoft's Kinect, and PlayStation's Move will combine with 3D monitors and inexpensive head mounted displays to increase consumer demand in the next few years as immersive worlds gain popularity (Blascovich & Bailenson, 2011).

In the converged media culture where communicators must be cognizant of all avenues of audience participation and engagement on the social Web, the potential of online 3-D virtual environments has yet to be fully grasped. One such environment is the Internet's most popular user-created virtual world, Second Life. According to Linden Lab, creator of Second Life, in 2010 and the first quarter of 2011, user hours in-world hovered at approximately 105 million per quarter, and more than 30 million in U.S. dollars were transacted quarterly in the virtual exchange (BK Linden, 2011). These dollars are most frequently spent on virtual land and virtual goods much the same as how people now download music rather than purchase CDs or LPs.

What distinguishes 3-D online environments from any previous medium are the elements of presence, immersion, and interactivity. The concepts of interactivity and presence are paramount to the powerful effects of new media. Presence has been defined as a personal, social, and/or environmental phenomenon (Heeter, 1992). *Personal presence* is experienced when a person believes that he or she is actually inside a virtual or remote environment. *Social presence* is the belief that a person is experiencing and interacting with other beings. *Environmental presence* represents how effectively the environment itself is able to acknowledge and interact with the user. In a 3-D virtual environment, presence can be more likely equated to Minsky's (1980) concept of *telepresence*, that is, when humans believe they are physically present at a remote location as a result of their: 1) interaction with a medium, and 2) their subsequent perception of the feedback they get as a result of their actions. It is this sense of presence that has dramatically influenced the growth of 3-D virtual environments. Compared to McLuhan's (1964) characterization of television as a "cool" medium for which viewers are required to give meaning to the images they see, the 3-D immersive world could be characterized as ice cold, as users are actually driving the action of their avatars as they interact with others. Participants in this environment are not passively participating. Rather, they are in essence creating the story of their virtual lives. For some, it is a mirror of their real lives. For others, it is a surreal mirror of the lives they would like to have.

Lee (2004) defined presence as "a psychological state in which virtual objects are experienced as actual objects in either sensory or nonsensory ways" (p. 27). *Physical presence* refers to the physical objects or environments that are typically experienced through visual and audio stimuli. Similar to Heeter (1992), *social presence* represents the social interactions that humans exchange, but in Lee's definition, they occur specifically through media and technology. Lee posited:

> Just as people pay special attention to other humans more than any other physical objects, technology users pay great attention to technology-generated stimuli manifesting humanness in both physical—and psychological (e.g., personality, reciprocity, interactivity, social roles, understanding language, etc.) ways. (p. 39)

Virtual presence as it relates to self presence is defined as either a state when "technology users do not notice the virtuality of either para-authentic representation of their own selves" (p. 45) or what might be considered an alter-self that is unique to the virtual environment.

Just as Lee differentiated social presence as it relates to media and technology, the role of interactivity also requires distinction. To best understand the concept of interactivity as it relates to 3-D immersive virtual worlds, Kiousis' (2002) definition best represents the measure as it relates to new media. In his explication of interactivity, Kiousis explained, "interactivity can be defined as the degree to which a communication technology can create a mediated environment in which participants can communicate (one-to-one, one-to-many, and many to-many) both synchronously and asynchronously and participate in reciprocal message exchanges (third-order dependency)" (p. 379). He also identified the importance of the technology user's ability to perceive this communication as a "simulation of interpersonal communication" (p. 372).

Kiousis (2002) also identified operational measures of interactivity that separate how it is experienced in interpersonal communication versus computer-mediated environments. These operational measures for computed-mediated environments relate to the medium's structure, including its speed, system range (or the system's ability to provide users with multiple actions such as sending and receiving information simultaneously), and technological complexity. The second set of measures is contextual, or how social presence is perceived in the media environment. The final set of measures represents "how well the communication experience simulated interpersonal communication" (p. 376). When interactivity measures are highest, the experience comes closest to that of a face-to-face exchange.

The simulation of interpersonal communication is far more enhanced in the 3-D virtual world than in text-based social media. Yee, Bailenson, Urbanke, Change, and Merget (2007) confirmed use of social norms and behaviors, including eye gaze and interpersonal distance (IPD), among avatars in Second Life. In other words, the social norms and interactions that occur in interpersonal communication in real life also apply to social behavior and communication in the virtual world.

Although most online virtual environments are strictly gaming cultures, virtual world environments have also become a tool for business, education, and research. Consider that in 2008 the New Media Consortium found that educators had moved from exploration to the use of Second Life for teaching and learning (New Media Consortium, 2008). At that time, more than 200 universities world-wide had "sims," or simulated campuses, in the virtual world. Although the highest volume of traffic and (as of this writing) the best technological interface remain in Second Life, new open source 3-D virtual worlds such as 3rd Rock Grid (http://3rdrockgrid.com/), InWorldz (http://inworldz.com/), OpenSim (http://opensimulator.org/wiki/Main_Page) and ReactionGrid (http://reactiongrid.com/) have recently emerged. In these environments, educators are not only teaching their regular courses, but also are providing distance education classes; developing bioterrorism preparedness programs; creating "safe" places where autistic and cerebral palsy patients can interact socially; creating interactive educational simulations about history, geography, art, literature, science, and environment; housing archives and reference materials that interact with multi-media; and collaborating with government agencies to develop multinational simulations for immersive cultural training prior to deployments.

Likewise, businesses are turning to the virtual world to host meetings and conferences. A powerful testimony to the changing business landscape in virtual worlds was in 2008 when IBM hosted its annual meeting in a secure Second Life environment. The website case study reported:

> IBM estimates the ROI for the Virtual World Conference was roughly $320,000 and that the Annual Meeting was executed beautifully at one-fifth the cost of a real world event. Many IBM staff were converted into virtual world advocates, paving the way for many future internal conferences and events to be held within the space. (Linden Lab, 2008, p. 1)

Non-profits are also benefitting. The American Cancer Society (ACS) began promoting Relay for Life in Second Life in 2004. Since 2008, ACS raised an average of nearly $250,000 annually among Second Life participants in 30 countries (American Cancer Society, 2011). Researchers, game developers, and corporations are also exploring the use of online games for social change. For example, the Korean Ministry of Justice launched "Adventures in Law Land" as a strategy to educate citizens about the importance of laws and to build respect for those laws (Games for Change, 2011). Additionally, the U.S. Department of Veterans Affairs developed "At Risk," an avatar-based game that is helping veterans address challenges associated with Post-Traumatic Stress Disorder (PTSD). The World Bank Institute funded a project called EVOKE. "EVOKE is a ten-week crash course in changing the world" with a goal of empowering "people all over the world to come up with creative solutions to our most urgent social problems" (Games for Change, 2011, *Case Study: Evoke*).

Online relationships: Developing digital social capital

Although research in these 3-D immersive virtual worlds is at a relatively early stage, there are many behavioral parallels between the online virtual world and other Internet venues. For example, while studying online forums Bargh and McKenna (2004) found that the "relative anonymity aspect encourages self-expression, and the relative absence of physical and nonverbal interaction cues (e.g., attractiveness) facilitates the formation of relationships on other, deeper bases such as shared values and beliefs" (p. 586). They found that communicating via the Internet was instrumental not only in maintaining close ties with friends and families, but also in forming new close and meaningful relationships in what is consistently interpreted to be a relatively safe environment. Bargh, McKenna, and Fitzsimons (2002) found that individuals had the tendency to project ideal qualities on the people they "meet" on the Internet. In this environment, people also let down traditional emotional and cultural barriers that may typically prohibit development of relationships, which in turn enhances the opportunity to create closer relationships.

In the 3-D immersive virtual world, these relationships appear to be occurring in much greater speed and intensity. Walther (1996) identified this trend in the development of the Hyperpersonal Model of Computer Mediated Communication (CMC). Walther found that users present themselves as their "best" selves online. Additionally, due to the reduced number of social cues in CMC (e.g., body language and eye contact) online interactions can lead to an idealized perception by the perceiver. He also found that the flow of communication could be more friendly and cordial than face-to-face communication because users are typically required to take more time to process their response to written words.

Likewise, the element of anonymity provides individuals the opportunity to explore their concept of self and express their true feelings in the "relatively safe environment" Bargh, McKenna, and Fitzsimons (2002) discussed. The 3-D environment is enhanced by the powerful sense of presence, combined with the ability to create a visual identity and experience levels of engagement and interactivity previously not possible in interpersonal or traditional Internet communications.

This virtual opportunity is of particular interest for public relations strategies that seek to engage a multitude of targeted audiences including employees, volunteers, community members, and constituents who otherwise may be hesitant to get involved. For example, when working offline, Sunstein (2006) reported that individual members of groups are often made to feel less than worthy when they attempt to participate. Whether they have truly been a victim of any form of bias or simply lack the self-esteem to confidently contribute, they are likely to follow authorities or defer to consensus, leading to Noelle-Newmann's (as cited in Shoemaker, Tankard, & Lasorsa, 2004) spiral of silence. This is especially problematic for low-status members of a group or individuals who have been made to feel marginalized, often as a result of their looks, gender, or social status. However, Gillmor (2006) found that online participation from citizen journalists has created participation "from people on the edges" (p. 100). He discussed how open source politics has provided a voice for people who otherwise would never have been heard. This has been particularly and consistently evident in the emergence of blogs, Facebook, and Twitter use during 2011 political uprisings in the Middle East and natural disasters around the globe.

Additionally, when individuals participate in online communities, including virtual worlds, without real life identifiers, the veil of anonymity empowers and emboldens them to interact in ways they would not likely consider in their physical realities. In a study of friendship development in multi-user virtual fantasy environments, Utz (2000) found that friendship development was "only weakly correlated with the general trait sociability" (*Discussion*, para. 2). She confirmed prior research (Parks & Roberts, 1998; Roberts, Smith, & Pollock, 1997; Zimmerman, 1987) that found shy people were less inhibited in CMC and were able to more easily form intimate relationships online than in real life. Valkenburg and Peter (2008) also found that adolescents experimented with their identities (i.e., pretended to be someone else) online, and those who communicated most often with people of different ages and cultural backgrounds were most likely to develop social competencies, which the authors defined as the "ability to effectively form and manage offline interpersonal relationships" (p. 210). Likewise, in a Pew Internet and American Life Project study, Boase, Horrigan, Wellman, and Rainie (2006) reported that not only are people "able to maintain active contact with sizeable social networks, even though many of the people in those networks do not live nearby" (p. 1), but also that media multiplexity (i.e., personal contact, phone use, and increased Internet activity) leads to stronger personal connections. These individuals were more likely to reach out to one another for help. Similarly, in a recent ethnography of the virtual world, Carter (2008a) explored the culture of one online virtual community, Cybercity, and the nature of human relationships within that culture. Carter concluded that online relationships are evidence that social relationships are in a state of transformation as they are no longer "anchored by the everyday social and cultural construction of gender, age or race" (p. 164).

This evolution in media use and relationship development provides further insight into a shift away from Shannon's (1949) linear communication theory—in which the message originates from one source and is distributed to many—toward new media communication in the more visionary circular model of Weiner's cybernetic theory (1948). In Weiner's model, the medium is at the center of the circle, and social networks are wrapped around and entwined with the medium. Barraket and Henry-Waring (2008) explained, "One of the most commonly cited culturally transformative possibilities of online technologies is their capacity to overcome the tyranny of distance, time and space to allow for the establishment of new networks and patterns of interactivity" (p. 156). They added, "More significantly—the dominant theoretical frameworks for understanding intimacy in the global era are predicated on a shared belief that we exist in a period of de-traditionalization, where socio-cultural traditions are being abandoned or reconfigured" (p. 161). Similarly, Bawin-Legros (2004) wrote, "the influence of traditional sources of authority and of social bounds has increasingly receded in favour of an endless and obsessive preoccupation with personal identity" (p. 241). In fact, peer influence has become the most trusted source of information with 9 out of 10 people globally reporting they trust the recommendations of the people they know more than information provided by traditional sources including media, clergy, parents, and political leaders (Smith, 2011).

This form of trust was also evident when Yee (2006) found that of almost 3,500 adults who participated in Massively Multi-user Online Role-Playing Games (MMORPGs), more than 20% of males and more than 30% of female players "had told personal issues or secrets to their... [online] friends which they had never told their real-life friends" (p. 320). Almost 40% of men and more than 50% of women surveyed also said that "their...[virtual] friends were comparable or better than their real-life friends" (p. 321).

This growing body of evidence of online trust and reciprocity suggests that although community social capital has been in decline in the past several decades (Putnam, 2000), digital social capital is on the rise. How people define their relationships and their communities is no longer bound to the physical world, but is embedded in the interaction they experience with individuals in a multitude of forms and from an equally complex blend of media channels. Through engagement in an online 3-D immersive space enhanced by presence and the ability to interact with individuals without many of the traditional barriers of geography, trait sociability, or perceived socioeconomic norms, organizations have new opportunities to explore innovative public relations strategies with both internal and external publics. As McGonigal (2011) explained, "as gamers are finding out, rebuilding traditional ways of connecting might not be the solution—reinvention might work better" (p. 93).

Conclusion

Social online 3-D immersive environments such as Second Life were initially met with media exuberance but subsequently suffered what many consider a rather significant fall from grace. Organizations and corporations dove into the virtual world expecting high volumes of traffic and invested millions of dollars in developing simulated products, shops, and showrooms expecting sales. However, due to a number of factors, such as technological challenges, limited access, and a complicated interface, the traffic simply wasn't there. In the theoretical lens of diffusion of innovation (Rogers, 1995), the technology experienced what is called the trough of disillusionment in the hype cycle (Fenn & Raskino, 2008). The hesitancy toward mass adoption has been further exacerbated by a sense of fear frequently found in media headlines of online addiction, infidelity, cyberstalking, isolation, and other maladaptive behaviors.

However, in the words of John Naughton (2008), "The First Law of Technology says we invariably overestimate the short-term impact of new technology while underestimating their longer-term effects" (para. 1). Although many public relations and marketing advisors predicted the demise of Second Life after the initial retreat of companies, the virtual environment continues to hold steady and evolve as new virtual worlds also enter the field.

Why is this important to the field of public relations? In a culture rife with apathy, alienation, boredom, and information overload, creating public relations strategies that reach targets and create impact has perhaps never been more challenging and yet more exciting. As individuals and industries scramble to determine how best to ride the tsunami of social media, public relations practitioners are grounded in a theoretical framework that understands the importance of attitudes, behaviors, and two-way symmetrical communication—the heart of social media. This chapter has focused on online games as a social medium that to date has yet to be fully understood, not only as a tool, but as a potential work environment or social community. As marketers begin to discuss "gamification" as a powerful way to engage consumers, how can the public relations field adopt this technology to help reach organizational goals?

To begin, public relations professionals need to approach 3-D online game environments not as a replacement for their current strategies, but as a complement and reinforcement. These environments have the ability to reach an expanded public sphere including those who are very limited in access to peers such as people battling life issues who seek support that might not be readily available in their physical world. When individuals can embrace the virtual environment through presence, organizations also have the potential to engage their publics in a way that can capture rapt attention for hours, as gamers often experience the state of intense engagement and heightened motivation that creates a diminished perception of time known as "flow" (Csikszentmihalyi, 1990; Nakamura & Csikszentmihalyi, 2002). This behavior could have tremendous value to team building and strategic planning activities.

As Walther (1996) posited the Hyperpersonal Model of CMC, considering the levels of trust and sense of safety of expression in NVEs, these environments could create a hyperpersonal model of Grunig's two-way symmetrical model. If individuals believe their participation has true meaning, wouldn't they be more likely to contribute to the communication process, and thus better enable us to validate Grunig's (2006) co-orientation process with our publics?

These 3-D environments have enormous potential for both internal and external communication. From an internal perspective, as previously discussed, companies such as IBM have discovered they could not only save significant dollars by holding meetings and training sessions in the 3-D environment, but also found increased engagement and interaction among their global workforce. Although this example primarily represents the functions of a social virtual environment, a more traditional gaming model could provide additional incentives for workplace engagement.

For example, Procter and Gamble created *Connect and Develop*, an online game that invited researchers, employees, and consumers to collaborate and guide product development with great success (Huston & Sakkab, 2006). The role of collaboration and team building in a game environment also fosters a strong sense of community, improves established relationships, rewards collaboration in performance reviews, and provides easy measurement of individual contribution to the team (Reeves & Read, 2009).

External audiences can also be targeted in 3-D online environments as evidenced by non-profits such as the American Cancer Society (ACS). As previously mentioned, ACS found an international audience willing to compete and contribute through events in the virtual world where individuals embrace a sense of community through their online presence. Likewise, the Smithsonian Institute is creating educational outreach for numerous programs in Second Life by recreating environments and providing video kiosks in-world. Publics whom the Smithsonian is committed to educating yet who may never have the opportunity to visit these places, or the Smithsonian, have access virtually (Smithsonian Institution, 2009).

As Reeves and Read (2009) noted, "In 2003, 55 percent of all U.S. workers used a computer at work" (p. 38) and that number has no doubt increased. As they explained, "Firms may own networks, but it is people who have relationships. A company can own a brand, but individuals own their own reputation" (p. 38) and that in today's workforce, companies must "create conditions in which they want to give you innovation, collaboration, and insight" (p. 38). As evidenced through the functions of immersion, interactivity, and presence, online 3-D virtual environments may offer excellent solutions for motivating and communicating with tomorrow's workforce.

However, it is also important to recognize that there remains resistance from a large population who fear the "Matrix" effect, or the fear that we will lose ourselves to technology, sacrificing real for virtual and giving up on real life challenges for fantasy. Likewise, non-gamers often see a gaming population as those who are socially inept or cultural outcasts. We cannot expect everyone to embrace this technology, just as the case of laggards in any technology adoption.

Also important to consider is that gamers are also immersed in a unique culture that they often embrace with strong personal commitment and are resentful of researchers or marketers who enter their domain without understanding the culture. Those interested in pursuing public relations opportunities in virtual worlds must research the environment just as they would any other culture before attempting to create communication strategies that will effectively create change.

PART II

Corporate Applications

Sandra Duhé

Overview

This section on corporate applications of social media adds to our understanding of how corporations attempt to engage with their publics online. Corporate entities have been found to be not as interactive as they could be on the Internet. Their hesitancy to fully engage in two-way communication can be related to regulatory, cultural, or risk factors, and/or a lack of understanding of technical capabilities or how online communication fits into an organization's overall strategy. These authors shed light on these important issues.

Tina McCorkindale starts by recognizing the dialogic challenges companies face in social media, noting how organizations embrace their capacity yet too often remain one-way in their communications with stakeholders. She provides specific guidance for social media use that is applicable to any type of organization, not just corporations. Her research-based advice includes ensuring a proper fit within an organization's overall strategy, being responsive, setting policy, and being "real" online.

Linjuan Rita Men and Wan-Hsiu Sunny Tsai offer an interesting cross-cultural comparison of how companies in the US and China build relationships online. They examined corporate pages in Facebook and Renren and found that communication practices related to low- versus high-context cultures apply to social networks as well. Whereas U.S. companies were more directly promotional in their messages, Chinese companies instead focused on being entertaining and informative while still promoting their brands. Hopefully, other researchers will follow their model and expand our understanding of how culture affects social networking.

Years ago, the Saturn car brand garnered attention when the company would sponsor events exclusively for Saturn owners. These were reunions where owners would get together, have a cookout, and share their stories of how much they enjoyed their vehicles. At the time, it seemed like such a quirky idea, but it was effective in enhancing emotional ties to the Saturn brand. Romy Fröhlich and Clarissa Schöller provide us with a contemporary view of online brand communities by drawing upon interdisciplinary research, proposing and defining an applicable communication model, and distinguishing strategic uses of brand communities for corporations.

Carol Ames is a former communication counselor and executive producer in the entertainment industry. In her chapter, she explains how social media are used in film and television promotions and what can happen when celebrities (think Charlie Sheen) get online without advice from their publicists. She also relays some celebrity-driven successes in the virtual world that are sure to be of interest to future publicists.

Alexander Laskin describes for us a world much more restrained than Hollywood. In his chapter, he describes how adoption of social media is slow among investor relations practitioners, and for good reason. Publicly held companies are strictly bound by regulations designed to protect current and potential investors from misleading information that could lead to unsubstantiated changes in stock prices, among other undesirable outcomes. But, he also explains that despite these restrictions, companies are finding ways to use social media to their benefit and within the limits of the law.

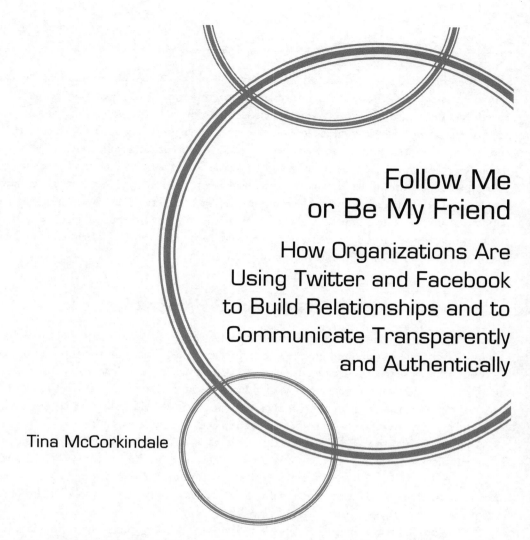

Follow Me or Be My Friend

How Organizations Are Using Twitter and Facebook to Build Relationships and to Communicate Transparently and Authentically

Tina McCorkindale

With the increasing popularity of social media, organizations need to properly and strategically integrate these channels into their day-to-day activities. Two of the most popular sites public relations practitioners use and monitor are Twitter and Facebook. Being able to communicate both transparently and authentically using dialogic communication on these sites may be a challenge for some organizations. This chapter explores the concepts of transparency and authenticity, discusses how organizations are using Facebook and Twitter to build relationships, and provides suggestions on how organizations can use these two sites.

Public relations practitioners use social media to generate awareness, manage their client's reputation, promote products and services, and listen to publics (Fathi, 2008). Two popular sites monitored and utilized by public relations practitioners are Facebook and Twitter. With more than 170 million unique visitors combined in March 2011 (Compete, 2011), the sites' high number of users and their ability to communicate with a wide range of stakeholders attracts public relations practitioners. Due to the speed at which information is diffused online, mastering how to effectively and quickly communicate using social media channels is important.

One of the ways public relations practitioners should engage on Twitter and Facebook is by creating and engaging in dialogue. As Jones, Temperley, and Lima (2009) attested, "in the social web, there is little space for monologue" (p. 930). Burson and Marsteller's (2011) study of the Fortune Global 100 found companies were more likely to devote resources to social media as well as to engage stakeholders through dialogue on Facebook and Twitter compared to the previous year's study. However, recent research indicates that while some organizations are embracing dialogic communication, others are still primarily using these channels for information dissemination (McCorkindale, 2010a; McCorkindale, 2010b; Rybalko & Seltzer, 2010; Waters, Burnett, Lamm, & Lucas, 2009). Not only do organizations need to engage in dialogic communication, but research suggests organizations need to be transparent and authentic in their communication (Rawlins, 2009; Tapscott & Ticoll, 2003).

Little research exists in terms of how organizations are using Twitter and Facebook to build relationships, and communicate transparently as well as authentically. Transparency and authenticity are frequently used to describe how public relations should be practiced, but understanding these terms as opposed to simply using them as buzzwords is essential for practicing public relations well. This chapter analyzes how organizations are using Facebook and Twitter in terms of relationship building as well as the importance of transparent and authentic communication.

Twitter and Facebook

The most popular online social networking site is Facebook. In March 2011, Facebook had more than 140 million unique visitors making the site the second most visited on the Internet (Compete, 2011). Facebook allows users to posts on walls, send messages, and update statuses in an effort to maintain close and distant relationships for the purpose of building social capital (Steinfield, Ellison, & Lampe, 2008). Organizations are using Facebook to connect with various stakeholders by posting updates in users' news feeds and encouraging stakeholders to "like" or join their online community.

Another popular site used by various stakeholders as well as public relations professionals is Twitter. The microblog was ranked 28th in popularity with 31 million unique visitors in March 2011 (Compete, 2011). According to the Twitter (2011a) website, Twitter asks "What's happening?" in a 140-character statement called a tweet. Church (2008) defined Twitter as a "conversation tool that allows organizations and customers to talk openly, directly, and quickly" (p. 10).

One of the challenges of using Facebook or Twitter is how organizations build relationships that do not violate the stakeholders' expectations, which may negatively impact the relationship. To be able to establish, build, and maintain positive relationships, public relations practitioners must understand how to effectively incorporate these two sites in an organization's strategy.

Building relationships

Hon and Grunig (1999) contended the strength of an organization-public relationship can be measured along three dimensions: commitment, satisfaction, and trust. Similarly, Bruning and Ledingham (1999) identified six conditions that should be present in an organizational-public relationship: awareness of influence; openness; trust; dialogue; understanding; and a willingness to negotiate. While little research has explored differences between offline and online relationships, consistency across the different media is important.

Although using social media to build relationships may be risky (Duhé, 2007), it also has rewards. According to Giles and Pitta (2009), while risks include inappropriate content or negative remarks, some rewards include promotional aspects, message dissemination, and stakeholder conversations. Goffman (1959) stated, "underlying all social interaction, there is a fundamental dialectic" (p. 247), which individuals use to assess a situation and make decisions. Therefore, the importance of engaging in two-way communication through dialogue and conversation cannot be overstated.

O'Neil (2008) found the single most important predictor for measuring the strength of relationships in nonprofits was communications that help donors understand how their donations will be used. By communicating transparently, nonprofits have the ability to broadcast this information on both Facebook and Twitter. Most public relations researchers agree dialogue is a necessary as well as ethical form of communication (Gilmore & Pine, 2007; Gilpin, 2010b; Henderson, 2010; Kent & Taylor, 1998; McCorkindale, 2010b). The dialogue must be honest and forthright in terms of transparency, as well as authentic (Gilpin, 2010b; Henderson, 2010).

Kruckeberg and Starck (2004) discussed ethics of consumer communities, and how the role of the public relations practitioner should be to foster and nurture these communities, thereby "making every attempt to encourage the positive benefits of such communities while minimizing or hopefully eliminating the potential negative outcomes" (p. 144). Social media comprise various communities, which are not limited to consumer communities but personal ones as well. This makes the public relations practitioner's role more challenging. However, Helm (2007) found regardless of the stakeholder groups individuals belong to, they have similar impressions of a firm's general reputation. Another challenge to building and maintaining relationships is crises, which disseminate rapidly due to the community nature of social networks.

The speed with which a crisis hits social media can cause irreparable damage. One example of an online crisis was the Motrin Mom's ad campaign, which suggested mothers wear baby slings as a fashion statement and to give the "impression" they are good moms. Although the issue was quickly disseminated throughout the Twitterverse, the crisis was also featured on Facebook and in the traditional media. Within 72 hours, Motrin canceled the ad campaign (Learmonth, 2008). Some critics argued Motrin did this prematurely, and should have instead talked with moms.

Research indicates stakeholders are more likely to participate in negative word-of-mouth communication when they have been angered by a crisis (Coombs & Holladay, 2007b). The rapid speed through which issues diffuse on social networks not only affects those directly impacted, but also the bystanders (Coombs & Holladay, 2007b), as they are the ones who witness the crisis unfolding on social networks and see the organizational response. Unfortunately, many companies are not using social media properly to deal with crises online (Cakim, 2007). Instead of engaging with various stakeholders by having conversations to build relationships, some organizations merely disseminate information, or even in some cases, spend their time creating a social presence that did not exist prior to the crisis. Regardless, stakeholders expect the organization to be transparent and authentic in their communication.

Transparency

How a company behaves online should mirror how they behave offline (Gilmore & Pine, 2007). Companies should be responsible citizens. Organizational stakeholders are demanding companies be more accountable and responsible for their actions, as well as more transparent and open. More companies are devoting resources to showcase their corporate responsibility efforts in online and offline reports (Global Reporting Initiative, 2010). Fundamental values of corporate responsibility include transparency and openness, and each is apparent in the dialogue process with stakeholders (Middlemiss, 2003). Professional codes of conducts and ethics, such as the Public Relations Society of America's (PRSA) Code of Ethics, have evolved to account for elements of transparency, openness, and authenticity.

Transparency research has become more popular in the past five years, attributed in part to the increase in social media use. Transparency has been linked to trust, corporate responsibility, authenticity, and ethics (Baker & Martinson, 2002; Fussell-Sisco & McCorkindale, 2011; Gower, 2006; Jahansoozi, 2006; Rawlins, 2009). Sixty-five percent of respondents to Edelman's (2011) Annual Trust Barometer reported transparency and honest business practices, as well as trust, were most important to corporate reputation. When there is a lack of trust in a relationship, transparency is needed to repair it (Jahansoozi, 2006). Transparency is a relational variable, which helps promote "accountability, collaboration, cooperation, and commitment" (p. 943).

Simply, transparency is the opposite of secrecy (Rawlins, 2009). Edelman was accused of not being transparent in their client Walmart's blogging campaign, which received criticism from both the blogging community and researchers (Burns, 2008). However, it should be noted that transparency is not the equivalent of disclosure, but encompasses a broader range of principles. Rawlins identified three important elements that are needed in the definition of transparency: truthful information; participation of stakeholders in identifying needed information; and objective, balanced reporting of activities and policies. One important concept in Rawlins' definition is the participation of stakeholders in the process of establishing and maintaining transparent communication by providing them with a satisfactory level of information. Based on his seminal study, Rawlins defined transparency as:

> the deliberate attempt to make all legally releasable information—whether positive or negative in nature—in a manner that is accurate, timely, balanced, and unequivocal, for the purpose of enhancing the reasoning ability of publics and holding organizations accountable for their actions, policies, and practices. (p. 75)

The issue is whether organizations can truly be transparent as well as balanced and objective.

Transparency is not an independent concept. Changing in an effort to be transparent requires a complete transformation of the organization's fundamental ways of thinking. Some organizations assume transparency requires the organization to give up secrets or proprietary information, thereby making the company vulnerable. But this is not the case. Management must support all efforts to engage in transparent communication, but should also offer social media guidelines or policies to do so effectively. Vargas (2011) suggested organizations should create a social media policy not just to protect themselves, but also to protect their communities. She believes these policies help build trust and credibility created by both online and offline interactions.

Even though certain best practices awards and regulations such as Sarbanes-Oxley are in place to increase transparency, the public is still experiencing low levels of trust (Dando & Swift, 2003; Edelman, 2011). Stakeholders are increasingly scrutinizing the financial, ethical, and social judgment of organizations (Duhé, 2007). According to Rawlins (2009), to be transparent, organizations should "voluntarily share information that is inclusive, auditable (verifiable), complete, relevant, accurate, neutral, comparable, clear, timely, accessible, reliable, honest, and holds the organization accountable" (p. 79). Gower (2006) contended stakeholders may not be as concerned with transparency as they are with being assured the organization is acting properly.

Transparency, though, should not be assessed on a continuum. Either an organization is transparent, or it is not. If an organization intentionally withholds information that is needed by stakeholders in the decision-making process, then the organization is not transparent. Stakeholders have the right to information that is not partial or distorted to be able to make decisions (Rawlins, 2009). According to Vargas (2011), "social media thrives on transparency. Nothing can be hidden…or at least, not for long" (para. 3).

Due to the communal nature of social networking sites, being transparent may pose more of a challenge especially in regulated industries (DiStaso, McCorkindale, & Wright, in press). Some organizations require all tweets or Facebook posts to be fielded by legal. Others establish social media policies, which may help guide or restrict the employee's communication. Some organizations have no policies or guidelines in place.

One way organizations can become more transparent is to disclose who is managing their social media accounts. In one study, organizations that disclose the person's name who manages the account had significantly more dialogic tweets, more "in reply to's" and more total tweets compared to accounts with unnamed tweeters (McCorkindale, 2010b). This finding may indicate a higher level of responsiveness by these organizations, and greater accountability to respond to tweets in addition to transparency. Also, providing a human voice compared to an organizational voice has been found to promote more positive relationships with stakeholders as well as positive word-of-mouth communication (Park & Lee, 2011).

Another way organizations can be transparent on Twitter and Facebook is to openly resolve stakeholder issues. If stakeholders observe the organizational response, then trust can be built and maintained (Jahansoozi, 2006). In addition, according to Jahansoozi, transparency enables both collaboration and cooperation between the organization and community.

Van Woerkum and Aarts (2009) proposed a concept of "visual transparency," which includes sharing pictures or factual images to demonstrate authenticity, increase trust, and educate stakeholders on how goods and services are produced. Organizations can post pictures and videos on social media sites to demonstrate transparency.

Transparency has also been linked to an organization's activity on Twitter and Facebook. Fussell-Sisco and McCorkindale (2011) found breast cancer nonprofits that tweeted more, had more likes, more followers, and more overall tweets were seen to be more transparent and credible by virtue of activity alone. Therefore, organizations should interact thoughtfully, but also do so frequently. According to Bulmer and DiMauro (2009):

> Companies should be mindful that a primary reason professionals participate in social networks is to collaborate, not to be sold to. Marketers should develop social media strategies that do not break or breach the social contract that professionals have when working within their social networks—by avoiding overt sales and marketing campaigns...Those that embrace transparency are the conversations that customers desire. (p. 5)

Underscoring the concept of transparency is legitimacy, according to Waddock (2008). She suggested even if a corporation is fully transparent, this may not satisfy stakeholders where legitimacy resides because of the imbalance of power between citizens and corporations. Legitimacy is related to the concept of authenticity, defined as "the level of knowledge and expertise" (Gilpin, 2010b, p. 261). Building trust, therefore, depends on both authenticity and transparency. A seamless connection between the organization's communication efforts and its authenticity is important in satisfying the needs of stakeholders' demands for transparency (Molleda, 2010). According to Gustafsson (2006), "changing to become more transparent is key for brands in order to acknowledge the change in what authenticity means to consumers, and to show consumers that they are trustworthy" (p. 527). Wakefield and Walton (2010) suggested in some circumstances transparency may not be the most ethical or emotional choice, but rather translucency. In some cases, translucency may be more beneficial for the interests and well-being of the stakeholders. For example, in issues communication if the issue is scientific and may be confusing for the stakeholders, then easy-to-understand language as opposed to full disclosure about the scientific method would be more appropriate.

Authenticity

Transparency is closely tied to authenticity even though Wakefield and Walton (2010) argued authenticity may be more important than transparency. Authenticity is described as something that is real, original, genuine, sincere, and not fake (Gilmore & Pine, 2007). Most definitions of authenticity include some reference to genuineness (Adorno, 1973; Alexander, 2009; Beverland, Lindgreen, & Vink, 2008; Van Leeuwen, 2001). In social media, authenticity relates to the way an organization interacts and has conversations with stakeholders. One description of authentic talk, according to Montgomery (2001), is dialogue that "does not sound contrived, simulated or performed but rather sounds natural, 'fresh', spontaneous" (p. 403).

Authenticity may also bear a "stamp of approval" from a third party verifying its genuineness (Van Leeuwen, 2001; Alexander, 2009). Some researchers argue whether one can actually have true authenticity as individuals have a true self hidden behind a public mask (Scannell, 2001; Tolson, 2001).

Grayson and Martinec (2004) suggested the distinction between authentic and inauthentic may be socially or personally constructed, which is also known as hyperauthenticity (Beverland & Farrelly, 2009). Therefore, stakeholders may become more skeptical of an organization's authenticity if expectations of behavior have been violated or manipulated in efforts to make a profit. Beverland (2005) extended this premise by suggesting authenticity must be non-commercialized, and developed by the organization's communities.

In order for organizations to be authentic on Facebook and Twitter, they need to have conversations that are real and not contrived, meaning a true conversation as opposed to a constructed conversation. However, one can argue that an organization's ability to have a truly authentic conversation may be difficult due to a perceived risk. Transparency and authenticity should not be considered mutually exclusive concepts, but rather necessary conditions for an organization's social media presence.

How organizations are using social media

While research has investigated the extent to which organizations are using social media, more research needs to determine how they are using it. In 2009, only a handful of Fortune 50 companies were using Facebook to respond to customer issues, recruit employees, and engage in dialogue with various stakeholders (McCorkindale, 2010a). However, more companies are using social media sites in B2B communication to talk with other businesses, including vendors and suppliers (Barnes, 2010b).

Research to date has found while many organizations are embracing the dialogic capabilities of social networking sites, others are using the sites to merely engage in one-way communication (McCorkindale, 2010a; McCorkindale, 2010b; Waters et al., 2009). Also, few have discussed corporate responsibility efforts (McCorkindale, 2010b). However, sites such as Facebook have experienced growth in grassroots campaigning and social activism (Zuckerberg, 2010).

Specifically, Facebook allows nonprofits to engage stakeholders in multiple ways, such as soliciting donations, recruiting volunteers, gaining support for its efforts (Zuckerberg, 2010), listening to stakeholders, and telling stories. In universities and colleges, Farrow and Yuan (2011) found when alumni participate in Facebook groups, they have stronger ties to the university, and that this relationship positively predicts behaviors such as volunteerism and charitable giving. Similarly, supporters used Facebook and Twitter after the Haitian earthquake to communicate issues, gain support, solicit donations, and communicate commitment (Smith, 2010b).

In terms of Twitter use by the Fortune 500 in 2010, slightly more than half (60%) of companies had accounts, but only 35% consistently responded with @replies or retweets within 72 hours (Barnes, 2010a). McCorkindale (2010b) analyzed Mashable's "40 best brands" and how they tweeted. Several ways organizations are using Twitter emerged: to provide updates, post information, respond to inquiries, interact with stakeholders, recruit employees, repost existing content, hold contests or giveaways, and provide opinions. Several companies regularly employed dialogic communication with their followers.

Rybalko and Seltzer (2010) found differences between Fortune 500 companies that engage in non-dialogic and dialogic communication in terms of their conservation of visitors and generation of return visits. Dialogic organizations made greater efforts to engage users but were less likely to tweet links to the company's website. The majority (60.1%) of those employing the dialogic loop responded to tweets, approximately one-third (30.1%) asked questions to generate conversation, and about one-quarter (26.9%) asked follow-up questions.

Organizations are now devoting more resources to engaging and monitoring social media. Bank of America (@BofA_help) and Whole Foods Market (@wholefoods) are just two examples of companies who have teams on Twitter devoted to responding to customer service issues, questions, and complaints. JetBlue (@JetBlue) has 17 staff members responding to requests, and Delta Airlines (@DeltaAssist) has 9 to 12 (Yu, 2011). The companies also acknowledged that although the accounts do not necessarily increase sales, they do provide another avenue for customer service. However, organizations should not merely use these sites for customer service, and should use them to interact with various stakeholders. Also, the public nature of social media should not be the impetus for giving customers and other stake-holders better attention.

Developing social media policies or training employees to properly use these channels is becoming more commonplace. Texas Instruments is training each of its 29,000 employees for a "conversation agent" certification in order to answer customer questions and gauge opinion online (Wilson, 2011). Dell requires each employee to be "certified" before they use social media channels. Today, organizations are savvier about how to both communicate and listen to various stakeholders, but more work needs to be done.

Conclusion

One important principle organizations must consider is consistency across media. Some organizations may be more concerned with online than offline presence due to social media's broadcast capabilities. Whaling (2011) offered suggestions for what public relations practitioners can do on Twitter, but these tips also apply to Facebook:

- Connect with reporters
- Offer a behind-the-scenes glimpse
- Crowdsource (open to community input) research and solve problems
- Leverage real-time professional development
- Find employment
- Strengthen crisis communication

As long as using Twitter or Facebook fits into the organization strategy, this chapter author suggests a balanced use of social media sites. These sites should be treated as communities where the organization engages in transparent, authentic communication. Letting stakeholders create a piece of the organiza-tion by inviting them to participate in the decision-making process also lends itself to authenticity, as suggested by Gilmore and Pine (2007).

Organizations must also respond to every *reasonable* question asked by a fan or follower. Moreover, they must listen. Even though much research has focused on the dialogic principles, there can be a bal-ance of one-way and two-way communication. Not every posting has to generate dialogue. The social media manager should also post articles and links to the company's website or blog to increase traffic and search engine optimization. Links could also include pictures and videos to round out the coverage.

Customer service is one of the most frequently cited reasons for why organizations engage on social media. A 2011 Harris study (as cited in Loechner, 2011) found more than half (68%) of the consumers who posted a negative review after the holidays received a response from the retailer, which resulted in 18% returning to the retailer as a loyal customer. Also, nearly a third deleted the original negative post, and the same percentage wrote a positive review. The importance of customer service cannot be understated, but there are many stakeholders who are important on social media sites. Therefore, social media should not be used only for customer service.

Although one can argue providing guidelines may restrict communication with stakeholders, the following non-inclusive guidelines should help organizations be transparent and authentic:

1. Set up a social media policy. Neef (2003) suggested in order to behave ethically, organizations should establish and follow both a code of conduct and value statement, the latter of which outlines expected organizational principles. The same can be said for a social media policy, which should provide guidelines for appropriate use of social media sites. However, these policies should not limit the transparency or authenticity of the organization.

2. Learn how to effectively measure social media. Some practitioners suggest a magic number of postings per day, which makes the organization seem inauthentic. Also, claiming there is a dollar value for each Facebook "like" or Twitter follower or using other methods that lack validity is ineffectual.

3. Follow those who follow you. To create dialogue and build relationships, an organization must listen. Find out who the followers or fans are, and converse authentically. Also, your friends and followers cannot send you a private message (direct message) on Twitter if you do not follow them.

4. Answer and respond to all inquiries/issues. If you have the capability of deleting a negative comment, don't. Instead, respond. Research indicates transparency in resolving issues in the public arena helps to maintain and rebuild trust (Jahansoozi, 2006).

5. Do not send auto direct messages (DMs) on Twitter. They are unauthentic and seem disingenuous.

6. Be open, honest, and forthright. This directly lends itself to transparency. If an organization moderates comments, then specific details of such moderation should be given. Organizations need to think carefully about how moderation should take place to not violate standards of authenticity (Henderson, 2010). Organizations must also disclose any information that may affect a stakeholder's decision-making process.

7. Be real. To be authentic, social media managers should be themselves. Tell stories, talk to people, and show empathy. Use a "human" voice and be professional.

Organizations must be aware of how they are communicating with and engaging various stakeholders. If the organization finds it difficult to adopt certain policies of transparency and authenticity, then there should be an evaluation, and possibly overhaul, of the organization's overall communication strategy. On the other hand, Facebook and Twitter may not be suitable platforms for all organizations. How useful they are depends on how well they fit into the organization's strategy. Finally, communicating transparently and authentically builds and maintains trust as well as credibility with various stakeholders, who determine the success or failure of an organization.

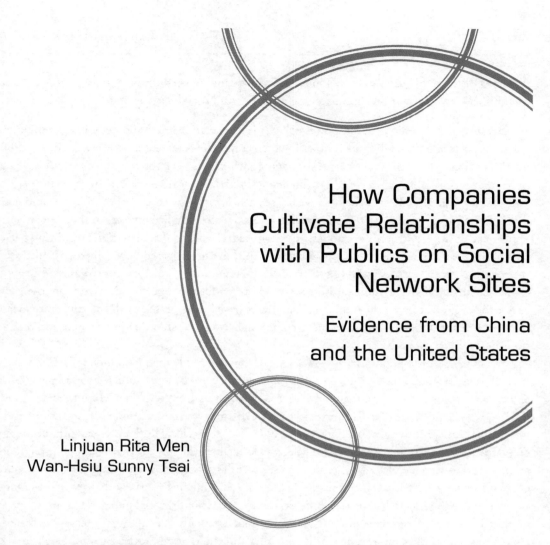

How Companies Cultivate Relationships with Publics on Social Network Sites

Evidence from China and the United States

Linjuan Rita Men
Wan-Hsiu Sunny Tsai

This chapter advances our understanding of relationship cultivation on social media from a cross-cultural perspective. We examined how companies use popular social network sites (SNSs) to facilitate dialogues with publics in two culturally distinct countries: China and the United States. We employed a content analysis of 50 corporate pages with 500 corporate posts and 500 user posts from each country. Overall, companies in both countries have recognized the importance of SNSs in relationship development and employed the appropriate online strategies (disclosure, information dissemination, and interactivity and involvement), but the specific tactics vary across the two markets. Furthermore, cultural differences among the types of corporate posts and public posts on SNSs indicate that culture plays a significant role in shaping the dialogue between organizations and publics in different countries. We discuss implications for corporate relationship management practice in the global market in the digital era.

In today's rapidly evolving media landscape, organizations use an extensive array of interactive media channels to engage with their stakeholders (Avery et al., 2010). Social network sites (SNSs) have become an integral part of many individuals' daily life, making SNSs an indispensable venue through which companies, nonprofit organizations, and even politicians stay connected with net-savvy citizens. boyd and Ellison (2007) defined a social network site as a Web-based service that allows individuals to construct a public or semipublic profile, build and maintain connections, and display their social connections to other members. SNSs integrate Web-based services and technologies such as blogs, bulletin billboard systems (BBSs), and social games. Thus, they constitute powerful multimedia platforms that provide such utilities as information sharing, entertainment, and social networking.

Because of their unique "viral" power in sharing information and building online communities, SNSs are an important public relations tool with which companies create dialogues with their publics. Social media are characterized by user-generated content, which has been found to be more effective than traditional marketing communications in influencing the attitudes and behaviors of other users (Thackeray, Neiger, Hanson, & McKenzie, 2008). As a consequence, organizations are now building and maintaining SNS public pages to improve their social network salience, enhance interest in their organizations, and build relationships with online publics. In response to hypes and speculations regarding the effectiveness of SNSs as a public relations tool, recent studies have investigated how companies cultivate relationships with their publics using SNSs (Bortree & Seltzer, 2009; Smith, 2010b; Waters, Burnett, Lamm, & Lucas, 2009). However, most of these investigations adopt a single perspective, from the viewpoint of either the organization or the public. Few analyses have incorporated both the corporate messages and the voices of the public to provide a more comprehensive understanding of organization-public interactions in the SNS context.

Furthermore, the popularity of SNSs has become a worldwide phenomenon. Although American sites such as Facebook have acquired a loyal following overseas, local audiences in many countries have embraced various culturally adapted native sites such as Renren in China, Mixi in Japan, and Orkut in India. Because members of different societies have distinct communication predispositions and Internet behaviors (Barnett & Sung, 2005; Kim, Coyle, & Gould, 2009), local differences can inhibit the application of Western public relations theories. However, existing literature on international public relations focuses mainly on traditional offline communication. Research has not adequately explored how cultural differences influence public relations professionals' use of social media to serve audiences in different cultures.

To bridge the research gaps and understand the dynamics of global public relations, we explored the implications of SNSs for corporate relationship management. We analyzed messages created by both companies and publics on leading SNSs in China and the United States, two important markets with huge Internet populations but with dramatic cultural differences. For this comparative study, we selected Facebook—one of the most popular SNSs worldwide (Zhang, 2010)—for the American sample, and we chose Renren to represent Chinese SNSs. Established in 2005, Renren remains among the "most popular, most open and best-financed social network" sites in China (Lukoff, 2010, para. 16). Nicknamed the "Facebook of China" (M. A., 2010), Renren is considered to be the Chinese equivalent of Facebook in terms of interface design (see Figure 1). Similar to Facebook, Renren gained its initial traction among college students and then began targeting a broader audience of young Chinese professionals.

This study is based on a multifold strategic and theoretical framework, including the online relationship cultivation strategies proposed by Kent and Taylor (1998), the widely adopted uses and gratification (UG) theory for understanding publics' media participation, and Hall's (1989) cultural context framework for understanding cross-cultural differences. In this chapter, we first examine the relationship cultivation strategies used by major companies in the two countries to build dialogues with strategic publics on leading SNSs, and then we explore how the patterns of corporate and public SNS posts in the organization-public dialogue project cultural differences.

Figure 1. Snapshots of Renren and Facebook Login Pages.

Literature Review

Relationship cultivation on social media

Because of the enormous popularity and the communal and collaborative nature of social media, recent studies have examined strategies for building relationships with online constituents (Smith, 2010b). Earlier studies provided empirical evidence that because of their interactivity features, blogs are more effective than static websites and traditional media for building dialogues with online publics (e.g., Kelleher & Miller, 2006; Kent, 2008). In recent years, sites like Facebook and Twitter—which are not only interactive but also inherently social and communicative (Avery et al., 2010)—have been advocated as optimum tools for online relationship building (Smith, 2010b). In addition, SNSs humanize organizations. On SNSs, users can become "friends" and "like" the organizations, facilitating relationship building at a more personal level (Kent & Taylor, 1998).

Recognizing the great potential of SNSs, scholars have recently begun to examine how organizations use SNSs for online relationship cultivation (Smith, 2010b; Waters et al., 2009). Existing literature has identified a variety of strategies for relationship cultivation in offline settings, including positivity, disclosure, assurances of legitimacy, networking, visible leadership, responsiveness, educational communication, and respect (e.g., Grunig & Huang, 2000; Hung, 2006). Kent and Taylor (1998) introduced the construct of dialogic communication, which has been adopted by several studies to examine the impact of various online strategies for relationship building (e.g., Kelleher & Miller, 2006).

Researchers have identified three essential strategies for relationship cultivation on the Internet. The first strategy, *disclosure* or *openness*, refers to the willingness of the organization to engage in direct and open conversation with publics. SNSs provide a convenient way to disclose organizational information to online publics. For full disclosure, organizations should provide a full description of the organization and its history, mission, and goals; use hyperlinks to direct users to its website; and use logos or other visual cues to provide intuitive identifications (Waters et al., 2009).

The second strategy, *information dissemination*, addresses the needs, concerns, and interests of publics while disseminating organizational information. Kent and Taylor (1998) argued that such information allows publics to engage with the organization as informed partners. On SNSs, organizations can post photos, videos, and announcements, and publicize information about products, promotions, or companies. They can also redirect users through hyperlinks to such external content as media coverage and microsites. According to Waters et al. (2009), including press releases and campaign summaries further maximizes the impact of the organization's presence on SNSs.

Last, *interactivity and involvement* plays an important role in cultivating relationships (Jo & Kim, 2003). McMillan, Hoy, Kim, and McMahan (2008) categorized three types of interactivity on the Internet, which can also be applied to the SNS context: human-to-computer interactivity (e.g., navigation); human-to-human interactivity (opportunities to contact the organization, to make a suggestion to a friend, or to share the content on one's own page on SNSs); and human-to-content interactivity (opportunities to comment on organizational posts and to respond to other users' posts). However, much remains unknown about the use of SNSs in public relations and in different cultures. Therefore, we propose the following research question:

RQ1: What strategies do companies in China and the United States use to cultivate relationships with publics on SNSs?

Understanding publics' voices on corporate SNS pages

In addition to building a corporation's connections with its publics, a corporate SNS page provides a platform upon which publics can interact with the organization and other important constituents, including fans, existing and potential customers, and opponents, who can collectively influence how the organization is perceived. To better utilize SNSs for building relationships, public relations professionals must understand the topics of publics' posts, and the publics' motives for involvement in corporate SNS pages.

Scholars studying online participation have applied uses and gratification (UG) theory to the Internet context (Ruggiero, 2000). According to the theory, audiences actively participate in media consumption to fulfill personal needs, or gratifications. Ginossar (2008) applied the UG theory to understand user participation in online communities where users function simultaneously as media audiences and content providers. Ginossar identified six categories of online activities: information seeking (questions to the organization or other users); information reply (responses to questions); unsolicited information (provision of information as an announcement, not a reply); emotional support (praise or encouragement, not information exchange); advocacy (calls for action); and conflict (expressing complaints or criticisms). Using Ginossar's categories of online participation, we explore publics' patterns of engagement in corporate SNS pages in the United States and China to understand publics' communicative needs and thereby help companies formulate dialogic relationships with their SNS members.

> RQ2: What are the patterns of communal participation as reflected in publics' posts on corporate SNS pages?

> RQ3: What are the differences in patterns of public participation between corporate pages on Facebook (United States) and Renren (China)?

Cross-cultural differences in global public relations

Verčič, Grunig, and Grunig (1996) argued that, although effective public relations across cultures shares general principles, strategies should be adapted to the political and media systems, levels of economic development, and cultural values of the local culture—a concept known as *specific application*. Cooper-Chen and Tanaka (2008) argued that culture often impacts public relations practice. Hall's (1989) typology of high- and low-context cultures has been widely adopted as a useful theoretical framework for determining differences between Western and Eastern cultures (An, 2007; Kim et al., 2009). According to Hall (1989), *context* refers to the situational information that one must acquire in order to understand the meaning of an event or subject. Hall categorized cultures into high-context and low-context, according to the degree of context dependence. In *high-context* communication, "most of the information is already in the person, while very little is in the coded, explicit, transmitted part of the message"; whereas a *low-context* communication or message is "just the opposite, i.e., the mass of the information is vested in the explicit code" (Hall & Hall, 1990, p. 8). Thus, whereas information in low-context cultures like the United States is conveyed in a clear and straightforward manner, marketing communications in high-context (e.g., East Asian) cultures tend to be more indirect and ambiguous (Cooper-Chen & Tanaka, 2008), demonstrating more emotional and harmony-seeking appeals (Miracle, Chang, & Taylor, 1992).

Prior studies indicated that high/low contextuality determines significant cultural differences in communication styles, and that these differences are reflected in various types of media content (An, 2007; Kim et al., 2009). Websites from low-context cultures tend to provide more product-specific information, whereas sites from high-context cultures tend to offer information about consumers' connections to their community (Lee, Geistfeld, & Stoel, 2007). As China exemplifies a high-context culture and the United States is characterized as low context, we hypothesize that

> H1a: Compared to the corporate messages on American SNSs, those on Chinese SNSs are less likely to convey direct and explicit information about products, promotions, or companies.

> H1b: Compared to corporate messages on American SNSs, those on Chinese SNSs are more likely to contain indirect and implicit information such as product-related educational or entertainment information or even messages irrelevant to the product or company.

Methodology

As public relations on SNSs comprises a relatively new development, an exploratory content analysis constitutes an appropriate method for understanding this phenomenon.

Sampling

The sample included 100 corporate profile pages: 50 from Renren and 50 from Facebook. We analyzed all 50 corporate pages available on Renren at the time of the investigation; 31 of these were Fortune 500 companies that had Facebook counterparts. To generate a comparable sample size for the Facebook data, we analyzed the 31 Fortune 500 counterparts on Facebook, plus 19 other randomly selected Fortune 500 corporate pages on Facebook. Then, using a systematic random sampling method, we chose 10 corporate wall posts and 10 public posts from each corporate profile page. User posts made for obvious advertising purposes were excluded from the sample. The final sample included a total of 1,000 corporate wall posts and 1,000 public/user posts: 500 from Renren and 500 from Facebook.

Coding scheme

Coding categories used in this study were adapted from previous research (Buis & Carpenter, 2009; Ginossar, 2008; McMillan et al., 2008; Waters et al., 2009). Consistent with Waters et al.'s approach of coding online relationship cultivation strategies, we examined the corporate profile page and wall posts to identify the presence of items representing organizations' strategies of disclosure, information dissemination, and interactivity and involvement. For strategies of disclosure, we evaluated whether the corporate SNS page provided detailed descriptions of the organization, organizational history, mission statement, URLs to the organization's website, or logos or visual cues to establish connections with the public. Regarding information dissemination, we examined whether the following items were present: news links to external media coverage; information in visual form (photos, illustrations, and videos); announcements; links to press releases; or campaign summaries. To measure interactivity and involvement, we combined several coding schemes from the literature on new media (Buis & Carpenter, 2009; McMillan et al., 2008; Waters et al., 2009) to capture SNSs' interactive features. Final coding for the interactivity and involvement strategy included organizational contact information (e.g., email, phone number, physical address); navigation (hyperlinks to external content); opportunities for commenting and sharing; action features for SNS engagement (polls, SNS applications, games, quizzes); and whether corporate communicators responded to user comments, thereby completing the dialogic loop.

To provide an in-depth understanding of how companies create dialogues with publics, we added six categories of corporate posts to the coding scheme: product specific, promotion specific, company specific, product-related educational or entertainment information, solicitation of responses, and non-brand-related messages. To analyze user comments, we adapted coding items from Ginossar's (2008) study examining Internet users' gratification. These included information seeking, unsolicited information, emotional support and expression, advocacy, conflict/complaints/criticism, and comments unrelated to the brand.

Coding procedure

Two bilingual coders familiar with both American and Chinese cultures coded the corporate pages on Renren and Facebook in early 2011. The units of analysis were the corporate profile page, corporate wall posts, and user posts. Following Waters et al.'s (2009) approach, the coders evaluated only whether each strategy and item occurred on the SNS pages. The coders scrutinized the first page of each corporate SNS profile. They then coded the randomly selected corporate wall posts and user posts into each category. The overall intercoder reliability ranged between .90 and .96 for the corporations' relationship cultivation strategies, and between .95 to 1.00 for the patterns of wall posts.

Results

RQ1: Relationship cultivation strategies on SNSs

All three online relationship cultivation strategies—disclosure, information dissemination, and interactivity and involvement—appeared on the corporate pages on Facebook and Renren, suggesting that corporations in both countries have integrated SNSs into their public relations campaigns. Yet, certain tactics were more commonly employed than others, and there were country-based differences. Table 1 shows a cross-tabulation between the American and Chinese SNSs and the relationship cultivation strategies employed therein.

Under the disclosure strategy, description of the company, URL to company websites, and logo/visual cues (i.e., product photos) were common. Mission statements and company histories occurred much less frequently, especially in China. In terms of providing information, companies on Facebook and Renren posted photos and made announcements equally frequently, but American companies featured significantly more videos and news links than their Chinese counterparts. Campaign summaries, which reported event turnouts and the results of promotion campaigns, showed lower levels of use. With respect to interactivity and involvement, the intrinsic attributes of SNSs—including navigation to external media content, and commenting and sharing features for publics—appeared on most corporate SNS pages. Corporate pages on Renren were more likely to provide organizational contact information such as email addresses, physical addresses, and telephone numbers ($c2 = 12.25, p < .01$). Yet, American companies incorporated significantly more action features on their Facebook pages, including online games and polls that engaged publics through online participation ($c2 = 33.28, p < .01$), while Chinese companies were more likely to redirect publics to external microsites. In terms of creating a complete dialogue loop, Chinese companies responded to users' posts more frequently than their American counterparts ($c2 =$

Table 1. Relationship Cultivation Strategies on Renren vs. Facebook

RELATIONSHIP CULTIVATION STRATEGIES	RENREN (CHINA)	FACEBOOK (U.S.)	C2
Disclosure			
Description	46 (92%)	44 (88%)	.44
History	25 (50%)	29 (58%)	.64
Mission statement	16 (32%)	25 (50%)	4.08
URL to website	43 (86%)	49 (98%)	4.89*
Logo/Visual cues	49 (98%)	47 (94%)	1.04
Information Dissemination			
News links	10 (20%)	19 (38%)	3.93*
Photo posted	49 (98%)	50 (100%)	1.01
Video files	43 (86%)	49 (98%)	4.89*
Announcements and press releases	48 (96%)	50 (100%)	2.04
Campaign summaries	12 (24%)	12 (24%)	.000
Interactivity and Involvement			
Organizational contacts	17 (34%)	3 (6%)	12.25**
Navigation	49 (98%)	50 (100%)	1.01
Commenting opportunity	49 (98%)	49 (98%)	.000
Sharing to one's own page	50 (100%)	50 (100%)	.000
Action features for online participation	17 (34%)	45 (90%)	33.28**
Response to user posts	47 (94%)	26 (52%)	22.37**

Note. Degrees of freedom (1), **p < .01, *p < .05.

22.37, p < .01). In many cases on Facebook, although the company may have initiated a series of discussions by posting a question or comment to solicit responses (e.g., "How much better would your day be with Coca-Cola?"), the company did not respond to or interact with users' comments.

H1: Cultural differences as reflected in corporate posts

Table 2 illustrates the mean differences of the numbers of various types of corporate posts on Renren and Facebook. Both American and Chinese companies were most likely to post information about promotions. The two countries did not differ significantly across the categories of product, promotion, and company-specific information. Therefore, Hypothesis 1a was rejected. However, corporate posts on Facebook tended to follow the hard-sell approach by frequently providing information directly related to the company and its offerings. In contrast, corporate posts on Renren were more likely to emphasize product-related educational/entertainment information ($t = 1.94$, $p < .05$) and to feature messages completely irrelevant to the company or its products ($t = 3.30$, $p < .01$). For example, a Chinese eyewear company offered tips for maintaining good eyesight, and Chinese companies were more likely to provide photos or videos of their celebrity endorsers or to simply post greetings, jokes, or human interest stories

Table 2. Corporate Posts on Corporate Pages on Renren vs. Facebook

CORPORATE POSTS (MEAN)	RENREN (CHINA)	FACEBOOK (U.S.)	T
Content of Corporate Posts			
Product specific	1.94	2.72	−1.82
Promotion specific	3.38	2.76	1.55
Company specific	.96	1.00	−1.14
Product-related educational/entertainment	2.46	1.76	1.94*
Solicitation of responses	.22	.86	−3.22**
Non-brand related	1.34	.52	3.30**

Note. Degrees of freedom (1), **p < .01, *p <. 05.

to engage with their SNS members. Such messages indicate an implicit and indirect communication style typical of a high-context culture, providing support for H1b. Additionally, corporate posts on Facebook were more likely to include direct questions soliciting responses from their members as compared to those on Renren ($t = 3.22$, $p < .01$), indicating a more explicit approach of communication.

RQ2 & RQ3: Publics' uses and gratification on corporate pages on SNSs

To inform companies' relationship cultivation strategies by providing an understanding of publics' uses of corporate SNSs in two distinct cultural settings, we analyzed a random sample of 1,000 public posts. RQ2 addresses the pattern of publics' participation on Renren and Facebook, and RQ3 explores the cultural differences as reflected in the different types of user posts featured on the two sites. The findings are summarized in Table 3. In general, the most common types on both sites included information seeking ("Where can I find Coca Cola Orange?"), emotional support ("I love Converse!"), and unsolicited information ("Wish myself a smooth new year!"). However, Renren users were less likely to post criticisms and complaints about the organizations' products or services ($t = -2.83$, $p < .01$), while Renren users were more likely to inquire about product information ($t = 2.27$, $p < .05$) and to engage in conversations that were not product or company related ($t = 3.44$, $p < .01$). No significant differences were found between publics' posts on corporate pages on Renren and Facebook across the categories of unsolicited information, expressing emotional support, or advocacy, although these types were slightly more prominent on Facebook.

Discussion

This chapter explored how companies incorporate social network sites (SNSs) to cultivate relationships with online stakeholders. We examined the content of communication created by both organizations and publics from a cross-cultural perspective. Findings suggest that companies in China and the United States commonly use the dialogic strategies of disclosure, information dissemination, and interactivity proposed by scholars (e.g., Kent & Taylor, 1998; Waters et al., 2009). Companies in both countries attempt

Table 3. User Posts on Corporate Pages on Renren vs. Facebook

USER POSTS (MEAN)	RENREN (CHINA)	FACEBOOK (U.S.)	T
Content of User Posts			
Information seeking	3.22	2.26	2.27*
Unsolicited information	1.46	1.60	-.47
Emotional support and expression	2.20	2.84	-1.64
Advocacy	.36	.56	-1.28
Conflict/Criticism/Complaints	.58	1.36	-2.83**
Comments unrelated to the brand/company	.96	.38	3.44**

Note. Degrees of freedom (1), **p < .01, *p < .05.

to be open and transparent by providing company descriptions and URLs for their websites and by using visual representations to establish brand recognition and identification. Notably, the high prevalence of corporate posts in the categories of product, promotion, and corporate activities, as well as publics' information-seeking posts, underscore the importance of SNSs in the dissemination of organizational information. Companies in both countries capitalize on the multimedia features (e.g., photos, videos, and interactive polls) available on SNSs. Thus, they are able to stay true to the relationship-oriented nature of SNSs by posting not only product and promotional information, but also brand- and product-related educational information and entertaining materials.

However, our findings suggest that companies have not taken full advantage of SNSs. For example, fewer than one-third of the Renren pages (only slightly higher for Facebook) provided links to external media coverage. Fewer than one-fourth of the pages provided campaign summaries, and only 6% of Facebook corporate pages provided organizational contact information to facilitate further dialogues. More importantly, much of the Facebook communication was one-way, as American companies were significantly less likely to respond to user posts and thus failed to truly engage with their publics. This underscores Wright and Hinson's (2009b) conclusion that practical gaps exist between what is happening and what should be happening in terms of corporate use of social media. Thus, our findings indicate that corporate communicators need to embrace SNSs with a "no post left behind" mentality to cultivate dialogic communication and meaningful relationships with their strategic publics.

This study also provides support for the global public relations theory of general principles and specific applications (Grunig, Grunig, & Dozier, 2002). Although companies on both Facebook and Renren made extensive use of the overarching relationship cultivation strategies of disclosure, information dissemination, and interactivity and involvement, we identified several significant differences in terms of specific tactics. For example, companies in the high-context Chinese culture featured more product-related educational information or entertainment news, or messages completely irrelevant to the brand, company, or product category, revealing a more indirect way to engage consumers and a relationship cultivation strategy based heavily on entertainment and socialization. A cosmetic company on Renren posted daily make-up tips without specifying brand name or product models, and a corporate post from an eyewear company evoked a highly personal tone to remind its SNS fans to drive home safely after a hard day's work. In a high-context culture like China, publics rely greatly on extended social networks for emotional exchange, and they value trust and the relationship with the company more than explicit product information. In such a context, corporate communicators also tailor their SNS messages to publics' social needs. Companies emphasize being personable and acting like a caring friend, capturing the essence of SNSs. In contrast, companies on Facebook are more likely to post messages directly and explicitly related their product, promotions, and corporate achievements, and are significantly less likely to engage in non-brand-relevant discussions.

Christ (2005) anticipated that the Internet would become a crucial avenue for stakeholders to learn about a company and its products, and this study found posts inquiring about the company's products, promotions, or after-sale services to be prevalent in both countries. However, cultural differences can be observed in publics' interactions with companies on SNSs. Renren users were more likely to use corporate pages for information seeking (inquiries directed to the company as well as to other users), revealing a greater dependence on social networks for information in the collectivistic Chinese culture. Also, complaints and criticisms appeared relatively infrequently on Renren, reflecting a cultural emphasis on group harmony. A closer examination of the numerous non-brand-related posts on Renren also reveals a high-context communication orientation that underscores users' socialization with the company. Many user-initiated posts greeted the corporate communicator as a friend ("Good night, Dell!" "Motorola, happy Valentine's day!"). After a corporate message was posted at a late hour, a user even advised the company's SNS communicator to "go to bed early." Moreover, there appeared to be a higher degree of interaction among Renren users, as many posts were responses to other user posts. In contrast, Facebook users were more likely to respond to company-initiated posts or to initiate individual posts. User posts on Facebook also tended to express complaints or criticisms about the brand, product, or company, reflecting the individualistic and low-context orientation of American culture, which favors competitive appeals and more straightforward modes of communication.

Conclusion

The innate social, communicative, and interactive characteristics of SNSs provide an advantageous tool for companies to build and maintain relationships with strategic publics. To better utilize SNSs to engage stakeholders and facilitate authentic conversations, public relations professionals must equip themselves with strategies, tactics, skills, and cultural sensitivities for social media relationship management. Overall, the principles of online relationship cultivation for Internet tools such as websites and blogs—disclosure, information dissemination, and interactivity and involvement—hold true for SNSs, the new genre of social media. However, online dialogic communications are culture bound, and global public relations professionals need to customize their SNS messages and tactics to local audiences when communicating to people with distinct needs and gratifications, as well as with people who have differing Internet behaviors and communication orientations.

Online Brand Communities

New Public Relations Challenges Through Social Media

Romy Fröhlich
Clarissa Schöller

With the advent of the social Web, online brand communities (BCs) have spread rapidly. Since their humble origin as rare, offline gatherings of fans of specific brands or products some ten years ago, companies have begun to deliberately initiate and sustain BCs, using social media platforms. However, little is known about the strategic purposes and potential uses of BCs within the context of corporate communications. In our chapter, we first discuss the role and characteristics of online BCs within a company's communication environment, drawing upon theoretical models from communication, business studies, and sociology. Consequently, we develop a model that distinguishes specific communicative changes raised by BCs as well as a typology of users in Web-based BCs. Based on a series of interviews with communication professionals experienced in the use of BCs, we subsequently distinguish four kinds of strategic uses of BCs in corporate communications, thus offering a starting point for future research.

Change of corporate communication through social media

Since Tim O'Reilly popularized the term "Web 2.0" early in the 21st century, numerous new applications have been created that allow common Internet users to publish their ideas and thoughts online. This new type of content was labelled *user generated content*, its applications are often summarized as *social software*, and Web 2.0 became the *social Web*. One interesting question that arises is what effects these changes have on corporate communications. In particular, how do they affect the day-to-day business of public relations professionals? To address these questions, we will first discuss several models of communication flows that help to explain the enormous transformation public communication has undergone in the past decade. Building upon this theoretical background, we will then introduce a communication instrument that may enable public relations practitioners to better face these new challenges and to profit from the shifts in the communication environment. We refer to this instrument as online brand communities.

From two-step flow of communication to peer-to-peer network communication

In 1948, Lazarsfeld, Berelson, and Gaudet published their famous study searching for mass media effects on people's voting decisions in presidential elections. Surprisingly, they found that voters hardly appeared to be directly influenced by media coverage in their decision-making. Rather, "informal personal contacts" (p. 150) emerged as a major driver of opinion formation. In particular, they identified a relatively small, highly influential group of people who functioned as so-called *opinion leaders* within their community. These "traditional" opinion leaders are trusted individuals within a community who typically access a wide range of mass media and other sources for information. Subsequently, they relay this information, mixed with their own opinion, throughout their community.

Based on these findings, Katz and Lazarsfeld (1955) developed their *two-step flow* theory of mass communication. Corporate communication professionals quickly discovered the practical relevance of this theory. That is, if a few opinion leaders can influence a much larger number of people in their voting decisions, persuading the opinion leaders might also provide an efficient route to affecting people's purchase decisions. Identifying and addressing opinion leaders became one of the main challenges in corporate communications (Muniz & O'Quinn, 2005).

In 1994, Eisenstein further developed the two-step flow theory into a *multi-step flow of communication*, which allowed for additional steps and mediators of communication. More importantly, whereas Katz and Lazarsfeld (1955) restricted the role of opinion leaders to individuals personally known to those following their advice, Eisenstein (1994) lifted this requirement. Her model also included *virtual* opinion leaders such as news anchors or experts shown on TV. Eisenstein argued that people can develop trustful relationships with individuals they know only from technically mediated communication, but have never met in person. One particular value of the multi-step flow model, hence, is that it can be easily transferred toward other technically mediated communication environments such as the World Wide Web. Following Eisenstein, a handful of virtual opinion leaders should be able to influence, via a cascade of multiple relays, a large number of people on the Web.

With the advent of Web 2.0 and its social software applications, the view of communication flows has finally been refined once more toward a notion of network communication (boyd & Ellison, 2007; Brown, Broderick, & Lee, 2007). The idea of networked online communication channels shares many similarities with the multi-step flow of communication. Chiefly, in both models, information is received and forwarded in multiple steps. They differ, however, in the assumed degree of hierarchical structure. In a multi-step flow model, the sender of information is easily identified and located on top of the communication chain. In network communication, by contrast, the source of information can be anywhere and is often hard to identify. Moreover, within Web 2.0, senders of information no longer need specialized skills to communicate. Social software enables millions of people to actively participate in networked communication. Communication professionals are confronted with an amazing number of stakeholders

who are now able to communicate their opinion on the Internet. This development raises several implications for public relations practice, which we will discuss in turn.

Symmetrical communication and cluetrain public relations

One consequence of social software enabling non-hierarchical, networked communication is that
companies and customers directly face each other at eye level on the social Web. This unprecedented
situation has been anticipated by Grunig and Grunig's (1992) excellence model, in particular the two-way
symmetrical model of public relations (see also Grunig & Hunt, 1984), which framed public relations
as an effort to build long-term, mutually beneficial relationships between organizations and their publics. Two-way communication, specifically two-way *symmetrical* communication, has been proposed as
a normative framework for effective and ethical public relations practice (Grunig & Grunig, 1992). The
means of achieving such relations is to discuss relevant issues in *emancipated dialogues* by way of how the
social Web and public relations models can interact.

For decades, the two-way symmetrical model had been considered a utopian ideal standard that can
hardly be implemented on a consistent basis in real public relations practice. Organizations were unable
to confer directly and dialogically with their multitude of stakeholders. Instead, organizations focused
on communicating with relatively few representatives of stakeholder groups and journalists to convey
their messages. With the emergence of the social Web, however, we now have a different situation. For
the first time in the history of public relations, the (technical) requirements for two-way symmetrical
communication can be met. Companies do not even need to develop their own technical tools, but they
can use available platforms like Twitter or Facebook for free. Not surprisingly, corporate communicators
have been eager to enter and leverage this online environment.

Pleil (2007) distinguished three main forms of online public relations: (1) *digitalized public relations*,
(2) *Internet public relations*, (3) and *cluetrain public relations*.[1] According to Pleil, digitalized public relations
simply means that conventional public relations information is provided digitally (e.g., on a company's
website). Communication remains asymmetrical, and stakeholders are addressed as largely passive audiences. Internet public relations takes communication one step further by providing channels for feedback
(e.g., contact forms), enabling collection of input from stakeholders. Cluetrain public relations, finally,
is "dialogue- and network-oriented" (p. 19) and defines stakeholders not only as *receivers* of communication, but also as *producers* of content and *senders* of communication. This view comes close to Grunig
and Grunig's (1992) two-way symmetrical model of excellent public relations. Cluetrain public relations
uses a variety of social software tools, featuring a mix of video, audio, and textual communication. By
enabling stakeholders to participate in an *emancipated* dialogue with an organization on a more or less
equal footing, social software is thus instrumental to achieving two-way symmetrical communication.

The most prominent social software tools used for public relations purposes are blogs,[2] microblogs[3]
(e.g., Twitter), podcasts and vodcasts (e.g., YouTube) and Social Network Sites[4] (SNS, e.g., Facebook).
SNS combine the advantages of other social software applications: (1) Their users feel committed to the
platform (since part of their social life takes place there), (2) information can spread quickly and easily
among relevant target groups, and (3) there is no need to attract users as they are already there.

Recently, some companies have attempted a very specific use of social software to create online communities around their products—so-called "brand communities" (BCs). We will define this unique and
promising social media instrument and delineate its opportunities and challenges for corporate communication activities that meld public relations and classic marketing communication.

Brand communities: Terminology, definition, and function for corporate communication/public relations

The origins of BCs lie ten years in the past, when they were rather uncommon, offline gatherings of fans that actively supported a certain brand or product. The term "brand community" was initially drawn from conceptual roots in both sociology and business studies.

The "brand" in "brand community": A business studies perspective

Despite its wide currency beyond marketing professionals and researchers, the term "brand" is rather inconsistently defined. There are multiple approaches, each of which stresses different aspects of brands. Haigh and Knowles (2004) examined different definitions and concluded that a brand is "everything from simple logos and trademarks up through the creation of a brand-focused company culture" (p. 24). Following Harris and de Chernatony (2001), there seems to be a trend in the literature to extend the brand concept from its initial focus on products to increasingly include organizations and employees in *corporate* branding. This shift followed from the observation that customers were no longer content with branded products only, but developed a keen interest in corporate attributes beyond products themselves (e.g., origins and circumstances of production, ethical standards, and information about other customers using the product). Harris and de Chernatony therefore advocated a more people-centered approach: "While management will still be required to initiate the process, staff should be encouraged to contribute to discussions" (p. 442). Dolak (2001) argued that a strong brand creates trust and an emotional attachment to a product or company. Accordingly, brand communication needs to move out of the exclusive responsibility of marketing communication and involve all corporate communication units, including public relations.

The "community" in "brand community": A social sciences perspective

The notion of community, although widespread across the social sciences, is a rather vague one. Cova and Cova (2002) claimed that in our postmodern society, individuals aspire to gather in "postmodern tribes" (p. 597). As deep running changes occur in society including the rise of individualism and dissolution of social bonds, people search for new forms of togetherness and belonging. Postmodern tribes address this need, forming communities that that are "inherently unstable, small-scale, affectual and not fixed by any of the established parameters of modern society" (Cova & Cova, 2002, p. 598). One person can belong to more than one postmodern tribe. Also, such communities are characterized by the possibility to leave the group without suffering major disadvantages.

As a result, marketing can address these communities directly, focusing on the customer-customer relationship rather than traditional company-customer relationships. In so-called *tribal marketing*, companies create platforms for customers and fans, allowing them to celebrate their brands as well as themselves using the brand, thereby sustaining the identity of the tribe. The duality of these platforms and the tribes using them is called a *brand community* (BC). BCs are thus not so much defined by the individual, but by the shared interest in one object of identification—the brand. Though BC members initially gathered offline, the rise of the social Web has enabled people to meet and share their common interests in buying, consuming, and celebrating a brand or product in an online environment.

The term "brand community" first appeared in the sociological literature around the year 2000. Muniz and O'Quinn (2001) defined a BC as "specialized, non-geographically bound community, based on a structured set of social relations among admirers of a brand" (p. 412). They understood BCs as "neo-tribes," (p. 414)[5] which are (in contrast to other forms of tribalism) *explicitly commercial*. The authors described three traditional markers of community that serve to define BCs as a specific kind of community in a sociological sense: Accordingly, BCs are characterized by: (1) *consciousness of kind*, (2) *shared rituals and traditions*, and (3) a *feeling of moral responsibility towards the community and its members*.

From a public relations point of view, these characteristics also describe what a company-initiated BC platform has to offer in order to flourish. That is, the platform must have an object of identification that is clearly defined, provide the space and possibilities to develop rituals and traditions, and encourage the formation of a sense of community by enabling its members to communicate directly (and on various levels) to get to know each other. The BC thus clearly departs from traditional, centralized top-down communication in which subordinate feedback channels are optional.

Brand communities, classic marketing communication, and public relations

In 2002, McAlexander, Schouten, and Koenig further developed Muniz and O'Quinn's (2001) idea of brand communities as a triad (with the brand and different customers on the corners of the triangle) into a *customer-centric model of brand community*. In doing so, they further differentiated the rather wide notion of brand used by Muniz and O'Quinn, distinguishing between *products*, *brands*, and *marketers*. This differentiation is crucial in that it now allows for the possibility of customers liking the brand but not all of its products, or enjoying a product while remaining critical of specific corporate policies. McAlexander, Schouten, and Koenig (2002) saw the main purpose of BCs in creating brand loyalty. The ideal BC member enjoys the product and trusts the marketer, which makes him/her loyal to the brand. At a first glance, this sounds like a classic aim of marketing. However, we argue that BCs go beyond marketing and tap into the domain of public relations because the formation of trust and loyalty is also heavily dependent on corporate reputation, which is classically the domain of public relations. Public relations and marketing communication need to go hand in hand. Although McAlexander, Schouten, and Koenig did not explicitly address this duality, they thus effectively described BCs as an online hybrid tool between marketing and public relations.

Because BC members are capable of critically reflecting (and communicating) their views on corporate behavior and reputation beyond the brand, we believe Muniz and O'Quinn's (2001) definition of BCs needs to be refined. BCs comprise not only supportive fans as "ideal members," but also other stakeholders who may hold different, and even opposing, views, as will be discussed. Based on Loewenfeld's (2006) definition, which introduced different types of BC members, as well as our own research, we therefore propose a new definition of a BC:

> A brand community is an interest-based, non-geographically bound community that is focused on one specific brand. A BC creates an environment that is characterized by high identification potential. It interactively unites adherents and admirers of the brand, customers with general interest in the brand, and/or a company's stakeholders sharing a professional interest in the brand. Brand communities exist in offline and/or online surroundings. Brand communities are initiated by users or companies and may be strategically used by companies as tools for corporate communication.

Specifically, the target groups noted in our definition underscore the need to involve public relations professionals when communicating with BCs. In our view, online (and offline) BCs could be an ideal tool to facilitate (or perhaps even force) a closer interplay between public relations and marketing communication. Combining different social software tools, the competencies of various communication units can be bundled and targeted to meet all relevant publics' interests. For public relations, BCs offer an ideal opportunity to enter a two-way symmetrical dialogue with clearly defined target groups. At the same time, social software applications (e.g., product-related vodcasts) allow for easy integration of marketing initiatives into communication with the BC. However, communication professionals are well advised to consider the communication and participation needs arising from the BC itself. That is, networked, community-conscious brand users may feel entitled to take part in *defining the brand* by bringing in their personal views on what the brand "really is about" (Arora, 2009, p. 15). Striking the proper balance between the strategic use of BCs and the needs of the community itself can be challenging. Even brand "evangelists" who greatly admire the brand may turn against the company if they feel that their input is not taken seriously or that the BC is nothing more than a disingenuous front for the company.

The application of BCs in corporate communications

The uses of BCs in corporate communications go far beyond spreading positive word-of-mouth (or negative word-of-mouth regarding competing brands) and "gain[ing] new customers without any additional attraction cost" (Arora, 2009, p. 15). Such views that limit the role of BCs to marketing communications fail to consider the broader public relations implications of communication with, and among, BC members. For instance, Sawhney, Verona, & Prandelli (2005) underlined the utility of collecting customers' opinions in virtual (Web 2.0) environments, involving them in a "joint experience of co-creation" (p. 6) as it occurs within BCs. By encouraging customers to talk about brands and products, companies gain valuable insights into possible problems, as well as potentials, for the development and improvement of both. Sawhney et al. specifically proposed that BCs can be instrumental in developing new products.

In summarizing different uses of BCs in corporate communications, it is useful to identify the strategic corporate goals that BCs can help achieve. By using BCs, companies can:

- address relevant publics at eye level
- increase brand loyalty
- increase brand equity
- win new customers by causing positive word-of-mouth
- build relationships among BC users and between the company and its publics
- use information from the BC to improve or even create products and services
- profit from BC users' knowledge

These goals can only be met, however, if BCs are employed *strategically* and based on *thorough planning*. This is particularly true for online BCs. In order to make full use of their potential, BCs need to be integrated with the company's overall communication strategy and complement other communication initiatives.

Communication in and around brand communities

Communication processes within and around BCs involve a variety of protagonists. Our model shown in Figure 1 attempts to clarify and describe these communication processes based on both theoretical and empirical findings.

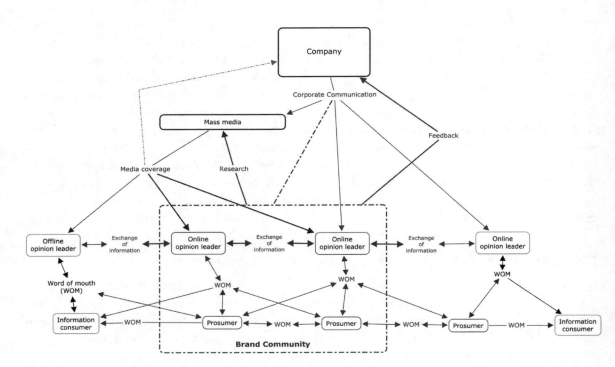

Figure 1. Communication Model of Brand Communities

Protagonists and target groups

The potential target groups for BCs comprise all publics relevant to a company. Although it is not necessary to always address all of them, BCs allow for a broad communication approach that targets multiple groups at once (e.g., by creating subgroups within the community). At the same time, BCs also interact with stakeholders outside the community. BC members utilize supplemental channels of communication to communicate with other users, journalists, and other interested publics. Specifically, journalists may be both targeted directly by the BC and use the BC for research on the company, its products, and services. Depending on the "openness" of the BC, journalists can also easily access information about BC members' perceptions of and ideas about the brand. Other stakeholders who do not actively participate in the BC themselves are likely to peek in and monitor information. The company itself is an important protagonist in the BC because it directly affects the community whenever it makes decisions about the brand and its products/services. However, within the social Web environment, the company can no longer claim to be the sole or perhaps even main communicator regarding the brand when BC members are able to voice their opinions in public. This is the epitome of two-way symmetrical communication as emancipated dialogue.

Within the BC (as shown in Figure 1), it is possible to distinguish three groups of members based on their interests in the BC and their extent of activity while using it: (1) opinion leaders, (2) prosumers, and (3) information consumers.

Opinion leaders perform precisely the role described by Lazarsfeld et al. (1948) by spreading (usually positive) information about a product among their acquaintances. Due to the trust they have earned among their peers, they are highly influential and may therefore be instrumental in persuading others to buy products as well. Obviously, this renders them extremely valuable for corporate communications, underscoring the relevance of the concept of opinion leaders even 60 years after its formulation (Muniz & O'Quinn, 2005). Within BCs, opinion leaders can be characterized as much better informed about a brand than average BC members. At the same time, opinion leaders also are much more likely than other BC members to share their knowledge with the rest of the community. Opinion leaders have potential to grow into true brand evangelists. They use the BC in the same way they would entertain a hobby by enjoying time spent on the platform, collecting and spreading the latest news about the brand and its products, commenting frequently on relevant media coverage, and spearheading discussions within the BC. They are very loyal to the brand, but also expect to be taken seriously by the company, as they often feel they understand the brand even better than the company does (Arora, 2009).

Prosumers refers to the possibility offered by social media for "consumers (audiences) to become producers" (Rennie, 2007, p. 25). Borrowing the term from sociology, prosumers are members who not only consume, but also "produce some of the goods and services entering their own consumption" (Kotler, 1986, p. 510).[6] They are not as deeply emotionally involved in the BC as the opinion leaders, but they likewise feel committed to the community and share their thoughts by taking part in conversations. Compared to opinion leaders, their use of the BC is more pragmatic and less driven by the urge to share their opinions. Rather, we characterize them as people who chiefly benefit from the pool of information the community provides them (e.g., when they have problems with a product or think about buying a new one).

While opinion leaders and prosumers actively take part in community life, the third type of BC members—or, to be more precise, BC users—is nearly invisible. *Information consumers* are best characterized as customers with a more general interest in the brand. Typical information consumers do not really see themselves as full members of the BC but rather use the platform to gather information on an ad hoc basis. Information consumers read through BC member discussions and reports. If they do not find the information they need, they may post questions and wait for them to be answered. Other than this, however, they are usually not active contributors to the group. For this reason, information consumers cannot really be regarded as *members* of a BC. However, from the company's point of view, they represent an important target group as potential new customers. Also, if a BC provides them with the desired information and enthuses them with positive evaluations of the brand and its products, they may eventually become prosumers or even opinion leaders themselves.

Communication processes in and around BCs

As a consequence of the BC's interactions with other publics, including potential customers, stakeholders, media, and offline target groups, communication professionals need to view BCs within their larger communication environment. Beyond catering to the needs of the different members within the group, they need to anticipate the possibility of further communication that involves the BC and its members. BCs neither depend on the company as their sole source of information about the brand (members also utilize, for instance, mass media for that purpose), nor do they necessarily keep their views contained within the BC (e.g., they may act as reliable sources of information for journalists). The company, its BC, and the media form a triadic communication relationship in which each actor may confer directly with the others, thereby circumventing gatekeeper functions traditionally performed by the company and the media. The same is also true in regard to other stakeholders who gain access to information from, and also convey information directly to, the BC. Discussions within the BC thus feed off and into a variety of other public discussions, rendering the use of a BC far more complex than most classic marketing communication and traditional (offline) public relations tools.

Types of online brand communities

Depending on both the company's strategic purpose and the needs of its BC members, BCs can serve a variety of corporate goals, among which public relations goals are merely one possitbility. In a June/July 2010 interview-based study of German corporate public relations practitioners,[7] we asked about the strategic uses and purposes of their BCs. Summarizing the results, there appears to be four main strategic approaches to the use of BCs as an emerging public relations tool. They are: 1) fan communities, which serve customer retention, feedback, and retention goals, for example, the fan page of software giant Microsoft on Facebook (http://www.facebook.com/Microsoft); 2) service communities, which serve customer retention and dialogue goals, for example, United Airlines' Twitter account (http://twitter.com/#!/United); 3) multi-purpose communities, which unite attributes of fan and service communities, for example, MINI's U.S.-based community (www.miniusa.com); and 4) special interest communities, which focus on what members wish to discuss more so than the brand itself, for example, Hobie's Facebook page (http://www.facebook.com/pages/Hobie-Fishing/347521991707?ref=ts) that focuses solely on fishing.

Conclusion

The development of social media has been rapid, causing several incisive changes relevant for corporate communications. Most of all, the shift away from familiar, established communication hierarchies has led to considerable insecurity among some communication professionals with regard to the social Web. As they are forced to abandon their heretofore comfortable positions at the controlling helm of communication chains linking the company to its environment, they suddenly find themselves eye-to-eye with their target audiences—with no journalists and mass media in between. This situation is particularly evident with regard to brand communities. Yet, is not this the situation that public relations professionals have always (claimed they) dreamt of? Is it not an ideal situation for public relations professionals to speak *directly* with their target audiences, cutting out the middle man? To work around censorious journalists who once had the power to decide what company-generated information would be conveyed to publics? Now, public relations professionals find themselves confronted with target audiences, that is, *numerous voices* of *numerous individuals,* who, using social media, are talking back to the company. The distant, anonymous mass toward whom press releases could be directed has morphed into a rout of opinionated, vociferous individuals.

This is also true for BCs, which represent a comparatively homogeneous group of individuals within this (often raucous) environment. By members involving themselves in topics once thought to be the exclusive domain of corporate communications, BCs (apparently) implicate incalculable challenges and risks—even the death of strategically planned, centrally controlled corporate communications. Against the background of the current professional discourse on "public relations and social media," one might easily be tempted to warn: Be careful what you wish for!

Threats of adverse rumors and compromising communications (possibly seeded by competitors) are included in Arora's (2009) list of disadvantages when working with Web-based BCs. Worst-case scenarios such as Greenpeace's attack on Nestlé's Facebook fan page[8] have communication professionals worried about security in their BCs and loss of control—even more so because social media tools accelerate communication across borders and around the clock.

Doubtlessly, positive word-of-mouth is what communication professionals strive for (Brown et al., 2007) but who is prepared to risk it turning negative and fanning out on the social Web? Companies need to be aware that company-initiated and -managed BCs are not only gathering places for fans, but also for everyday customers who might express their anger about shortcomings of products and services. Gone are the days when angry customers expressed their dissatisfaction only to the customer relations department in the offline world. Such voices now spread quickly on the Web and may severely affect a company's communication environment. However, of important note is that these risks exist on the social Web *with or without* the presence of company-sponsored BCs. Attributing these risks to BCs misrepresents the productive role that these venues may play in channelling, addressing, and even countering harmful communications spreading online. In a company-sponsored environment, communication professionals at least have a platform from which to respond, and respond quickly, not to mention the significant upside of building genuine relationships with publics through emancipated dialogues. BCs enable companies to have a voice in social Web communications. Although online BC activities are easy to monitor, companies need to react quickly to developments.

In an ideal world, companies would create perfect products and services and flawlessly deliver them to the marketplace, consistently delighting each of their stakeholders. There would be no need for dissatisfaction to spread across the social Web. Alas, no such utopia exists, and even the "best" companies meet opposition in their interaction with publics. Companies are best positioned when they engage with their stakeholders online and address misunderstandings in communication or negative word-of-mouth with bold, open, and trusted dialogue.[9] BCs provide a useful tool for communicating in this new environment—albeit under changed conditions.

ENDNOTES

1 See Levine et al. (2000) and http://www.cluetrain.com/

2 e.g., Boulos & Wheeler (2007)

3 e.g., Zhao & Rosson (2009)

4 e.g., boyd & Ellison (2007)

5 Neo-tribes are very similar to the postmodern tribes described by Cova & Cova (2002).

6 Typical prosumers are members of the do-it-yourself (DIY) community who sew their own clothes, grow their own vegetables, bake their own bread, etc. By using the social Web to share their ideas, communication becomes a DIY process.

7 See Schöller, 2010

8 See http://www.smh.com.au/technology/enterprise/handling-bad-pr-turns-sticky-for-nestle-20100326-r0t2.html and http://www.greenpeace.org/international/campaigns/climate-change/kitkat/

9 Numerous authors (e.g., Langheinrich & Karjoth, 2010; McLennan & Howell, 2010) recently highlighted the need to involve employees in social Web communication.

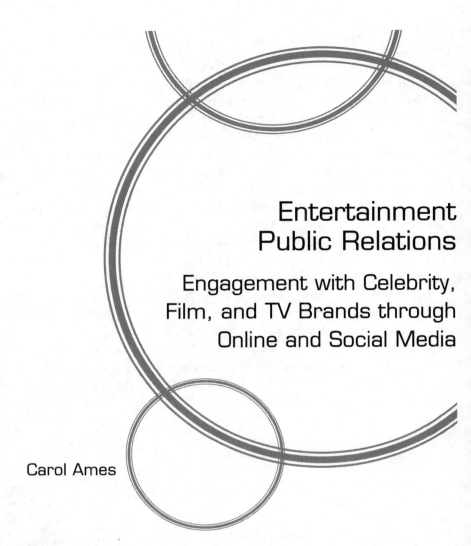

Entertainment Public Relations

Engagement with Celebrity, Film, and TV Brands through Online and Social Media

Carol Ames

Entertainment companies have been in the forefront of discovering ways to use online and social media. In 1999, online postings propelled the super-low budget The Blair Witch Project *to commercial success. The audience voting mechanism of "American Idol" beginning in 2002 also represents a milestone. More recently, when fired by NBC, Conan O'Brien went from no Internet presence to a pro presence and used Twitter to sell out his live tour within hours (Stelter, 2010a). Currently, "Tosh.O" on Comedy Central draws on viral videos as content, includes audience tweets within the show, and tweets live during the initial cablecast of each episode.*

This chapter will look at some special ways social and online media are used for public relations for celebrities, motion pictures, and television. For example, Twitter's most followed accounts are almost exclusively entertainment celebrities, and entertainment's traditional fan clubs now congregate and disseminate every new tidbit on fan sites, Facebook, and Twitter. Actors and even fictional characters have Facebook pages and blogs. In addition, advances in live streaming of video now make it possible to watch full episodes of television shows or specially created webisodes on company websites or dedicated YouTube channels, among others. Entertainment brands also have apps for the iPhone, iPad, and other mobile devices.

The star power of celebrities makes their live chats and Twitter accounts especially useful for publicizing entertainment brands, as various entertainment public relations practitioners, called publicists, try to leverage the two-way audience-interactive possibilities of online and social media to augment and sometimes replace traditional public relations tactics for creating buzz and excitement. Social media can facilitate fan engagement, but the entertainment business also presents some warning signs of social media usage gone awry, even to the point of damaging a brand beyond repair.

The paradigm is two-way communication—a "conversation." As at any dinner party, some conversationalists online are better than others. The better ones listen, respond, and share appropriately, while the bores just drone on about their own concerns without a thought for the other participants. Whatever public relations tactics are used, in entertainment, publicists and their campaigns are evaluated, as always, by bottom-line results: opening weekend box-office grosses for films and ratings for television shows that equate to higher prices paid by advertisers.

Celebrities lead the way—both up and down

How much of the general public even knew what Twitter was before April 15, 2009? That is the date Ashton Kutcher posted a Web video challenging CNN's breaking-news feed to a race to become the first Twitter account to have a million followers. At the time, Kutcher was third with 897,000 followers, Britney Spears was second with 906,000, and the CNN feed was first with 938,000 (Sutter, 2009). Kutcher said, "I found it astonishing that one person can actually have as big of a voice online as what an entire media company can on Twitter" (Sutter, 2009, para. 4).

Two years later (at the time of this writing), Twitter is at the center of our celebrity-obsessed culture. Entertainment celebrities count as 17 of the top 20 Twitter accounts, while CNN breaking news comes in at 21 with 4.2 million followers. Two of the non-celebs are Twitter itself and Twitter en espanol at 13 and 14, respectively. The other non-entertainer is President Obama at 4 with 7.4 million. The President of the United States is surpassed by Lady Gaga (9.5m), Justin Bieber (9.0m), and the enduring Britney Spears (7.5m). Numbers 5, 6, and 7 are Kim Kardashian (7.2m), Katy Perry (6.8m), and Ashton Kutcher himself (6.6m). Numbers 8, 9, and 10 are Ellen DeGeneres (6.5m), Taylor Swift (6.1m), and Oprah Winfrey (5.6m) (http://twitaholic.com/). Winfrey made her first tweet live on the air on April 17, 2009, the finish line of the race in which Kutcher beat CNN Breaking News to one million followers, only two days after his improbable challenge ("Oprah, Ashton Kutcher," 2009).

Since 2009, Twitter has become the preferred medium for direct communication between celebrities and fans. Lady Gaga has used her number one Twitter ranking not only to sell music, but also to mobilize her fans, as in the case of her campaign to help repeal the Don't Ask, Don't Tell law that prohibited gays from serving openly in the military. Her campaign included traditional public relations tactics such as: speeches at the National Equality March in Washington in October 2009 and in Maine in September 2010 that were widely covered in broadcast and print; and TV appearances on *The Ellen DeGeneres Show* and on the red carpet of the MTV Video Music Awards, where she was escorted by four members of the Servicemembers Legal Defense Network. That organization's website received almost 100,000 first-time visitors within 72 hours after Lady Gaga referred her fans to it (Parker, 2010). The speech was tweeted more than 1,000 times and "liked" on Facebook by more than 30,000 (http://www.ladygaga.com). This case demonstrates how social media can be used to supplement traditional celebrity-driven event publicity.

Recent incidents, however, demonstrate a darker side of celebrities' direct access to the public via new media. Charlie Sheen, formerly of *Two and a Half Men*, has a long history of drug, relationship, and legal problems. In January 2011, CBS and Warner Bros. put the series on hiatus, while Sheen supposedly once again entered rehab. This time, Sheen eschewed the typical celebrity crisis template, which is usually to: apologize through a publicist's statement to those one has hurt and humbly ask the fans' forgiveness; go into rehab and go silent in the media; after a quiet period, reemerge into public life as a "new person," complete with a serious commitment to a worthy charity, non-profit, or non-controversial issue; and finally, prove by ongoing good works and positive actions that the former alcoholic/drug addict/DUI offender/ spouse-abuser/minority-hater no longer exists. *Hancock*, the 2008 feature film starring Will Smith as a down-and-out superhero, presents an amusing take on celebrity image rehabilitation (Ames, 2010).

This time, Sheen went on the offensive in the mainstream media, including radio call-in shows, and others, on which he claimed to be healthy and ready to work. Eventually TMZ reported that Sheen's publicist Stan Rosenfield resigned minutes after Sheen told TMZ that Rosenfield had previously lied in a press statement about the reason for one of Sheen's hospitalizations ("Charlie Sheen's," 2011). Without the restraint of his publicist, Sheen's media binge continued with ABC's and NBC's national morning shows, which "have been relentless in recent days in aiding and abetting the epic meltdown of a celebrity who happens to be the biggest star on the biggest comedy hit at rival CBS" (Rainey, 2011, p. D1).

Shortly thereafter in early March, Sheen debuted on Twitter with the help of, "Ad.ly—a Beverly Hills firm that writes messages on Twitter or Facebook for celebrities who, for a fee, endorse products or brands," per the *Los Angeles Times*, which also reported that Sheen's first tweet said, "Winning..! Choose your Vice…" and linked to a photo of Sheen holding a bottle of chocolate milk, and porn star Bree Olsen—one of his two girlfriends—holding a Naked Juice fruit smoothie" (Olivarez-Giles, 2011, p. B3). A day later Sheen had 910,000 followers. Only days after this astounding social media debut, on March 7, Warner Bros. sent Sheen an official letter of termination from *Two and a Half Men* (Littleton & Weisman, 2011).

While threatening a lawsuit, Sheen next mixed new and traditional media approaches by creating a Web show and "My Violent Torpedo of Truth/Defeat Is Not an Option Show," a live tour of a new one-man show, which had a disastrous debut in Detroit on April 2 (McNulty, 2011; Scott, 2011), despite advance press coverage and feature articles placed by his new publicist (Flint, 2011).

Other celebrities have confronted the downsides of the digital media: their ubiquity and ease of use, their permanence, their speed, and their viral spread. For example, since cameras are now in almost every cellphone, members of the public, not just aggressive paparazzi, can post embarrassing candid photos online. In April 2011, online photos quickly and permanently appeared of the dressing room window that Chris Brown broke after a tough interview on *Good Morning America*.

Most cell phones now also have video/audio capability that provides members of the public with digital proof of celebrity misbehavior that might previously have been denied—the glamorous and famous celeb's word against that of the ordinary person. In March 2011, Dior designer John Galliano was taped in a drunken rant: "'I love Hitler'—caught, as so many career-enders are these days, on video and circulated on the Web" (Horyn, 2011, p. E1). The rant happened on a Thursday night and immediately made the viral video rounds. Galliano was suspended on Friday.

In another digital mishap in mid-March 2011, within two days of the catastrophic Japanese tsunami, caustic comedian Gilbert Gottfried unthinkingly tweeted 10 jokes about the Japanese victims. In less than 48 hours, he was stripped of his long-standing gig as the voice of the spokesduck for Aflac, which earns 75% of its revenue in Japan (Elliot, 2011). Gottfried's publicist, Steve Honig (as cited in Holson, 2011),

> first heard about it from reporters contacting him for comment. When he looked up the news online, he said he initially saw the postings mentioned 300 times. Seconds later it was 600 times, and within a few minutes, more than 1,000 times. "You could watch it spread in real time," (para. 14, 15)

and it was too late for the apology Gottfried wanted to make.

Many celebrity publicists are frustrated by their clients' refusal to understand the inherent dangers of new media. Veteran Hollywood marketing executive Terry Press (as cited in Holson, 2011) explained:

"If someone does not understand the value of a public image that has to be managed or considered there is nothing you can do...." A career is built on a certain public perception, she added, "and there is no advantage in showing people the reality is different." (para. 9)

The "different reality" travels fast online and is permanent.

On the other hand, a celebrity's understanding of social media can lead to outstanding public relations results, as demonstrated in the case of Blake Griffin. In February 2011, he won the slam dunk contest during the NBA All-Star Game by jumping over the hood of a Kia car to make the shot, accompanied by a gospel choir. Posted on YouTube by the NBA and numerous others, the views went parabolic. The viral video, along with "Blakegriffin" as a trending topic on Twitter, raised the profile of Griffin from a talented rookie playing for the perennially struggling Los Angeles Clippers to a magnet for product endorsements (Block, 2011).

The publicity teams for Griffin and the Clippers did much of the advance work and the follow up on Twitter and Facebook through postings, contests, photos, and video. Griffin also tweeted to stir excitement. Griffin's Facebook page doubled in popularity and gained more than 300,000 "likes" between the All-Star weekend and March 15 (Baca, 2011).

These cases—both positive and negative—demonstrate that it is important for celebrities to understand the ramifications of social media efforts and to take responsibility for helping to manage their brands toward a positive outcome. Celebrity uses and misuses of Twitter demonstrate a fundamental change in the relationship between a celebrity brand and the public, as well as a change in the role of the publicist. Previously the celeb's personal publicist negotiated and vetted the details of appearances and contacts with the press and thus the public. Now celebrities can evade both media questioning and the guidance of the publicist because Twitter offers a direct, unmediated line of communication to millions of followers. They can broadcast their every thought and whim in 140 character bursts, with no outside restraints, media training by the publicist, self-censorship, or even common sense.

What is the publicist to do? Some become the voice of the celebrity or, as in other businesses, oversee a social media staff that creates digital content and handles the celebrity Twitter, Facebook, and other social media accounts. Entertainment tweets may include a combination of promotional messages (e.g., concert dates) and ghost-written mini-insights that could plausibly come directly from the celebrity. Some publicists write under their own names on the celeb's account. Others valiantly try to rein in an egomaniacal celeb tweeter or serial interview grabber and gabber. Still others like Stan Rosenfield find that they can no longer do damage control, especially when new damage to the brand keeps going viral.

Film publicity, digital adaptations of time-tested public relations tactics

Films fall into two categories—studio films and independents—and their public relations and marketing campaigns have always varied accordingly. Because new media are often cheap to use, digital public relations tactics are likely to become more common for both categories.

For blockbuster studio films—the kind with $100 million plus production budgets—another 30–40% is spent on the marketing campaign. Publicity, advertising, and promotion have to be synchronized to create the maximum buzz and interest for the film's opening. In his outstanding analysis of film marketing, Friend (2009) wrote:

Modern campaigns have three acts: a year or more before the film debuts, you introduce it with ninety-second teaser trailers and viral Internet "leaks" of gossip or early footage, in preparation for the main trailer, which appears four months before the release; five weeks before the film opens, you start saturating with a "flight" of thirty-second TV spots; and, at the end, you remind with fifteen-second spots, newspaper ads, and billboards. (p. 44)

About three-fourths of the budget is spent on TV commercials, but trailers are key to attracting filmgoers, so disseminating them through new media has become crucial to film publicity. Ever since "real" footage posted on the Internet turned out to be a trailer for *The Blair Witch Project*, a dedicated film website has often "debuted" first the teaser and then the full-length trailer, along with other sneak-peeks and extras. For example, fans of a film franchise such as the Star Wars saga haunt the film website for new tidbits and flood the site's message boards. More recently, studios have used other media, including social media, to premiere trailers before they go onto the website and into theatres (e.g., use of TV for *Transformers: Dark of the Moon* during the Super Bowl and for *Cowboys and Aliens* on *American Idol*, Amazon.com for *Water for Elephants*, and Facebook for *X-Men: First Class*) (Stewart, 2011).

In late 2010, innovative publicity tactics began to generate awareness and buzz from fans of director J.J. Abrams and producer Steven Spielberg for *Super 8*. Tactics included an official website, not with a teaser trailer, but with a flash animation that when decoded frame by frame sent fans to another website (http://www.scariestthingieversaw.com/) that, in turn, led to a complicated scavenger hunt with complex clues to engage fans' interest and emotional investment. Meanwhile, the producers and the publicity team stood back and let fans engage with one another through blogs and other social media (Marani, 2011). On March 11, 2011, Paramount made news by starting a *Super 8* Twitter account and announcing the industry's first deal to release a trailer on Twitter so that, "Fans can visit Twitter.com/Super8Movie to get the first look at the trailer and use #Super8Movie to share information about the movie with friends" (Friedman, 2011, para. 4). Links to Facebook further enhanced communal viewing options. By March 15, the trailer had almost 4 million views on Twitter (Marani, 2011). Altogether, the marketing tactics resulted in a $35.5 million opening weekend for *Super 8* (Pandya, 2011). As Landmark Theaters CEO Ted Mundorff said, "Distributing online is faster than getting it [a trailer] out to the theaters.... Therefore, the faster you can create buzz, the better" (Stewart, 2011, p. 18).

YouTube is another place where trailers and "extras" can be posted with links to the film website, although a trailer on YouTube must fight through the clutter of cat-chasing-laser-light and laughing-baby videos. A more film-friendly, online environment is available on IMDb.com, the Internet Movie Database. IMDb is an elaborate and commercially adept film and television industry wiki that posts trailers and promos, cast lists and bios, technical information, and publicity information, such as the hidden *Super 8* URLs of scariestthingieversaw.com and rocketpoppeteers.com, as well as fan discussion boards. In addition, studios can buy home-page IMDb placement for "featured HD trailers" that link directly to ticket sales at local theaters (via movietickets.com), thereby converting a search for information not only to awareness or interest, but also to purchase intention and buying. Because of its wealth of archived information about thousands of films and TV shows, IMDb is usually one of the top search results for a current film title (along with the film website).

Its search visibility also makes IMDb a crucial and inexpensive publicity tool for the second category of films—the independents. These low-budget movies are produced and edited with no distribution deal in place (i.e., no way to get into movie theaters). The films "earn" distribution through the buzz and publicity they can garner, usually by first being selected for prestigious film festivals such as Sundance, Toronto, and New York, where they hope to be well received and thus earn an offer for a distribution deal. Because no advertising or marketing budget exists until a distributor steps in, low-cost, creative publicity is essential.

Indie film publicists create inexpensive press kits usually packaged in a colored, two-pocket folder with a sticker of the film's logo on front and all the typical movie press kit essentials inside: a media release about the film's basic elements including its stars, director, genre, and story; cast and credits; bios; production information, including story hooks and human interest anecdotes that can stimulate press coverage; and the availability of production photos and video clips. These elements are often also made available on a low-cost website, when possible. Everything for indies is done on the cheap at a tiny fraction of the cost for elaborate studio kits and websites. The filmmaker (who often functions as producer-director-editor and sometimes star) can also try to function as an amateur publicist at the pre-festival through distribution deal phase.

Low-budget film publicity needs to be creative and original to get noticed. For example, in early 2011, *Connected: An Autobiography about Love, Death and Technology* made the industry trade press because the filmmaker assembled a "discussion kit" for Sundance that included a book on a similar topic, a widely available phone app, "and a deck of conversation cards (i.e. 'What would the world look like when there are 12 billion people?')" (Kaufman, 2011, p. 4).

The biggest current weakness in recent publicity campaigns seems to be the inability to integrate new media with convergent marketing communications. That is, there seems to be a failure to optimize the use of new public relations tactics, especially the two-way capabilities of social media. Film publicists send out information in the same way they have always sent out press releases, but now the stories go directly to customers using Facebook and Twitter. What publicists do not do, with a few notable exceptions, is answer back when fans and consumers ask questions and try to have a conversation. This is probably due to how time-intensive, and therefore cost-intensive, it would be to execute two-way conversations with thousands of individuals. So, typically, the publicist may attempt to launch a discussion by asking an interesting question, and then letting consumers talk to one another, as *Super 8* did, thus creating a sort of online "discussion kit" and hopefully prompting the kind of discussion online that used to take place around the water cooler.

New media—Activating the couch potato

Television viewing has been a social activity since the days of the first TVs, when groups of neighbors would get together to watch. Later families would select one of the limited broadcast network choices to view together. Now that new media such as Facebook and Twitter have enabled social interactions among larger groups of distant viewers, the television networks, producers, and publicists are striving to find engaging ways to enter the conversations, whether about live broadcasts, reality shows, or scripted series.

As with sports programming, live broadcasts of award shows, such as the Academy Awards, the Emmys, and the Grammys, have unpredictable outcomes. Social media become virtual water coolers, and everyone can have an opinion. Each year there seem to be reports that the ratings of the Oscar Broadcast have fallen, but it is consistently the year's number two-rated broadcast (after the Super Bowl). Almost yearly, these awards shows change hosts, tweak their formats, and cut down on the number of awards presented and the length allowed for the winners' speeches. Now the shows are also trying to involve the audience via social media.

Audience engagement is the key goal, which is why CBS, the Grammy Awards, and Bon Jovi used Web voting to decide which song the band would perform live during the 2010 broadcast ("Fans Invited," 2010). MTV held a fan contest to select its first TJ (Twitter Jockey) and used its Twitter tracker for the September 12, 2010, Video Music Awards. The TJ read fan tweets during the pre-show and show. Twitter tracker, which was viewable both on the TV screen and in the theater during commercials, tracked who and what about the show was trending, and allowed fans and celebs in the audience to respond and influence the rankings ("MTV's VMA Twitter Tracker," 2010). For dual-screen viewing (TV and computer or hand-held device), the 2010 Emmy Awards set up eight cameras backstage to webcast behind-the-scenes footage, as well as a "Thank You Cam," for uninterrupted thank you speeches. In addition, host Jimmy Fallon invited viewer tweets about presenters to use on air, and he also tweeted live (Stelter, 2010b).

The 2011 Oscars also got into the social action, with co-host James Franco tweeting live as part of a "techno-wide…approach, offering unprecedented coverage along the red carpet and backstage, leaking details of the ceremony and soliciting fan requests and 'mominee' secrets (in which the mothers of some nominees got Twitter handles and let loose)" (Ryzik, 2011, p. C6).

Live broadcasts and Twitter seem made for each other. Everyone can participate, and the 140-character limit means that no one can hog the conversation. Advertisers love live shows, including sports, because the immediacy of the show and the engagement of the extra live social media elements discourage DVR time shifting. DVRs allow viewers to skip commercials, defeat the purpose of companies paying for expensive commercial airtime, and ruin the business model that sustains broadcast and basic-cable TV.

The networks, cable channels, Nielsen ratings service, and advertisers struggle to resolve how to evaluate and monetize time-shifted viewing on computers, tablets, and smart phones. Reality series, dramas, and situation comedies that are not broadcast live are trying to find different uses for social media that will engage viewers with both the show and the advertisers' products. Particularly with reality series, product integration into the story line of the show is hugely important, sometimes seeming to make *Survivor* and its reality offspring the contemporary equivalent of *The Price Is Right* or *Let's Make a Deal*, the enduring game shows that are actually non-stop commercials. New media public relations focuses on the broadcaster's website and the show's dedicated website. For *Survivor*, these sites include previews, contests, and *Survivor*-themed games, giveaways, and discussions. Twitter is used mostly for promotional messages, with links to either the *Survivor* website or the Facebook page, which was "liked" by more than one million people, as of October 2010 (Edmondson, 2010).

Facebook has become a primary publicity tool for television series, and as of January 25, 2011, fan-pagelist.com listed these five series as leading in "Likes": *Family Guy* (Fox); *South Park* (Comedy Central); *The Simpsons* (Fox); *House* (Fox); and *SpongeBob SquarePants* (Nickelodeon) as reported in *Emmy*, the member magazine of the Academy of Television Arts and Sciences ("Face Time," 2011).

Now networks, producers, and marketing departments are drawing on the wisdom of the crowd by using Facebook fans as vast focus groups. For example, ABC "asked fans their opinion on proposed key art for the current season of *Grey's Anatomy*" (Garron, 2011, p. 9). At Bravo, the reliance on social media for internal decisions is even more pervasive, with the channel creating new shows such as *Bethenny Ever After* based on the social media buzz about Bethenny Frankel's appearances in *The Real Housewives of New York City*.[1] Although Bethenny and her social media team do not respond to fan comments on Facebook at the time of this writing, she tweets and retweets a number of times a day on Twitter, and she answers fan questions (James, 2011), putting her in the rare company of the few entertainment brands that participate in two-way communication.

Evaluating online public relations efforts

Like other businesses, entertainment brands are wrestling with how to evaluate online public relations efforts. The bottom line for entertainment is still about the ultimate metric of money: sales of tickets for the all-important opening and second weekends of feature films, and the Nielsen ratings that determine the rate advertisers pay for a commercial on a specific television show. The film business has always valued "buzz," that intangible "sound of lots of people talking" created by advance publicity that companies try to quantify and evaluate. Firms including National Research Company conduct surveys and solicit feedback at free advanced screenings. And, on opening night, CinemaScore does exit testing of paying customers to assign a film a score that closely and positively correlates with word of mouth and the eventual worldwide box office gross (Goldstein, 2009). Clips of print placements and copies of TV and radio coverage, of course, still play a part in evaluating the effectiveness of traditional media placement efforts.

Many of the new media display built-in metrics: Twitter (followers and retweets), Facebook (likes), and YouTube (views). Tracking the metrics is time-consuming, however. Therefore, numerous film business and film-fanatic websites compile evidence of both buzz and box office results. For example, for current films, boxoffice.com compiles numbers of Facebook fans and tweets. Also, starting a month before opening, boxoffice.com tracks daily trailer views and their viewer ratings, as well as number of tweets and percentage of positive versus negative twitter buzz.

To evaluate Web tactics, entertainment publicists are using Google Analytics to study traffic sources (such as search, referral, direct, or campaign traffic) and navigation (page views, unique page views, average time on page). Search traffic can be broken down by keyword, and referral traffic can be analyzed by source, such as google.com or twitter.com (Claiborne, 2011).

Other entertainment practitioners use hootsuite.com, a dashboard for managing and linking social media (Twitter, Facebook, LinkedIn, WordPress, Myspace, Foursquare, and mixi, the Japanese site), and for analysis, such as active monthly Facebook users, total likes, and daily post views. Additional metrics attempt to evaluate online influence, rather than just count raw numbers, such as the millions who follow celebrities. The Klout Score encompasses 35 variables, such as retweets, @mentions, comments, and likes (Klout, 2011). In another example, *The New York Times* worked with research firm Twitalyzer to develop a Twitter Influence Index. The results showed, for example, that Kim Kardashian scored highly because she actually answers and interacts, links, and follows others (Leonhardt, 2011), rather than dominating the conversation with one-way, outgoing-only promotional messages.

Conclusion

Entertainment brands have been in the forefront of using digital technology for public relations purposes, because celebrities, high profile film marketing campaigns, and television shows drive so many water-cooler conversations. When a celebrity such at Ashton Kutcher begins to talk about the power of Twitter, the public hears about it via all the traditional media, as well as digital media. In recent years entertainment publicists at film studios, television networks, and entertainment publicity companies such as Rogers & Cowan have had their junior staff members and interns experimenting with every digital innovation that comes on their radar and often allow interns to make exciting contributions to publicity campaigns and to the field.

Analyzing the current entertainment uses of digital media shows that "two-way" and "conversations" need to become more central to entertainment publicity efforts. Also, celebrities and entertainment brands need to be aware of digital media's pitfalls, which include over exposure, the impulse to comment before thinking, and the permanence of digital content.

To achieve future best practices, digital campaigns will have to include budget lines for efficient social media management tools and analytics, as well as for the manpower to undertake and maintain labor-intensive, two-way communication and content tactics.

ENDNOTE

1   As stated by Jennifer Fader, vice president of e-media at Rogers & Cowan, at the Entertainment Publicists Professional Society conference May 4, 2011, Los Angeles, CA.

Social Media and Investor Relations

Alexander V. Laskin

Adoption of social media by investor relations professionals lags behind other communication specializations such as marketing or customer relations. The delay can be explained by investor relations' complex legislative landscape, absence of financial audiences on social media sites, and lack of understanding about the proper protocols of engagement in social media. Today, however, the situation is likely to change. The Securities and Exchange Commission issued an interpretive guidance that allows companies to communicate with investors online. More financial professionals are joining social media, and specialized investor-oriented new media sites, such as StockTwits and SeekingAlpha, have started to appear. Finally, investor relations officers are increasingly using social media to monitor investor chatter, conduct issues management, and combat rumors and misinformation.

What is investor relations?

Public Relations Society of America (PRSA) recognizes investor relations as one of seven specializations of public relations along with media relations, employee relations, consumer relations, community relations, government relations, and fundraising/donor relations (PRSA, 1988). The definition of investor relations clearly has parallels to other public relations practices. For example, National Investor Relations Institute (NIRI, 2003), a leading investor relations professional association, defines investor relations as "a strategic management responsibility that integrates finance, communication, marketing and securities law compliance to enable the most effective two-way communication between a company, the financial community, and other constituencies, which ultimately contributes to a company's securities achieving fair valuation" (para. 2). The definition focuses on two-way communication between a company and its publics—one of the cornerstone concepts in public relations.

At the same time, some argue investor relations is not a public relations functions at all, but rather a financial function. For example, Petersen and Martin (1996) discovered that most investor relations professionals have a financial background rather than a communication/public relations background, and that most investor relations professionals work in treasury departments and report to a chief financial officer rather than to a chief communication officer. The study concluded that investor relations "is most frequently treated as a financial function, both in terms who is in charge, and what are the qualifications for the job" (Petersen & Martin, 1996, p. 204). Laskin's (2009) research reported similar results with investor relations professionals proposing that "finance should take the lead" (p. 218) in investor relations. Laskin (2009) investigated whether investor relations really belongs to public relations or finance. He concluded that despite the differences in tactics employed, investor relations is part of the strategic function of managing relationships between an organization and its publics. In other words, it is part of the public relations function. Laskin (2010b) proposed that successful investor relations professionals must "combine the expertise of both communication and finance to devise sophisticated two-way symmetrical programs to facilitate dialogue between company's management and the financial community with the purpose of enhancing mutual understanding" (p. 24).

The ultimate measure of investor relations is a fair price of company's shares (Laskin, 2011c). Fair price does not mean the higher the better. Rather, it should be an accurate reflection of the company's actual value. Fair pricing is based on the deep and accurate understanding of the company. Such understanding requires investor relations professionals to provide both positive and negative information. In addition, to enhance understanding, companies have to expand their communications with shareholders from obligatory financial disclosure to include information beyond U.S. Generally Accepted Accounting Principles (GAAP), especially information about non-financial and intangible assets. In fact, the Centre for the Management of Environmental and Social Responsibility (as cited in Hockerts & Moir, 2004) explained, "Investors increasingly consider non-financial aspects in their assessment of companies" (p. 85). Thus, the quantity and quality of information investors require today are constantly growing.

Laskin (2010b) suggested that these demands place extra pressure on investor relations professionals. It is not enough to be good communicators or good financiers. Investor relations practitioners must know and understand the company's business as well as or even better than the company's CEO:

> Demands on the investor relations officers are increasing as well—professionals are required to understand every aspect of their company's business, understand its position in the industry and be able to discuss this information with investors and analysts in compliance with variety of laws and regulations that govern the financial markets. (p. 27)

The obligation for a constant dialogue between the company and its financial publics flows from the need to help investors and financial analysts understand the company well. In fact, for years companies have engaged in interactive communications with investors: one-on-one meetings, conference calls, interactive webcasts, investor conferences, and even in-person roadshows to meet and have a dialogue with shareholders, investors, and financial analysts (Laskin, 2009). Such dialogue can be greatly enhanced by relying on new media technologies, and so it is logical to expect investor relations to benefit greatly from new media and be on the forefront of social media adoption. Yet, this is not the case. In fact, adoption of social media for investor relations purposes lags behind other uses such as marketing, customer service, and community relations, among others. The key reason behind this delay is legal pressure.

Indeed, investor relations is a highly regulated activity in most countries. In the US, the primary agency responsible for oversight of the stock market is the Securities and Exchange Commission (SEC). On the agency's website (www.sec.gov), the SEC (2011) stated its main mission is "to protect investors, maintain fair, orderly, and efficient markets, and facilitate capital formation" (para. 1). To achieve these goals, the SEC oversees federal securities laws, maintains the disclosure of financial information by publicly traded companies, and brings enforcement actions against violators of securities laws. Today, the SEC employs almost 3,500 people. The SEC also maintains the EDGAR (Electronic Data Gathering, Analysis, and Retrieval) system. All publicly traded companies are required to submit their financial information to EDGAR, and that information becomes available to anyone with an Internet connection.

The key pieces of legislature that govern securities markets today are the Securities Act of 1933, Securities Exchange Act of 1934, Regulation FD, Sarbanes-Oxley Act, and the Dodd-Frank Act.[1] The Securities Act of 1933 requires any original interstate sale or offer of securities to be registered. The company must file a document, called a prospectus, that describes the specific types of securities offered; information about the company, its business, and its management; and financial statements certified by independent accountants.

The Securities Exchange Act of 1934 aims at regulating secondary trade of securities. The Act provides regulation of brokerage firms, transfer agents, clearing companies, stock exchanges, and so on. The Act established guidelines for periodic reporting by major corporations. However, the reporting was often not fair. Large institutional investors received information before small retail shareholders.[2]

As a result, the SEC adopted a new rule, Regulation FD (Fair Disclosure), in October 2000 (Securities and Exchange Commission, 2000). The key stipulation of Regulation FD was to eliminate the practice of "selective disclosure," or, disclosure of information to some select parties (largely, institutional investors). Instead, Regulation FD requires the following:

> The regulation provides that when an issuer, or person acting on its behalf, discloses material nonpublic information to certain enumerated persons (in general, securities market professionals and holders of the issuer's securities who may well trade on the basis of the information), it must make public disclosure of that information. (Securities and Exchange Commission, 2000, *Executive Summary*, para. 2)

Such disclosure must also be done simultaneously to securities market professionals and everyone else, thus eliminating the opportunity for professional investors to "beat the market" by receiving information earlier.

The Sarbanes-Oxley Act, officially called Public Company Accounting Reform and Investor Protection Act and often referred to in industry as "SOX," was enacted July 30, 2002. President George W. Bush, when signing the law, stated that Sarbanes-Oxley is "the most far-reaching reforms of American business practices since the time of Franklin D. Roosevelt" (Bumiller, 2002, p. A1). Other observers called SOX "a most welcome gift to shareholders" (Bloxham & Nash, 2007, p. 14). SOX primarily focuses on further improving the quality and quantity of financial disclosure. To some extent, SOX was a governmental response to a wave of corporate scandals that shocked corporate America at the beginning of the 21st century. Many of these scandals were directly related to senior management's manipulation of information disclosed to investors, and, as a result, the inability of investors, both private and corporate, to properly understand the company's business and its value. SOX significantly expanded the scope of disclosure by public companies. SOX emphasized the accuracy and completeness of the disclosure by public companies and introduced personal responsibility of senior corporate managers for such disclosure.

Finally, the Dodd-Frank Wall Street Reform and Consumer Protection Act, signed into law July 21, 2010, promised a fundamental change in financial regulations, including improved investor protection and changes to securities regulation. The Act gave additional powers to the SEC, required creation of an Office of the Investor Advocate, and set to further level the playing field for retail investors, thereby enabling them to better compete with large institutional investors.

Although these regulations focus on enhancing disclosure, they do not specifically recognize social media sites such as Facebook, Twitter, or any others as legitimate vehicles for disclosure. Now that senior executives are held personally responsible and liable for the information disclosed, not many companies are willing to take the risk of using social media. Initial regulatory language made it unclear whether publishing an earnings release on a website would constitute public disclosure.

State of social media in investor relations

In August 2008, however, the SEC published guidance on the use of corporate websites for investor relations. This guidance was a long-awaited response to many inquiries that the SEC received from companies and investors about using the Internet for communicating financial information. Although the SEC (2008) still did not recognize all websites as legitimate vehicles for disclosure, the agency noted:

> In the context of a company web site that is known by investors as a location of company information, the appropriate approach to analyzing the concept of "dissemination" for purposes of the "public" test as it relates to the applicability of Regulation FD to a subsequent disclosure should be to focus on (1) the manner in which information is posted on a company web site and (2) the timely and ready accessibility of such information to investors and the markets. (p. 19)

In other words, the SEC suggested that some companies can rely on their websites for disclosure *if* such websites are commonly known to investors, information is easy to find on the website, and investors are notified of such posting. In fact, the SEC proposed a so-called "notice-and-access" release that would be issued through traditional wire services to simply notify investors about the disclosure and provide a link to the actual information on the company's website.

More importantly, the guidelines also recognized that in addition to websites "there are now many different channels of distribution of news and other information which account for the rapid dissemination of news today" (SEC, 2008, p. 19). This opened the door for companies to use new media channels for communicating with investors. In fact, the SEC directly encouraged such interactions between companies and investors:

> Companies are increasingly using their web sites to take advantage of the latest interactive technologies for communicating over the Internet with various stakeholders, from customers to vendors and investors. These communications can take various forms, ranging from "blogs" to "electronic shareholder forums." (p. 40)

The SEC launched its own Twitter feed in 2008 that, as of this writing, has almost 170,000 followers and provides updates on SEC activities, changes in regulations, and notices of enforcement actions. Once the agency clarified how online tools can be used for investor relations, companies carefully started experimenting with new media channels.

Many of the pioneers using social media in investor relations were Internet companies. It is not surprising that executives running online businesses and using new media for customer service, marketing, and other applications would also employ social media for communicating with investors, shareholders, analysts, and other financial publics. In 2008, eBay Inc. launched a corporate blog and Twitter feed. During eBay's earnings calls,[3] eBay corporate blogger Richard Brewer-Hay tweeted financial information from the call in real time, expanding the reach of the conference call to the Twittersphere. However, to comply with legal requirements, some eBay tweets must include a link to regulatory disclaimers, such as Safe Harbor statements[4], making communication seem unnatural. A *Wall Street Journal* article about corporate usage of social media also observed difficulties with eBay's tweets, noting they lacked originality and simply restated the words from earnings releases: "These days, Mr. Brewer-Hay is more restrained around financial matters. He posted around 75 tweets from his computer during eBay's quarterly earnings call Wednesday. Most repeated verbatim comments by eBay executives" (Tuna, 2009, p. B4).

Dell was among the first to launch a blog, Dell Shares,[5] dedicated exclusively to investor relations. The blog included slides from corporate presentations, YouTube videos of executive speeches, and webcasts of conference calls. CiscoSystems, Microsoft, Sun Microsystems, and other technology companies were also among the pioneers.

Some of the most common uses of new media in investor relations are still based on offline information. For example, it is common for companies to post presentation slides used at industry or investor conferences on SlideShare (http://www.slideshare.net/). Companies often re-broadcast conference calls after earnings releases or webcast other important events, such as shareholder meetings. Increasingly, it is becoming popular to post videos of senior executives' speeches on YouTube, as well as to post pictures of products, facilities, or meetings on Flickr.

Some companies also create RSS feeds to incorporate push-technologies into their content-distribution efforts. Twitter is often used to post links to this content with the goal of promoting it. Facebook is not used as often for investor relations purposes as it is for communicating with consumers. However, some companies use it to publicize their top executives.

Specialized new media venues have appeared that are dedicated exclusively to investing. One of the most popular is SeekingAlpha (http://seekingalpha.com/), a platform for professional and retail investors to share their opinions on a variety of stocks and their relative performances. SeekingAlpha aggregates content from hundreds of professional investors, creating non-stop conversation about companies and their investment value. Another one is StockTwits (http://stocktwits.com/), which utilizes Twitter to allow investors to discuss best and worst investment opportunities. StockTwits allows companies to create a verified investor relations account so that a company's investor relations department can have an official voice in this conversation. Wikinvest (https://www.wikinvest.com/account/portfolio/regx/start) is a wiki-based platform that allows individual investors to upload, analyze, and review their portfolios in one convenient online location. Needless to say, companies have to monitor these and many other social media channels and engage in conversations when appropriate (e.g., to correct a factual error or address a dangerous rumor).

One example of a long-standing company using social media for investor relations purposes is Citigroup. Citi is on the forefront of technological innovations. Its Twitter account established in 2009 had almost 6,500 followers by mid-2011. Although most of its Twitter usage focuses on consumer-related issues, Citi also announces its financial results over Twitter, attaching links to full earnings releases on the company's website. In addition, CFO John Gerspach has discussed earnings in YouTube videos. Citi's presence on Facebook includes CEO Vikram Pandit's personal page, where he has posted slides from various presentations, videos, links to his media mentions, and thoughts on Citigroup's future. One of the videos posted on Pandit's Facebook page was from Citi's annual shareholders meeting, making it available to the general public rather than limiting access to shareholders as is often the case.

Dominic Jones (2011), an investor relations consultant, proposed that using social media in investor relations is "now fast becoming standard practice for scores of companies—and many more will likely join the parade in the coming weeks" (para 2.) He even suggested that social media may soon replace other channels companies use to communicate with investors, shareholders, and financial analysts:

> In fact, company disclosure channels that once seemed innovative—such as investor relations websites, webcasting and PR wire services—are struggling to stay relevant as investors grow accustomed to receiving information from companies in real-time on their favorite social networks in formats that are easier to access and use. (para. 3)

It becomes vital to understand why social media channels are taking such an important place in investor communications. Investor relations professionals have always focused on direct, ongoing communications with financial publics. Talking to investors through media releases is less important in investor relations than direct conversation (Laskin, 2009). Investor relations officers often focus their attention on tens or hundreds of institutional investors rather than millions of retail shareholders. Social media can help investor relations professionals maintain conversations with a broader range of investors. In fact, the use of social media makes such communications more Regulation FD friendly because the conversation is online and visible to everyone rather than privately held via phone or email.

New media investment venues like StockTwits and SeekingAlpha facilitate conversations with multiple investors. Investors and financial analysts discuss companies, industries, and earnings releases simultaneously. Companies also participate in the conversation to make sure the information is accurate. These social media channels become important for investor relations professionals not only because of the number of visitors to these sites, but also because these visitors are current and potential investors.

Heaps (2011) reported that engaging in social media can also create savings for a company. Instead of sending hundreds of emails to hundreds of investors addressing similar concerns, an investor relations officer can create one Facebook post or one Twitter update to reach audiences using these channels. Follow-up questions and answers, when posted publicly, are visible to all users.

Another important feature of social media is its interactivity. Investor relations is a specialization in which the input of shareholders is highly valued. L. Grunig (1992) argued that due to their ability to constrain an organization, stockholders are one of three priority publics for corporations, and public relations practitioners should respond to them first. Laskin (2010a) explained that collecting feedback from investors and communicating it to the company's management is vital for investor relations. Because social media enable two-way communication and the SEC encourages it, investor relations professionals are well positioned to benefit from these online venues.

Social media provide more than two-way communication. These emerging technologies can help companies engage their shareholders. It is one thing for shareholders to read information, it is another thing for shareholders to comment on it, but it is even better if shareholders re-tweet or share this information with their networks of friends. Social media allow investor relations professionals to have an influence not only on the communications between a company and its shareholders, but also on the communications between shareholders and other publics as company content is shared and discussed. Laskin (2011a) proposed that we no longer live in the Information Age, but rather are transitioning to a Community Age in which personal networks become individuals' most valuable assets. Thus, companies must produce investor-related content that is engaging, innovative, and relevant to stimulate sharing between various publics and penetrate a variety of networks.

One more benefit of new media for investor relations is impossible to overlook, and that is access to information about investors. Research is an integral part of any investor relations program. Laskin (2009) reported that research is one of the most common activities of investor relations officers. They must carefully monitor what financial publics are saying about the company, its management, performance, and future prospects. It is equally important to monitor competitors and the industry at large. Investor relations officers often engage in formal investor perception studies. Monitoring social media can serve as daily investor perception research, which helps professionals better understand who is buying and selling their stock, changes in trading volume, and the best prospects for future contacts.

Despite its benefits, there are still many factors limiting the usage of new media for investor relations. Regulations, although allowing companies to interact online, still hold companies accountable for every word they say or type. As the SEC (2008) explained:

> Companies are responsible for statements made by the companies, or on their behalf, on their web sites or on third party web sites, and the antifraud provisions of the federal securities laws reach those statements. While blogs or forums can be informal and conversational in nature, statements made there by the company (or by a person acting on behalf of the company) will not be treated differently from other company statements when it comes to the antifraud provisions of the federal securities laws. Employees acting as representatives of the company should be aware of their responsibilities in these forums, which they cannot avoid by purporting to speak in their "individual" capacities. (pp. 42–43)

Every time investor relations officers or other company executives log on to a social network, they risk violation of SEC guidelines. For example, an investor relations professional checking in at Foursquare (https://foursquare.com/) *could* suggest that person is meeting with a potential target for a merger or acquisition. This check-in could prompt increased trading volume from those who are listed as friends on such a social network and decide to act on this assumption. Yet, shareholders outside of the network would not have access to such information and, from the SEC's perspective, would be at a disadvantage, thus creating a potential selective disclosure violation. Even bigger problems can be caused by an executive "liking" or "re-tweeting" a financial analyst's report on the company. This can be perceived as an endorsement of analyst's report by the company, once again putting the company in violation of SEC requirements.

This legislative restraint often creates a situation in which all tweets, Facebook updates, and other new media messages simply restate sentences from earlier press releases (or other official statements) because investor relations officers must avoid the risk of saying something that could violate a number of regulations. This also limits opportunities to actually engage in a dialogue with financial publics by answering questions in an informal, conversational manner.

Another problem with social media is that most online investor relations efforts are limited to private retail investors. Laskin (2011b) suggested that institutional investors often have policies that prohibit them from logging on to social media sites at work and thus do not have access to Facebook updates or Twitter feeds. It is unlikely they would rely on this information anyway because they have access to wire services and Bloomberg terminals, where information tends to first appear. For example, a recent survey of European institutional investors ranked social media as the least important source of information behind business media, analysts' research, primary market research, and information received directly from companies (Deutsche EuroShop, 2011). In fact, the study concluded that "the majority of respondents is skeptical about the importance and reliability of the information on social media platforms" (p. 21).

Because most shares of a company's stock are managed by institutional investors, it is not cost-effective for investor relations professionals to focus the bulk of their communication efforts on retail shareholders. Thus, it is unclear how investing in social media programs can contribute to the bottom line and whether such efforts are justifiable from a return on investment perspective.

Furthermore, even those institutional investors who have access to social media at work are less likely to engage with investor relations officers publicly. They are more likely to call or email their company contact. For some, it is a matter of corporate policy, for some, habit, but for many, it is also a reflection of their industry. The investment industry is highly competitive, and financial analysts are often not willing to publicly discuss specifics of a company's performance out of fear that they would expose their proprietary analytical processes and thus lose an opportunity to beat the market (i.e., other analysts). This, in turn, limits new media usage to distributing information rather than engaging financial publics in two-way communication and utilizing the very strengths of social media.

Future developments of new media in investor relations

New media already have a presence in investor relations, and it seems clear that new media will continue to gain importance. Undoubtedly, investor relations professionals will continue using social media to monitor conversations related to their stocks. In light of severe regulatory restrictions, listening, for now, is perhaps the greatest benefit of engaging in social media.

Although many investor relations professionals treat social media as a "large microphone for pushing out information" (Morgan, 2009, p. 1), such an approach is not sustainable in the social media landscape. Laskin (2010b) explained that "social media is a tool for dialogue and using it for pushing out information can cause more harm than good as the audiences will resist the push, challenge the information and provide critical feedback" (p. 71). Social media are an excellent monitoring and diagnostic tool, but not necessarily a microphone. Monitoring social media enables companies to constantly scan shareholder forums, blogs, and microblogs to identify and respond to issues before they spill over into newspapers, TV news, and the professional investment community. Indeed, companies have many opportunities to send information out, but significantly fewer opportunities to draw information back in. Social media can help mitigate this imbalance.

Companies will continue to participate in conversations, whether it is via traditional or new media channels. Executives want to have their voices heard, report their company's successes, and explain its failures. The SEC has encouraged use of electronic communications between companies and their shareholders by enacting a number of amendments about blogs, proxies, and electronic shareholder forums. Thus, the share of electronic communications is expected to grow. Laskin (2010b) proposed that social media can expand the scope of dialogue between corporations and their shareholders, particularly in regard to social responsibility. Investors, consumers, suppliers, local communities, and other publics are increasingly concerned with environmental sustainability, fair treatment of employees, and social accountability. Since these issues do not always make it into traditional corporate disclosures, companies may use social media channels to discuss related initiatives.

Both investor relations and public relations professionals must remain mindful of a spillover effect that frequently occurs in social media. For example, someone concerned with a company's pollution record may engage with a Twitter account or Facebook page reserved for investor relations issues. Conversations on SeekingAlpha, a site dedicated to investing, often include discussions of environmental issues, employee issues, or international bribery issues, all of which can affect stock price. Collaboration among various communication professionals is required to address the appropriateness of company policies and manage issues that can be damaging to corporate reputation.

As such, social media will likely accelerate the process of integrating all communication functions into one department or at least under the leadership of a chief communication officer. Investor relations, public relations, media relations, community relations, marketing, government relations, and similar functions must have a unified voice in new media. Increasingly, there are fewer opportunities to isolate messages for consumption by only one public. Today, information transcends over a variety of publics almost instantaneously.

New media's focus on engagement will continue to expand. Producing content that is unique, creative, and engaging, and having conversations about it with investors and financial analysts, will remain important, but how investors, analysts, and other relevant publics share, discuss, and otherwise use the content will gain even more importance. Integrated communication departments will be better suited for producing such innovative content. Every reader, then, has the potential to become an opinion leader by bringing this content into her or his network, thus truly utilizing the web-like infrastructure of the Internet.

Conclusion

Federal regulations, especially the introduction of Regulation FD, along with the availability of new media technologies, enable corporate financial information to be shared with current and potential investors, large and small, simultaneously. This regulatory push for equal access to information was influenced by corporate failures and financial abuses by large institutions. Both Sarbanes-Oxley and Dodd-Frank Acts represent government responses to the market's failure to regulate itself as well as a compelling interest to protect individual consumers and shareholders.

The scope of relationships investor relations officers have to manage is broadening. Companies are no longer permitted to limit their relationship-building activities to a handful of large institutional investors and analysts. New media can help. Webcasts, podcasts, and Twitter updates, among other options, can be designed to reach interested parties. Social media provide opportunities to listen to and engage in conversations with various stockholders and stakeholders, within legal limits. As the new media landscape continues to evolve, contacts will be evaluated not only by their direct involvement with the company, but also by the reach of the social networks behind them.

These changes will continue to push for various communication departments to speak with one unified corporate voice to consumers, investors, local communities, and others. This does not mean that customer service and investor relations will (or should) have the same Twitter account. Rather, this means that all communication functions will be located in one integrated public relations department. In a new media environment where messages cannot be segregated for consumption by just one public, such integration is a necessity.

ENDNOTES

1   Information about each of these regulations can be found on the SEC's website (www.sec.gov).

2   Institutional investors are entities that accumulate money from individuals and organizations and invest on their behalf. Some of the most common types of institutional investors are banks, mutual funds, pension funds, and insurance companies. Alternatively, people can invest individually—buying and selling securities on their own. Then, they are called retail shareholders.

3   Following public release of quarterly earnings, company executives will hold conference calls to discuss their earnings report with journalists, financial analysts, and other relevant parties.

4   For example, companies making any forward-looking statements or projections of future performance must include a disclaimer that such goals may not be realized.

5   http://en.community.dell.com/dell-blogs/dell-shares/b/dell-shares/default.aspx

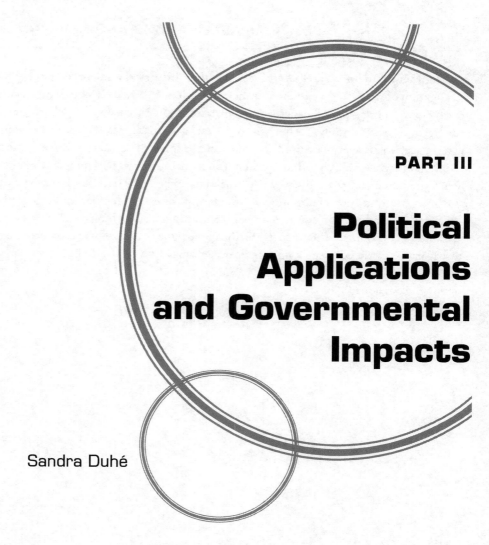

PART III

Political Applications and Governmental Impacts

Sandra Duhé

Overview

The extent to which governments have an impact on the daily lives of citizens and the degree to which governments are considered active publics for organizations differ around the world and are highly dependent on the political economy of established market structures. This section offers an interesting array of perspectives on the influence governments can have on the public relations process.

In the first chapter, Jordi Xifra offers an insightful look into how stateless nations, such as Catalonia, rely heavily on new media and social networking as part of their nation-building process. He additionally suggests a new approach to public diplomacy studies based on the concept of soft power. Soft power is closely tied to the theory and practice of public relations because of its focus on relationship building with foreign publics. Jordi's framework offers ample opportunities for future research.

Brooke Fisher Liu and Rowena Briones are the first, to my knowledge, to discuss how new media can be used to strengthen relationships between government entities and publics who are generally unprepared to protect themselves in the event of a terrorist attack. Preparation, they explain, diminishes the effect of a primary weapon in terrorist activity: fear. Thoughtfully planned two-way communication tactics are required. Brooke and Rowena offer best practices for public relations practitioners and issue calls to action for researchers to further investigate this important and timely topic.

Mahmoud Eid and Derek Antoine drew upon historical and contemporary foundations in their investigation of whether Canadian Members of Parliament used Facebook to provide constituents with opportunities for access and debate or, instead, to simply further their own information dissemination and electoral goals. Findings from their interviews are amazing and indicate politicians have a long way to go if Facebook is ever to be used as a channel to support and advance democratic ideals.

Yi Luo shares with us her findings from extensive interviews with public relations practitioners working in China. She accurately highlights the need for public relations scholars to examine the use of social media in communication environments curbed by government regulations. She explains the fascinating Chinese social media landscape and, drawing on the input of her interviewees, reveals the extent to which the Chinese government affects social media campaigns. Those of us accustomed to communicating in a relatively unrestricted social media environment may find her findings somewhat surprising, but they are particularly intriguing because they come from professional sources inside the country rather than from media outside of China.

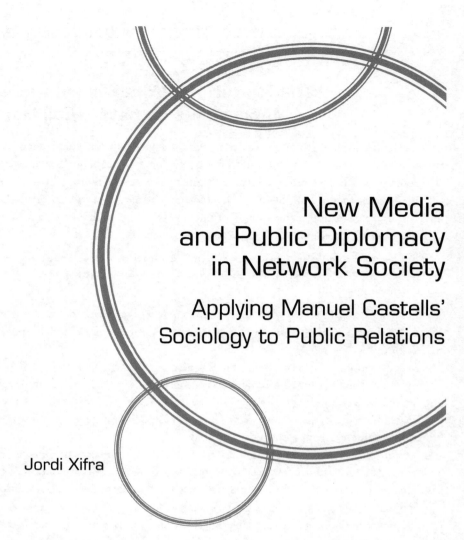

New Media
and Public Diplomacy
in Network Society

Applying Manuel Castells'
Sociology to Public Relations

Jordi Xifra

Nation-building has been a major topic of the public relations research agenda in recent years. However, these studies have been conducted from a limited perspective. In particular, the research has focused on nation-states, ignoring the role of stateless nations. Using the case of Catalonia as an example, this chapter first links with work on the network society by contemporary Catalonian-Spanish sociologist Manuel Castells, and then with the concepts of soft power and noopolitik coined respectively by Nye, and Arquilla and Ronfeldt, to offer a new public relations and public diplomacy approach to nation-building.

In this approach, the role of new media and social network services and websites is crucial. Arquilla and Ronfeldt coined the term "noopolitik" to mean a form of governance, to be undertaken as much by non-state as by state actors, that emphasizes the role of soft power in expressing ideas through all manner of media. In the noopolitik, non-governmental actors and the transnational networks woven by civil society and stateless nations are more successful than nation-states. The use of social media is a key factor for this success. From this standpoint, the idea of noopolitik offers an approach to managing relationships with foreign and national publics through new media and social networks. Furthermore, noopolitik offers a new model for the theory and practice of public relations as a governance tool for any organization. The aim of this chapter is to analyze this new model starting from a broader public relations/diplomacy approach to nation-building.

The limitations of public relations approaches to nation-building

One of the fields in which public relations scholars have focused their research in last two decades is the role of public relations in nation-building (e.g., Taylor, 2000a, 2000b; Taylor & Kent, 2006). Nevertheless, this research has been conducted from a limited perspective, mainly for three reasons.

First, it has focused on the idea of law-nation (the nation-state), ignoring the idea of cultural nation, and therefore excluding other forms of nation-building, such as those in nations without state.[1] As Castells (2008a) stated:

> Any detached observation shows that in modern times there are nations, there are states and there are different types of relationship between both: nations without state, nation states, multinational states and imperial nation states that absorb different nations by force. (p. 15)

To these, we can add states with shared nations (e.g., Korea), states without nations (e.g., Andorra or Singapore), and nationalist movements[2] to provide a more comprehensive public relations approach to nation-building.

Second, public relations scholarship has disregarded crucial issues of a new international order and, in particular, globalization. Third (and accordingly), it has attached little importance to public diplomacy and other forms of building relationships with foreign publics, such as those developed by territories other than nation-states. By not considering this situation, the public relations approach to nation-building has been conducted from a limited perspective.

Kymlicka (2001) suggested that national minorities (stateless nations) tend to respond to the nationalism of the majority (nation-states) by adopting their own alternative nationalism. They often utilize the same techniques that the national majority uses to promote these nation-building efforts (e.g., control over language). In consequence, both stateless nations and nation-states are being continuously built and rebuilt, particularly in the current globalized world. That is, the formalized governance characteristic of nation states, while not disappearing, is in the throes of major change as states face increasing political and economic pressures from outside their geographical borders. Keating (2001) discussed the effects of this ongoing political reorganization on the broader social order:

> This is providing opportunities for the construction of new systems of social regulation and collective action below and beyond the state, but these new systems take on a variety of forms. One of these is revived minority nationalism in territories where there is a historic sense of identity, an institutional legacy and a political leadership able to construct a new system. (p. 41)

This process of reorganization is what Keating (1997) dubbed *stateless nation-building*, that is, the building of new or transformed communities (what Anderson, 1983, referred to as imagined communities) and collective action systems. One type of these communities is represented by stateless nations (Guibernau, 1999). Catalonia, like Scotland, Wales, Flanders, and Quebec, is "a stateless nation" (Castells, 2004c, p. 45).

It is important to distinguish between nation-building and institution-building. Whereas institution-building focuses on the establishment of fundamental political and economic systems within a society, nation-building focuses on the creation of national identity and national unity (Taylor & Kent, 2006). The stateless nation-building process requires a great deal of institution-building but does not require nation-building in the conventional sense of the term because there is no state to impart unity (Keating, 2001). Rather, the stateless nation-building process is based on three major elements: identity, the creation of institutions, and global dimension. These elements are underpinned by the fact that stateless nations claim the right to be acknowledged as political and international actors and to have a voice in different forums, access to which hitherto has been limited to nation-states.

Hence, diplomacy is no longer the privilege of nation-states. Since 1945, international politics has become much more complex. Gradually, new non-state actors have entered the international scene (Arts, Noortmann, & Reinalda, 2001). Some of these non-state actors are of a non-territorial nature, such as non-governmental organizations (NGOs), multinational corporations, and sports organizations. Others, including stateless nations, have a territorial nature.

In keeping with the purpose of this chapter, we will now analyze the role of public relations and public diplomacy in the stateless nation-building process using the works of Castells (2000a, 2000b, 2001, 2004c), Nye (2002, 2004), and Arquilla and Ronfeldt (1999) regarding the network society, soft power, and noopolitik, respectively. Like Castells (2004c), we will draw from the public diplomacy efforts of the stateless nation of Catalonia to overcome problems derived from its lack of status, build an international reputation, and convey its national identity abroad. The Catalan model will also inform an approach to public diplomacy and public relations based on the notions of soft power (Nye, 2002, 2004) and noopolitik (Arquilla & Ronfeldt, 1999).

Building a national reputation abroad as a critical effort of stateless nations: The example of Catalonia

Following head of state Franco's death in 1975, Spain evolved into a democratic and autonomic state, which is formally defined in the Constitution of 1978. In 1977, the *Generalitat* (the official name given to the autonomous political institutions in Catalonia, made up by the Parliament, the President, the Executive Council, the High Court and the Ombudsman) was restored. In 1979, The Statute of Autonomy of Catalonia was approved, making the restoration of self-government possible. In 1986, Spain joined the European Union, where Catalonia proposed the recognition of autonomous regions as a driving force for economic development and social welfare.

Today, Catalonia has one of the highest levels of self-government in Spain, and the *Generalitat* is basically sovereign in health care, education, regional security, trade, industry, tourism, and agriculture. The region has gradually been allowed to use tax revenues as a source of autonomous funding. Use of these tax revenues has been the subject of controversy among Catalan politicians given that Catalan taxes amount to one-third of national income tax, and this revenue has been redistributed by the central administration in Madrid. In addition, Catalonia has, along with a few more regions, a special status in terms of language and culture. The regional language, Catalan, differs from Spain's official language, Spanish.

As Guibernau (2010) acknowledged, Catalonia's funding is the main problem of its condition as a stateless nation because it affects the action capacity of its government. However, this is not Catalonia's only problem. One must also highlight the lack of cultural and political recognition of the Catalan nation as problematic. Catalonia is not recognized as a nation, neither in Spain, inside the European Union, nor in the international arena. From an international perspective, Catalonia is in a position of inferiority and one of automatic exclusion "because its representatives are not fully represented on international forums" (Guibernau, 2010, p. 149). This situation is compounded by another fact, and one of special relevance: the complex nature of an international system framed within what Castells (2000b) called the "network society" (p. 21). Network (or knowledge) societies are those where the manipulation of symbols, and therefore of the realities inherent in them, becomes the core of existence in its multiple facets—social, economic, political and cultural (Castells, 2000b, 2001). This network society, Castells explained, conditions the international emergence of states by changing traditional sources of power.

In this context, it is crucial for stateless nations to enact policies that establish and maintain relationships with foreign publics and effectively position a region's identity in the international arena. In this regard, Catalonia has made significant strides in its public diplomacy efforts. As Keating (1996) recalled, Catalonia has demonstrated a commitment to the promotion of the Catalan nation as a different national society inside and outside Spain, although the separatist feeling of the Catalans is weak. Generally speaking, Catalans do not wish to be separate from Spain.

In 2010, Catalonia's outreach efforts resulted in the Catalan government's adoption of a Foreign Action Plan in which public diplomacy plays a main role. This strategy highlights new media as a crucial channel for building a Catalan image and reputation abroad. Indeed, the international sphere is like a spider's web, in that a small tremor at any of its ends is eventually transmitted throughout the web. Global interdependence explains how multiple facets of reality (e.g., political, economic, social, cultural) are united through a loose and mobile connection, which means that the sources of wealth and social power are constantly changing hands, places, and origins (Castells, 2000a). The basis of this global network is the noosphere, or knowledge (Arquilla & Ronfeldt, 1999). The management of this knowledge and its related elements (i.e., channels for the reception and dissemination of information along with "factories" and "intermediaries" of ideas, image, and opinion creation centers) will decide who will occupy the main nodes of this grid-like system constantly traversed by flows of knowledge, tangible and intangible capital, goods, and people (Castells, 2001).

Public diplomacy, soft power, and noopolitik: The triad of a new public relations model

Soft power is the ability to get what you want through attraction rather than coercion or payments (Nye, 2002, 2004). It arises from the attractiveness of a country's culture, political ideals, and foreign and domestic policies. In contrast to hard power, soft power strategies emphasize common political values, peaceful means for conflict management, and economic cooperation in order to achieve common solutions. The practice of soft power is similar to one of the forms in which power can be exercised in network society: "by the construction of meaning on the basis of the discourses through which social actors guide their acts" (Castells, 2009, p. 10).

Indeed, advances in new communication technologies have given rise to a knowledge-based society, where the appropriation, management, and transmission of ideas and know-how are a source of power. In this context, along with the notion of soft power, it is clear that political leadership is based, in large measure, on a race to achieve attraction, legitimacy, and credibility. Nye's (2004) approach to this topic is that although both hard power and soft power are necessary instruments for implementing a country's foreign policy interests, the tactical use of attraction is less costly than coercion. In the information age, a well developed soft power provides a competitive advantage to a country or a stateless nation because soft power indicates that a nation is adapting its culture and values to prevailing global standards, gaining access to flows of information and communication, and achieving greater influence in the management of national and international affairs (Nye, 2002).

In this setting, the network society differs from the information society, in that, as Castells (2000b) explained, information societies have always existed, since the very first *homo sapiens* until the human being of the industrial civilization. In the information society, knowledge was applied to technology. But nowadays we encounter a totally new phenomenon. In the network societies, the opposite is true. It is technology that is applied to knowledge. The role of public relations in the network society thus becomes essential by stimulating and provoking, from a symbolic relational perspective, active participation in the social construction of meaning.

The noosphere, a term coined by French anthropologist and theologian Teilhard de Chardin, has contemporary relevance for public relations. The term, however, must not be confused with cyberspace or infospace. Cyberspace fundamentally refers to the information that flows through the virtual network, while infospace combines this network-based information with information that circulates in the mass media. The noosphere is not just information, however. It is also the sum of the ideas, myth, beliefs, and attitudes that man produces through the collection and analysis of data. Noosphere refers to the extension of infospace to the mind, that is, all information provided on the Internet, plus any other information available in other formats (not necessarily electronic) and, finally, all available information in human minds (Arquilla & Ronfeldt, 1999). As an example of how information evolves across these various levels, Arquilla and Ronfeldt (1999) described the tenets of cyberspace as being interconnectivity and democracy, the tenets of the infosphere as prosperity and interdependence, and, ultimately, the tenets of the noosphere as sharing ideas.

This analysis has given rise to the term *noopolitik*, which underlines the role of soft power in expressing ideas, values, procedures, and ethics through the mass media. According to Arquilla and Ronfeldt (1999), this new paradigm has emerged in the information era and is affecting the prevailing paradigms of international relations. In this regard, Arquilla and Ronfeldt (1999) prefer to talk of "global interconnectivity" (p. 29) rather than interdependence. Like soft power, noopolitik seeks to attract, persuade, and influence public opinion.

From this standpoint, Castells (2001) warned that there is an increasingly greater need to design a new public diplomacy with a virtual-digital profile suited to the new communication context. This new model must go beyond the traditional hierarchical diplomacy in which information flows from top to bottom, and the state is the sole transmitter. Because the aim of public diplomacy "is not to convince but to communicate, not to declare but to listen" (Castells, 2008b, p. 91), public diplomacy now has to target society (Castells, 2001) and the players that emerge and act in it. In Castells' view, public diplomacy has to act as a soft power facilitator, promoting the capacities of persuasive discourse and using suitable technological resources to do so.

Thus, the noosphere is one of the planes on which international activity unfolds. Knowledge is the key. Knowledge may be achieved not through a one-way, linear process, but rather through multiple processes and stages, which may be summarized as information, perception, analysis and dissemination (Arquilla & Ronfeldt, 1999). New communication technologies act on the first horizon (information); soft power, in combination with public diplomacy, acts on the latter two (analysis and dissemination), and public diplomacy, as a public relations area, on perception and reputation.

For non-state actors, the international promotion strategy of a country's brand becomes increasingly important in the design and implementation of its economy and foreign policy. As such, the Catalan government has woven international promotion of Catalonia as a tourist destination into its strategic tourism plan. In this plan, use of new communication technologies is considered a competitive imperative. One of the most relevant efforts for promoting Catalonia through new media has been creation of the "9+1 Catalonia Experience" social website (http://experience.catalunya.com/?lang=en), in which interactivity with users is required to experience the Catalonia brand. This initiative uses different social nets, such as Flickr (http://www.flickr.com/photos/catalunya_experience) and Twitter (http://twitter.com/#!/catexperience), and also has a YouTube channel (http://www.youtube.com/user/CatalunyaExperience). Another governmental initiative is the social site *Fans de Catalunya* (*Catalonia fans:* http://www.fansdecatalunya. com/en/home/home), which gives users the opportunity to meet and share tourism in Catalonia, become a fan and participate in groups, and create their own groups, along with other interactive tools related to Catalonia as a desirable tourism destination. These technologies "herald an opportunity for Catalonia to increase the awareness of the Catalonia brand, together with the leverage of tools for the creation of knowledge and the improvement of competitiveness and promotion" (Generalitat de Catalunya, 2006, p. 195). Indeed, as Castells (2004c) stated:

> By not searching for a new state, but fighting to preserve their nation, Catalans may have come full circle to their origins as people of borderless trade, cultural/linguistic identity, and flexible government institutions, all of them features that seem to characterize the information age. (p. 54)

The noopolitik's implications for nation-building, such as the Catalan one, are observable. One of the main characteristics of Catalan nationalism (together with the distinctive language) is its relationship with the nation-state. The father of modern Catalan nationalism, Prat de la Riba stated in 1906: "Catalonia is at the same time European, Mediterranean and Hispanic" (as cited in Buffery & Marcer, 2011, p. 274). That is, Catalonia rejects separatism from Spain, but looks for a new kind of state. According to Castells (2004c), it would be a state that incorporates respect for the historically inherited Spanish state and the growing autonomy of Catalan institutions in conducting public affairs. Likewise, it would integrate both Spain and Catalonia in a broader entity, Europe, which translates not only into membership of the European Union, but also entry into various networks of regional and municipal governments, as well as civil society. These are the essential associations that multiply horizontal relationships throughout Europe under the tenuous shield of modern nation-states. Castells (2004c) expanded on this idea:

> Only a Spain that could accept its plural identities—Catalonia being one of its most distinctive ones—could be fully open to a democratic and tolerant Europe. And, for this to happen, Catalans have first to feel at home within the territorial sovereignty of the Spanish state, being able to think, and speak, in Catalan, and thus creating their commune within a broader network. (p. 53)

This differentiation between cultural identity and the power of the state is a historical innovation in most nation-building processes. As such, it fits well within the realm of network society in that nation-building is based on flexibility within and adaptability to a global economy, and the networking of media. In other words, the nation-building process is adapting to the noosphere.

In the progressive extension of the noosphere, stateless nations will find the management and use of soft power to be an asset in the implementation of public and virtual diplomacy policies as well as in their positioning in the world economic and political arena. Arquilla and Ronfeldt (1999) commented on the forthcoming role of soft power in global politics:

> Our discussion of the noosphere anticipates the next key proposal: At the highest levels of statecraft, the development of information strategy may foster the emergence of a new paradigm, one based on ideas, values, and ethics transmitted through soft power—as opposed to power politics and its emphasis on the resources and capabilities associated with traditional, material "hard power." Thus, *realpolitik* (...politics based on practical and material factors —those of, say, Henry Kissinger) will give some ground to what we call *noopolitik* (politics based on ethics and ideas, which we associate with many of those of George Kennan). (pp. 4–5)

The idea of soft power is highly relevant for public relations in that it renders possible the analysis of different forms of power positions constructed in the international arena between state and/or non-state actors. Furthermore, public relations can be analyzed as a form of soft power, or soft power can be studied as the ontological power of public relations practice.

Table 1. Contrast between realpolitik and noopolitik, as described by Arquilla and Ronfeldt (1999)

REALPOLITIK	*NOOPOLITIK*
States as the unit of analysis	Nodes, non-state actors
Primacy of hard power (resources, etc.)	Primacy of soft power
Power politics as zero-sum game	Win-win, lose-lose possible
System is anarchic, highly conflictual	Harmony of interests, cooperation
Alliance conditional (oriented to threat)	Ally webs vital to security
Primacy of national self-interest	Primacy of shared interests
Politics as unending quest for advantage	Explicitly seeking a *telos*
Ethos is amoral, if not immoral	Ethics crucially important
Behavior driven by threat and power	Common goals drive actors
Very guarded about information flows	Propensity for info-sharing
Balance of power as the "steady-state"	Balance of responsibilities
Power embedded in nation-states	Power in "global fabric"

Having reached this point, noopolitik offers a new public diplomacy perspective of managing relationships with foreign publics. Arquilla and Ronfeldt (1999) identified five trends that promote noopolitik: the growing fabric of global interconnection, the continuous consolidation of global civil society, the increase in soft power, the new importance of cooperative advantages, and the formation of the global noosphere. These trends do not render the paradigm of political realism (which focuses on hard power) obsolete, but they do make its use more complex, by rendering its limitations patent. The authors explained how soft power is inherent to noopolitik:

> Noopolitik is an approach to statecraft, to be undertaken as much by nonstate as by state actors, that emphasizes the role of soft power in expressing ideas, values, norms, and ethics through all manner of media. This makes it distinct from realpolitik, which stresses the hard, material dimensions of power and treats states as the determinants of world order. (p. 29)

Arquilla and Ronfeldt (1999) distinguished the aspects of *realpolitik*, which is based on hard power, and noopolitik, which relies on soft power. Table 1 draws from their work to illustrate the differences. Both paradigms are used in the present political realm, though each is appropriate for certain circumstances. For example, realpolitik works better when diplomacy can be managed mainly in the dark, far from the public eye, and under strong state control without the need to share information with other players. However, the information revolution has made this type of diplomacy very difficult and is favoring the actors who operate transparently and leverage the advantages of information sharing.

Conclusion

The characteristics of noopolitik described by Arquilla and Ronfeldt (1999) provide us with common concepts of the prevailing paradigms of theory and practice of public relations: win-win, harmony of interests, cooperation, shared interests, ethics, common goals, and info-sharing, among others. Hence, the noopolitical model confirms Hiebert's (2005) idea "that the new communication technologies can save democracy by restoring dialogic and participatory communication in the public sphere, thus reserving a role for public relations as two-way communication rather than propaganda and spin" (p. 1).

L'Etang (2008) applied Wight's (1994) diplomacy traditions to public relations, thus emerging three public relations models: real public relations, rational public relations, and revolution public relations. As L'Etang (2008) stated:

> Applying the diplomatic models to PR highlights various intentions that may underpin representational work on behalf of organizations, not just states. This little exercise also shows that it is possible to derive alternative perspectives about PR from other disciplines and that it is possible to build different typologies...from those that currently dominate the field. (p. 240)

This chapter suggests a new outlook for the discipline, based on the phenomenon of stateless nation-building and public diplomacy in the new global context, in which social movements, NGOs, and particularly stateless nations have acquired an enormous capacity of influence intervening in the noosphere, a system of communication and representation where behavioral models are constituted (Castells, 2001). From the concepts of soft power and noopolitik, a perspective on public relations can be culled that is more realistic than the prevailing theories of the discipline. Concepts of soft power and noopolitik have a clear relevance for understanding and analyzing public diplomacy and public relations theory and practice. The public relations noopolitical model includes the three diplomatic models established by L'Etang (2008), but they go beyond the international arena.

From a nationalist perspective, both nation-states and stateless nations have to move from classical communication strategies to knowledge management in the noosphere. Public relations may play an important role in those strategies, building relationships between the noosphere actors in different ways. One of those ways is projecting national identity using public diplomacy, soft power, and noopolitik tactics in the stateless nation-building process. As Chernilo (2007) argued, Castells' three-volume work on *Information Age* connects the thesis of the rise of the network society with that of decline of the nation-state. From this perspective, and through the use of new media, public relations helps non-states actors to participate actively in the reverse process outlined by Castells (2000a, 2000b, 2004c).

Accordingly, noopolitik offers an optimal model for public relations to control, listen, and, primarily, influence one of those new symbolic spaces—the noosphere. The noopolitical model of public relations emphasizes public relations as a knowledge management function in the information age, not only applicable between governments and their publics in the international arena, but also to the relationships between any organization and its publics.

Table 2. Contrast between government and governance

	GOVERNMENT	GOVERNANCE
Field	Public affairs	Collective affairs
Horizon	War	Peace
Spirit	Vertical Hierarchic	Horizontal Democratic
Decisions	Order	Negotiation
Goals	Maintenance Unity	Creativity Diversity

Source: Adapted from Moreau-Defarges (2008).

When we talk about public diplomacy, soft power, and noopolitik, we are talking about governance instead of government (i.e., diplomacy, hard power, and realpolitik). In this proposed model, the flows and networks between organizations and their publics provide a more suitable description of a type of governance and, as an extension, corporate governance, where power is more mobile and unstable, and where the metaphor of hard power makes way for that of soft power (see Table 2). Certainly, effective governance implies or entails a soft power that acts through the persuasion and commitment of men and women linked by reciprocal obligation networks. In this context, public relations become an essential function as a source of the soft power needed for effective political and corporate governance.

In conclusion, a noopolitical model of public relations describes more accurately the role of public relations as a two-way communication process and is likewise applicable to corporate governance. Even more so than new communication technologies, the model also expands the traditional practice of public relations (and, by extension, public diplomacy) to the relationship-building efforts of non-state actors. The influence of new media in these endeavors is crucial because they form the structure through which noosphere, knowledge society, and, therefore, noopolitik are shaped. Thus, from both theoretical and practical standpoints, this chapter has described a new public relations approach to nation-building and has exposed how noopolitical public relations manages relationships with publics across space, time, and new media, specially in two domains: 1) the relationship management function of non-state actors in their nation-building efforts; and 2) the governance function of related actors and corporations.

In sum, this chapter has described public relations as relational management of influence, that is, as a form of soft power, with power understood as "the relational capacity that enables a social actor to influence asymmetrically the decisions of other social actor(s) in ways that favor the empowered actor's will, interests, and values" (Castells, 2009, p. 10). Castells' definition can also be used to describe public relations because public relations is an inherent function of governance in the current network society.

ENDNOTES

1 An organized political community living under a government with a sovereign political entity in international public law.

2 e.g., the National Federation of Students of Catalonia (FNEC, *Federació Nacional d'Estudiants de Catalunya*), founded in Barcelona in 1932, is an organization of Spanish students with Catalan nationalist ideology.

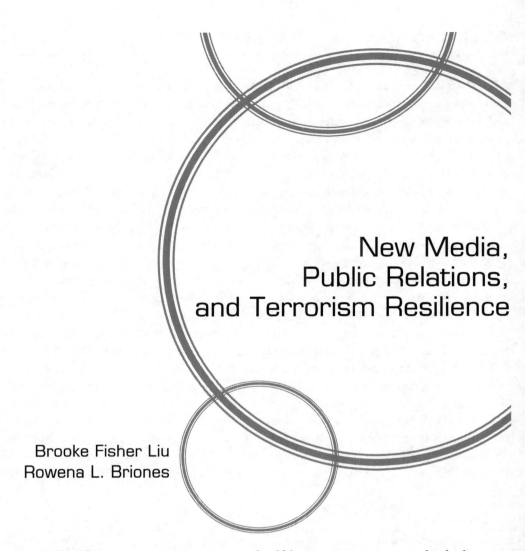

New Media, Public Relations, and Terrorism Resilience

Brooke Fisher Liu
Rowena L. Briones

Though Americans believe that counterterrorism measures should be a top priority, most individuals are not taking the proper precautions to protect themselves in the event of a terrorist attack. Traditionally, governments have relied on one-way media to communicate to publics about terrorism, but new media can offer a venue for more two-way, dialogic communication, strengthening relationships with publics. Given that no previous comprehensive review of new media and terrorism resilience exists, this chapter provides valuable insights by bringing together applied and theoretical findings from diverse fields including public relations, sociology, and terrorism studies.

Strengthening publics' resilience to terrorism is one of the most effective methods for countering terrorism because it diminishes terrorists' abilities to instill fear (Weimann, 2009). Americans agree that defending the country from terrorist attacks should be a top policy priority ("Public's State of the Union," 2011), yet only 4% of Americans take responsibility for fully preparing themselves for disasters including terrorism ("America's Preparedness," 2007). Traditionally, governments relied on one-way media to educate publics about terrorism, but new media allow governments to engage in customized, two-way dialogues (DiMaggio, 2008; Peterson, 2002). For example, after 9-11 the U.S. Department of Homeland Security (DHS) launched a color system that signaled the relative threat of a terrorist attack from low danger (green) to severe threat (red). Since the system's onset, the nation was never below the third threat level (yellow)—a significant risk of a terrorist attack –leading to the conclusion that the system taught the public to be scared not prepared ("U.S. Ditching," 2011). Consequently, in 2011 DHS replaced the old system with a simpler two-level threat advisory system featuring messages selectively disseminated via Facebook and Twitter (Sullivan, 2011).

This progression from using a one-way color system to a campaign that encourages two-way dialogue is promising for enhancing publics' terrorism resilience. Yet, more can be done to curate two-way dialogue through integrating new media. To this end, the next section defines key terms to lay the chapter's foundation. We then discuss factors that foster publics' disaster resilience, subsequently highlighting the unique role of new media. We conclude with discussing implications of and future directions for integrating new media into counterterrorism communication.

Definitions

Terrorism

There is no agreed upon definition of terrorism, likely because many definitions raise "questions about whose interests or agendas are being served by doing the defining" (Tuman, 2010, pp. xi-xii). Due to governments' inability to consistently define terrorism, Schmid and Jongman (1998) conducted a seminal review of 22 terrorism definitions and identified 16 overlapping components of these definitions:

> Terrorism is an [1] anxiety-inspiring method of repeated [2] violent action, employed by (semi-) [3] clandestine individual, group, or state actors, for [4] idiosyncratic, criminal, or political reasons, whereby—in contrast to assassination—the direct targets of violence are not the main targets. The [5] immediate human victims of violence are generally chosen [6] randomly (targets of opportunity) or [7] selectively (representative or symbolic targets) from a target population, and serve as message generators. [8] Threat- and violence-based [9] communication processes between terrorist (organization), (imperiled) victims, and main targets are used to [10] manipulate the main target (audience(s)), turning it into a [11] target of terror, a [12] target of demands, or a [13] target of attention, depending on whether [14] intimidation, [15] coercion, or [16] propaganda is primarily sought. (p. 28)

Thus, in this oft-cited definition, terrorism essentially is violence as communication, which is supported by other leading terrorism researchers (e.g., Nacos, 2012; O'Hair & Heath, 2005; Tuman, 2010).

Counterterrorism and counterterrorism communication

There is less debate in defining counterterrorism, perhaps because there is arguably less research on counterterrorism (Lum, Kennedy, & Sherley, 2006) as well as the ambiguity in defining terrorism itself. Nacos (2012) defined counterterrorism as offensive measures that prevent, deter, and/or respond to terrorist acts. Nacos (2012) further distinguished counterterrorism from anti-terrorism, the latter of which focuses only on defensive actions. Similar to how leading scholars define terrorism as communication, scholars also emphasize the central role of communication in defining counterterrorism (e.g., O'Hair & Heath, 2005; Sparks, Kreps, Botan, & Rowan, 2005). Based on their review of the extant counterterrorism communication research, Liu and Levenshus (2011) offered the first definition of counterterrorism communication: "counterterrorism communication is a communicative process that includes the exchange of symbols between sender(s) and receiver(s) where the focus, intent, and/or outcome is on deterring, preventing and/or or reducing vulnerability to, preparing for, responding to, and/or or recovering from the risk, threat, or actions of violent extremists" (p. 26).

Disaster and community resilience

Metaphorically, the term resilience deals with the notion of "bouncing back" or returning back to a state of equilibrium. Holling (1973) first used the term resilience to describe a "measure of the persistence of systems and their ability to absorb change and disturbance and still maintain the same relationships between populations or state variables" (p. 14). Like terrorism, there are multiple definitions of resilience, with no one broadly accepted definition (Manyena, 2006).

In relation to disasters, resilience comprises four factors (Paton, 2006). First, institutions that make up a community must have the proper resources necessary to ensure the safety and continuity of that community when a hazard has the potential to disrupt societal functions. Second, a community must possess the skills and competencies to deal with the hazard. Third, the community must utilize planning and development strategies to facilitate resilience and to integrate the resources at every level of the community. Finally, the adopted strategies should ensure the sustainability of the available resources and the overall competencies of members over time.

New media

In this chapter we use the terms new media and social media interchangeably. The Pew Internet Research Center defined social media as "an umbrella term that is used to refer to a new era of Web-enabled applications that are built around user-generated or user-manipulated content, such as wikis, blogs, podcasts, and social networking sites" (Pew Internet & American Life Project, 2010b). These Web-enabled applications include a range of tools instrumental for public relations professionals that encompass "a number of different forms including text, images, audio, and video" (Wright & Hinson, 2010b, p. 1). New media can facilitate or trigger disasters (Gonzalez-Herrero & Smith, 2008), but can also facilitate publics' disaster resilience (Sutton, 2010).

Factors that foster terrorism resilience

To understand how to create a more resilient society, it is important to take into consideration the various factors that foster terrorism resilience. This section will discuss how community resilience and risk communication affect resilience to terrorism.

Community resilience

Although community resilience can be observed and practiced as a result of many different types of hazards, it is most frequently studied in the context of a natural disaster, such as hurricanes, floods, or earthquakes (Paton, Millar, & Johnston, 2001; Tobin & Whiteford, 2002). Community resilience in a natural disaster context provides best practices that can be adapted to a terrorism context, keeping in mind that terrorist events are distinct from natural disasters.

There are nine primary characteristics that constitute what makes a particular community resilient (Subcommittee on Disaster Reduction, 2005; Tobin, 1999):

- Relevant hazards are recognized and understood by the community;
- The community at risk knows when a hazard event is imminent;
- Plans to alleviate or eradicate the hazard are made at the appropriate scale;

- Community members are safe from the hazard in their homes and at work;
- A less hazard prone community with strengthened networks emerges;
- Reduced levels of vulnerability for all members of the community surfaces;
- Planning for sustainability and resilience continues and is ongoing;
- There are high levels of support from federal agencies and political leaders; and
- There is a minimum disruption to the life and economy of a resilient community after the hazard event has passed.

Communities should also be aware of their particular approach to disaster resilience, as approaches over the years have shifted and changed. Static and linear approaches are now giving way to strategies that perceive hazards in a larger context, addressing the resilience issue while ensuring the community's overall sustainability and better quality of life (Bruneau et al., 2003; Mileti, 1999). Once the community has identified its key resilience factors and approach, it then considers five steps to enhance community resilience in the event of a disaster: (1) communities must develop economic resources, reduce risks, and attend to areas of greatest social vulnerability in order to increase their resilience to a disaster; (2) local people within the community must be meaningfully engaged in every step of the mitigation process to increase social capital; (3) communities should tap into pre-existing organizational networks and relationships to rapidly mobilize support services for disaster survivors; (4) interventions need to be developed that will help boost and protect naturally occurring social supports in the aftermath of a disaster; and (5) communities should plan, but also should plan for not having a plan to remain flexible (Norris, Stevens, Pfefferbaum, Wyche, & Pfefferbaum, 2008).

Risk communication

In public relations research, community resilience most often has been explored within risk communication. Risk communication is the "opportunity to understand and appreciate stakeholders' concerns related to risks generated by organizations, engage in dialogue to address differences and concerns, and carry out appropriate actions that can reduce perceived risks" (Palenchar, 2005, p. 752). Like community resilience, at its inception risk communication also took a very linear, source-oriented approach based on experts' risk assessments (Palenchar, 2010). However, as the field progressed, risk communication evolved into the study of how individuals can be more knowledgeable about certain risks to feel more confident about their health and safety decision-making. This evolved focus emphasized a relationship-building approach to addressing community residents' and employees' concerns and perceptions (Palenchar, 2010). Thus, risks are constructed out of a combination of social and cultural factors that underlie the relationships built within a community, or between an organization and its publics.

Understanding risks can be seen as a cognitive process that depends on the involvement of the individual to the imminent issue or problem. A variety of key variables has been tested to determine their impact on risk communication management and processing. First and foremost, *cognitive involvement* is vital, as people are more likely to communicate about issues they feel are personally relevant to them (Grunig & Hunt, 1984). Cognitively involved people are also more likely to seek information and are more easily able to support and/or refute arguments surrounding the issue (Heath & Douglas, 1991).

Another factor that is important in the risk communication process is *trust*. Distrust increases among the general public as government officials take opposing viewpoints on various estimations of risk (National Research Council, 1989). This can become problematic as risk estimations conflict with each other, resulting in more complex decisions that need to be made (Palenchar & Heath, 2002). Therefore, for risk communication to be fully effective, stakeholders must believe that they are relying on a source of information that is trustworthy.

The factors of *uncertainty* and *control* are related as publics seek to reduce uncertainty by way of control. Risk communication deals with gauging how uncertain stakeholders are to the risk at hand (Heath & Nathan, 1990). The communicator must then attempt to reduce the uncertainty by offering information that can help further explain the risk. The notion of control, then, is the natural response to the uncertainty that is implicit in risks (Palenchar & Heath, 2002). Individuals are more likely to accept risks if they feel that they have some sort of control over the situation (Nathan, Heath, & Douglas, 1992).

Finally, the notion of *support-opposition* has played a role in risk communication research and practice. Organizations will build relationships with communities that will support, rather than oppose, their perceptions of risk. Both the organization and its publics must collectively decide on risk estimates in the surrounding community. If a risk is perceived as having an adverse effect on the community-at-large, it is less likely to be tolerated and more likely to be opposed (Palenchar & Heath, 2002).

New media and terrorism resilience

During disasters, communities often turn to media not only to receive information, but also to assist in helping the community return to a semblance of normalcy after disasters end. However, in most cases, media typically cover disasters episodically, focusing on the immediate aftermath instead of the broader context (Nacos, 2012). In spite of this, the media can move communities toward resilience. For example, the media can deliver information to publics quickly and efficiently that can be easily processed and interpreted (Paton & Johnston, 2006). Additionally, the media can increase collective efficacy within the community, which in turn can increase the adaptive capacity of individuals (Paton, McClure, & Bürgelt, 2006). New media in particular have expanded the options publics have to assist and involve themselves with disaster preparation, response, and recovery (Vieweg, Palen, Liu, Hughes, & Sutton, 2008). In this section, we first summarize the limited research on best practices for organizations using new media to enhance community disaster resilience. We then turn to the larger body of research on best practices for publics using new media to enhance their disaster resilience.

Organizations, new media, and disaster resilience

There is relatively minimal research on how organizations use new media to enhance publics' terrorism resilience, perhaps because public organizations historically have been absent from new media (Sutton, 2010). In addition, there are relatively few terrorism incidents, limiting opportunities for researchers to examine how new media affect publics' resilience after an attack. Even when there are terrorism incidents to examine, public authorities may not integrate new media into their responses because of a lack of awareness of new media's value, a lack of familiarity with new media, and/or a lack of resources (Liu, Jin, Briones, & Kuch, in press). Further, traditional command and control disaster response strategies may limit the flexibility of public sector officials to act creatively by using new media (Kendra & Wachtendorf, 2003).

To encourage public officials to more fully integrate new media into their national security missions, government agencies should (Drapeau & Wells, 2009):

- Empower some individual government employees to be authentic;
- Envision citizens as communities of conversations;

- Unlock the government cognitive surplus so that resources and knowledge are more easily shared across agencies and levels of government;
- Create a return on engagement through indirect influence such as connecting with journalists before crises;
- Develop modern brands for agencies and market them; and
- Answer questions within specific areas of responsibility.

Given that new media have decentralized crisis communication, publics are now more responsible for their own level and quality of knowledge (Bucher, 2002). Consequently, it is essential for public officials to adopt an engagement approach to countering terrorism that involves relationship building to increase "customized, two-way dialogue in place of conventional one-way, push-down mass communication" (Peterson, 2002, p. 81).

Publics, new media, and disaster resilience

Adopting such an engagement approach requires that public officials understand why and how publics use new media to learn about terrorism. Crises typically result in increased information seeking and sharing (Coombs, 2012; Jin & Liu, 2010). Particularly after crises, publics converge online to share information and collectively solve problems (Sutton, 2009). To foster positive forms of post-disaster communication, public officials need to create online environments, tools, and features tailored to publics' specific information needs. Disaster sociologists have identified seven types of publics that converge after disasters: the anxious, returners, curious, helpers, exploiters, mourners, and supporters (Fritz & Mathewson, 1957; Kendra & Wachtendorf, 2003). For example, public officials could support mourning activity by allowing users to easily create, customize, and maintain their own online memorials (Hughes, Palen, Sutton, Liu, & Vieweg, 2008).

Public officials also need to also understand why publics use new media after disasters. Research examining why publics used new media after 9-11 provides valuable insights. For example, after 9-11, 64% of Americans used the Internet to find more detailed information, to find information they could not get elsewhere, to find more up-to-date information, and to find out what was happening while they were at work (Bucher, 2002). New media, however, co-existed rather than replaced traditional media. For example, individuals who communicated with family and friends about 9-11 via the phone were also more likely to communicate with family and friends about 9-11 on the Internet (Dutta-Bergman, 2004). Further, individuals who participated in online communities were significantly more likely to attend a meeting to discuss 9-11, volunteer in relief efforts, write about their views in the news, and sign a petition regarding the attacks compared to individuals who did not post their thoughts in online communities (Dutta-Bergman, 2006). Researchers have further noted that publics use new media to provide insider information during disasters, especially when traditional media are perceived as "outsiders looking in" (Shklovski, Palen, & Sutton, 2008, p. 6).

Increase in social support after disasters plays a critical role in disaster survivors' abilities to cope with and recover from disasters (Procopio & Procopio, 2007). Therefore, new media can play an important role in publics' disaster resilience, especially if new media are an integral part of publics' pre-disaster communicative behaviors. The first source of crisis information typically is diffused through existing social ties and familiar communication modes (Perry & Greene, 1982; Sutton, 2010). Further, new media channels are often the primary means by which time-sensitive disaster information first reaches the public (Bucher, 2002; Jin & Liu, 2010). New media also can raise awareness and hold the traditional media accountable for not adequately covering disasters. For example, Sutton (2010) found that Twitter was particularly instrumental in holding the traditional media accountable for not adequately covering a 2008 coal waste spill in Tennessee.

Further, new media allow for publics who are distributed after disasters to continue sharing important information (Sutton, 2010). For example, after 2005 Hurricanes Katrina and Rita strangers from across the county worked together to unite separated families and used social media to organize relief efforts (Majchrzak, Jarvenpaa, & Hollingshead, 2007; Scaffidi, Myers, & Shaw, 2007). Finally, while misinformation and rumor have the potential to spread quicker through online social networks, citizen editors can employ crowd wisdom to self-correct misinformation. For example, after the 2007 Virginia Tech mass shooting incident, Vieweg et al.(2008) found that students who were told to stay inside most of the day used social media to determine the safety of their friends, engaging in highly distributed problemsolving to determine who the victims were.

Conclusion

As indicated in the research review, the majority of the new media and disaster communication research focuses exclusively on disaster responses, similar to the majority of the research on emergency management (McEntire & Myers, 2004). Research testing and expanding the Social-Mediated Crisis Communication model (see related chapter in this book) is beginning to fill in this research gap as well as research applying the transtheoretical model to disaster preparation (Paek, Hilyard, Freimuth, Barge, & Mindlin, 2010). As indicated in the transtheoretical model, people change their behavior through five stages of action: precontemplation, contemplation, preparation, action, and maintenance. Thus, communication messages should be different for people in different stages of change. Social media are one method for fostering unique messages (Paek et al., 2010), but more research is needed, particularly in a terrorism context.

Research also has concluded that self-efficacy, perceptions of what one's friends and family think, and emergency news exposure are positively associated with possession of emergency items and stages of emergency preparedness (Paek et al., 2010), but again more research is needed in a terrorism context, particularly given the rare and unique nature of terrorism incidents. As research on optimistic bias concludes, publics consistently underestimate the likelihood of risks affecting them, and this is particularly true for terrorism that results in high levels of fear (Lerner, Gonzalez, Small, & Fischhoff, 2003). We now discuss the chapter's implications for public relations professionals.

Best practices for public relations professionals

First, **incorporate new media into training and education.** Resilient communities must have the skills and competencies to deal with a disaster or threat (Paton, 2006). One way to ensure that individuals are properly trained in a terrorist event is to offer online training sessions for those who are unable to participate in full-scale drills. These sessions can be livestreamed through a secure online space to guarantee that information is kept private and confidential. Videos of best practices can also be shared among first responders so that they are aware of the proper actions to take during a terrorist attack.

Second, understand how **publics use new media, specifically in a terrorism context.** As previously mentioned, publics use new media during a crisis event to share information (Sutton, 2010), collectively solve problems (Sutton, 2009), and provide insider information (Shklovski, Palen, & Sutton, 2008). Governmental agencies need to further explore *how* publics use new media tools in a terrorism context, and whether or not this usage is similar or differs from other types of crises. In other words, *do people use new media more, less, and/or differently in a terrorism context?*

Third, **increase empowerment, collaboration, and involvement.** After a terrorist incident, it is even more crucial to increase collective efficacy, create a sense of empowerment, and form partnerships with governmental agencies, local organizations, and community members. By building strong relationships with all parties involved, important decisions can be made efficiently, and a stronger network emerges (Tobin, 1999).

A call to action for public relations researchers

In addition, this review calls for more research in these particular areas:

- **More pre-disaster research.** As it currently stands, the majority of the new media and disaster communication research focuses exclusively on disaster responses, similar to the majority of the research on emergency management (McEntire & Myers, 2004).
- **The effects of new media on policy makers.** More research needs to be conducted that explores the effects of new media for policy makers. Relationships need to be built with this public to garner trust and to more easily demonstrate the value of new media for policy work.
- **The global environment.** New media is no longer about the local community. Research thus needs to explore how new media affects the global environment, especially in terms of terrorist events that can impact more than one community on a much grander scale.

In conclusion, with advances in new communication technologies crisis managers have greater opportunities for fostering terrorism resilience. Through greater collaboration between crisis managers and researchers we can better harness the full potential of new media to increase publics' terrorism resilience.

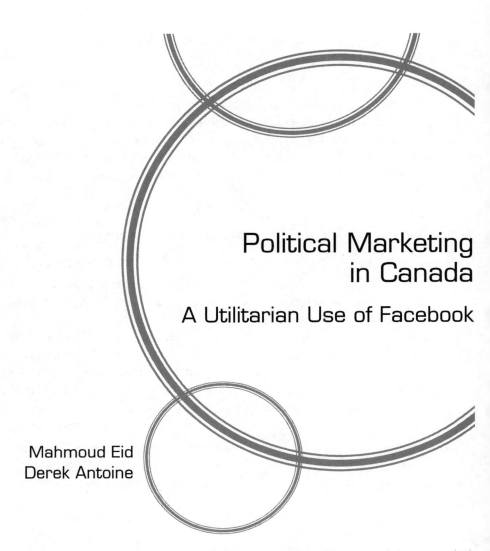

Political Marketing in Canada

A Utilitarian Use of Facebook

Mahmoud Eid
Derek Antoine

Social network sites are increasingly playing a large role in politics. This chapter examines the ways in which Canadian Members of Parliament (CMPs) use Facebook as a communication tool, through the lens of Mill's (1867) liberal utilitarianism, Habermas' (1962/1989) public sphere theory, Castells' (1996) network theory, and Lievrouw's (2002) social shaping of technology theory. It looks at attitudes, choices, and personal views, in order to explore whether CMPs use Facebook to empower grassroots Canadians and follow the public sphere elements of access, universality, and rational debate; or, for political marketing purposes to further their own electoral success. In-depth interviews with CMPs and qualitative content analysis of their Facebook profiles revealed that CMPs used Facebook individualistically and utilitarianly as an easy, inexpensive, and effective communication tool to disseminate information for political marketing purposes. They posted content that reinforced their and their party leaders' positions, while limiting user interaction and the tenants of public sphere.

Introduction

Through social network sites, which have evolved as a result of Web 2.0 concepts and new media technologies, millions of people around the globe are recently building online local, regional, and global communities to communicate their shared interests and activities, disseminate information, and interact through a variety of Web-based tools that have implications for society, culture, and politics (Eid & Ward, 2009). In Canada, social network sites play a large role in politics. Virtually every current Canadian Member of Parliament (CMP) uses these sites in one form or another. From Twitter to Facebook to YouTube, many CMPs are active users.[1] Canadians are also very active social network users. In the spring of 2009, *CTV News* learned that social network sites and blogs overtook email in popularity, growing at a rate twice as fast as more traditional forms of communications ("Social Networking Sites," 2009). This increase in the use of social network sites has led some to believe that they may be used as a tool to help strengthen Canada's democracy.

Due to the fact that voter turnout in Canada is at a record low, some researchers have studied the relationship between the Internet and political participation. Polat (2005) and boyd[2] (2008a), for instance, have argued that technology on its own cannot change social institutions or democracy. Polat (2005) in particular argues that if the reason for low voter turnout rates is inconvenience, then the Internet can likely make a difference. However, she says it is more likely that voter participation is in decline due to an unwillingness to participate or systemic hindrances, and the Internet has little effect. On the other hand, Coleman, Lieber, Mendleson, and Kurpius (2008) argue that well built websites can have an effect on political participation. They assert that if a website is designed in a way that makes information easier to access and understand, then it is likely that those who visit the website will become engaged and more politically active.

Given this debate, it is important to examine the ways in which the Internet is being used for political purposes. Specifically, the increasing use of social network sites and the popular discourse on how they affect politics in Canada make it essential to explore the ways in which Canadian politicians use such social network sites as political communication and marketing tools. More precisely, and given that Facebook[3] is one of the most popular social network sites used by Canadian politicians, this chapter explores the ways in which CMPs use Facebook as a political communication and marketing tool. Therefore, their use of Facebook will be compared to the tenets of liberal utilitarianism, public sphere, network theory, social shaping of technology, and political marketing.

Liberalism and utilitarianism

With the erosion of a sense of community, citizens are becoming less engaged in their democracies because they feel that they are losing control of the public space that governs them. Sandel (1996) argues that contemporary liberalism, a philosophy that emphasizes and confers a great deal of respect to the individual, is not only the dominant school of thought in Western democracies, but also the reason that society is losing its sense of community. The focus on the individual causes people to act in their own self-interest rather than as a collective. The result is an erosion of community and increased distrust in public institutions.

The origins of individualism can be traced back to John Locke (1689) and the second book of his *Two Treatises of Government*. Locke argues that "man's state of nature" (sect. 4) is a state of true equality where all of the power is equally distributed among individuals. Locke believed that as part of a collective under a government, the individual was the most important unit, and actions of the state should work to preserve freedoms and protect property. To Locke, the only reason individuals are willing to join together with others is "for the mutual preservation of their lives, liberties and estates," which he then refers to in its proper term as "property" (sect. 123). To Locke, that is the framework by which individuals relinquished some of their freedoms to the state.

Similarly, Jean-Jacques Rousseau (1762/1957) traces the individual's state of nature, arguing that "the first law of man is to provide for his own preservation" (ch. 2). He famously wrote, "Man is born free; and everywhere he is in chains" (ch. 1). Rousseau also believes that individual liberty should be held in great esteem, saying that "to renounce liberty is to renounce being a man, to surrender the rights of humanity and even its duties" (ch. 4). Rousseau argues that governments are necessary to enforce contracts that protect property. Even today in modern democracies, the vast majority of laws concern property. Such laws, Rousseau argues, must reflect what he calls "the general will," which is when individuals assemble and collectively make laws that both promote individual well-being and facilitate self-preservation (chs. 5, 6). Again, decision-making is about individuals coming together to advocate on behalf of their private interests in a public space to form a general will.

Alternatively, John Stuart Mill (1867) introduces the idea of liberal utilitarianism in which he argues that individuals are morally obligated to pursue their own happiness. In doing so, people maximize the happiness of everyone around them. He says that "the only freedom which deserves name is that of pursuing our own good in our own way, so long as we do not attempt to deprive others of theirs, or impede their efforts to obtain it" (ch. 1). In *Utilitarianism*, Mill deepens his argument by saying, "The creed which accepts as the foundation of morals, Utility, or the Greatest Happiness Principle, holds that actions are right in proportion as they tend to promote happiness, wrong as they tend to produce the reverse of happiness" (ch. 2). To Mill, it is not only a right but also a moral responsibility to act in a way that maximizes one's happiness. The benefits are both individual and societal because the pursuit of one's own happiness translates into happiness for the entire community. Utilitarianism is fundamentally about individuality and the supremacy of individual rights. The community benefits from this. However, liberal philosophy ignores the inherent contradiction in this reasoning.

Public sphere and network theory

Jürgen Habermas' (1962/1989) public sphere theory emphasizes the individual over the community. Public sphere is primarily about individuals coming together to form a collective. The individual advocates on behalf of the issues that matter to her/him through rational and critical discourse, and is thus able to make decisions that will enhance her/his happiness and protect her/his freedoms. But as individuals come together to discuss their private interests, it is possible that the sense of community suffers and causes, as Sandel (1996) argues, individuals to lose trust in their public institutions and democracy.

However, researchers such as Boeder (2005) and Sassi (2001) believe that Habermas' (1962/1989) public sphere theory establishes the parameters of an environment ripe for rational and critical discussion in a democratic society in a way that advances the human cause. The public sphere has never been a physical place. Rather, it is an abstract concept where rational/critical debate flourishes. The theory of communicative action also contributes to the understanding of how communication shapes social structures. Sassi (2001) suggests that theories of public sphere and communicative action serve as important starting points when examining the Internet.

From these two theories, parallels can be drawn between the public sphere salons and coffee shops that Habermas originally conceived of and today's online social network sites. At the time of his writing, Habermas conceived of the public sphere as taking place in coffee shops, pubs, and other public places where people congregated—"the sphere of private people coming together as a public" (Habermas, 1962/1989, p. 27). Today, the Internet has become one of those places (e.g., Bohman, 2004; Corrigan, 2008; Keane, 1995; McClurg, 2003; McGary, 2008; Papacharissi, 2002; Rasmussen, 2008; Slater, 2001; Sparks, 2001).

Facebook in particular has the capabilities, features, and constituency to act as a virtual public sphere. Kluth (2007) argues that social network sites are "like sitting around campfires again, only now with vastly superior tools" (para. 8). On Facebook, there are tools that encourage discussion, tools for sharing information, and tools for networking like-minded individuals. Facebook, as a communication tool, has the ability to become a public sphere, but only if it is used for such purposes. Similarly, Facebook can be used to strengthen democracy, but only if it is used to achieve democratic goals.

To Habermas (1962/1989), a public sphere must include three basic elements: *accessibility*, *universality*, and *rational/critical discussion*. For accessibility, he describes an environment in which a discursive culture exists, and the issues being debated are general in both their significance and participation because everyone is allowed to, and does, participate. Topics relevant to everyone are being discussed and debated because they are chosen at a grassroots level. Universality cannot be reduced to individuals viewed as equals. Rather, the public sphere is a place where status is completely disregarded, and people view one another as members of a "common humanity" (p. 36). Finally, Habermas explains that a public sphere should be based on reason and rational/critical debate. Without reason, one cannot expect to advance what is right and what is wrong, which Habermas further explains by developing the theory of communicative action. Habermas (1981) argues that communication is a key element to humanity's survival. Humans must be rational if they are to continue preserving their species. What makes them a dominant life form is the ability both to interact with one another in a social manner and work cooperatively to further their interests and to carry forward historical teachings through reason.

In addition to these Habermasian elements, Bennett and others (2004) introduce the idea of mediated public spheres. They argue that a mediated public sphere should be assessed on three qualities—*access*, *recognition*, and *responsiveness*. Identical to Habermas, access examines who is allowed to participate in discussion. To determine this, it should be asked whether the public sphere is inclusive or exclusive: Who is allowed to join the network, and who is not? For recognition, the question of access goes one step further: For those who are included in discussion, how much space is given to them to express themselves and share their opinions? Is there enough discourse space allotted to individuals or groups and their ideas? As for responsiveness, the question is whether there is a dialogue between those who have access: When one side comments, does the other side respond? Is there opportunity to respond? Social network sites form a mediated public sphere due to the sites' framework and ability for users to filter information. Therefore, the added criteria of a mediated public sphere are considered in the discussion of the Habermasian public sphere because these three elements provide a concrete and measureable interpretation of the modern public sphere.

Internet and social network sites can form a mediated public sphere (e.g., Bakardjieva, 2005; Hjavard, 2006; Hutchby, 2001; Silverstone, 1999). Social network sites can be communication tools that accelerate this process by connecting decision-makers to ordinary citizens in a way that provides an environment of mutual respect and recognition, ultimately leading to collaboration.

To understand this, however, one must also understand the changing nature of society and social networks as outlined by Manuel Castells (1996; 2004a), who frames the notion of networks in society. Castells (1996) defines a network as a set of "interconnected nodes" that share similar communication codes. Nodes can be understood as communication intersections much like street intersections. Codes are the common language or similarity that connects a network. Networks are open structures that are able to expand without limits as they integrate with new nodes. Most importantly, Castells argues that in a network society, power rests in the switches that connect each network. The switchers are the power holders linking together multiple networks and changing the dynamics of these relationships. Switches are the instruments of power that become the fundamental sources shaping societies. Political power, for

Castells (2004a), rests in the networks that structure society rather than the institutions, states, or large corporations that seem to exercise it. Society's organization around networks is powerful because networks are not bound by borders, are flexible and continually expanding, are more effective and efficient than other forms of organization, are continuously restructuring civil society at local and global levels through networks of activists, and are communicating in public space to define reality. Castells' network theory looks at the power relations that shape society. From Castells' perspective, networks are a source of power in society, and certain technologies can make them more effective and efficient. Social network sites, and the way they are used, have the potential to form a powerful switch that connects networks to government officials or public institutions if they are used for such a purpose by both sides.

A network society can be argued to be an evolution from Habermas' view of the public sphere, as original conceptions of the public sphere continue to change. It is a "complex mosaic of differently sized overlapping and interconnected public spheres" (Keane, as cited in Boeder, 2005, *The Future of the Public Sphere*, para. 2). A network society is essentially a decentralized public sphere where people come together around communication codes and engage one another in specific issues that affect them. It is impossible for each individual to implicate oneself in every issue or group that exists, but one can take part in several networks and has varying degrees of influence in each of them. As well, the public sphere is fundamentally about networks, and social network sites can play a role in developing, maintaining, reinforcing and switching—if not becoming switches in and of themselves. This leads to the importance of the role of technology in society and whether or not communication tools such as social network sites hold deterministic elements.

Social shaping of technology

The network theory conceptualizes society "as a fundamentally interactive and intersubjective web of affiliations rather than a static, separate superstructure overshadowing human action" (Lievrouw, 2002, p. 184). Like Castells' (1996), this theory rests on the idea that networks shape society and work to determine the relations between individuals. However, Castells (2004a) argues that although technology does not determine society, some social structures could not develop without specific technologies. The potential for social network sites to change social networks for better or worse depends on how they are used. The social shaping of technology (SST) theory explores this idea more deeply.

SST researchers reject technological determinism as an approach to analyzing technology and prefer to look at the complex relationships between society and technology. SST emphasizes "the importance of human choices and action in technological change" rather than the effects of technological change on human choices (Lievrouw, 2002, p. 185). This reinforces the point that the effects of technologies are a result of how the users choose to use them. But even within the SST camp, there are two schools of thought that seek to analyze the society-technology relationship: social construction of technology (SCOT) and actor network theory (ANT). SCOT analysis looks at the social context of innovation and change. It explores the "choices available to designers, developers and users in the course of technological development" (Lievrouw, p. 185). ANT rejects both strong technological determinism and strong social constructivist approaches, opting instead to study technology in terms of the actors such as people, the technology itself, and institutions that "have equal potential to influence technological development" (Lievrouw, p. 186). ANT researchers believe technologies are not infinitely flexible and become "embedded and stabilized within institutional and social structures and influence or even determine subsequent technological choices" (Lievrouw, p. 186).

The SST approach provides a strong foundation for analyzing social media as it gives perspective to the role of users in shaping technology. Without users, technology has no purpose or function. The Internet can be a tool for democracy, but only if its users use it for such purposes. The tool itself does not hold deterministic elements that necessarily impose stronger democratic principles. This is why it is important to analyze the ways in which CMPs are using social network sites to determine whether these sites are being used for engagement purposes and to build a sense of community.

Political marketing

Political marketing is a professionalization and market-oriented version of political communication, primarily focused on election results. The difference between political marketing and political communication is that the former is concerned with election results, whereas the latter with civic engagement. Political marketing has created political consumers rather than citizens and perhaps undermines the very nature of the public sphere, political party structures, and representative democracy. (Scammell, 1999). Political parties have learned to use communication technologies not to engage citizens in rational public discourse, but to mobilize and demobilize different segments of the electorate. They do this by breaking audiences into narrow segments and communicating differently and directly to each of them. (Bennett & Manheim, 2001).

Facebook and other social network sites can be very effective political marketing tools. They have features that allow users to broadcast messages to wide networks as well as provide tools to gather feedback on how users react to those messages. They also work to categorize users by network in a way that allows them to segment and tailor messages for specific constituencies. Users can categorize members of their networks by age, geography, interest, school, and other demographics.

As an increasing number of politicians log on to social network sites, the potential for engagement and interaction with citizens increases as a result of the nature of the space. boyd (2008b) argues that there are two ways in which political decision-makers can use social network sites. They can use the sites to digitally handshake citizens by creating an "open channel for communicating" with constituents (boyd, 2008b, p. 93). Politicians can also use the sites to engage citizens in public discourse within an online public sphere by allowing citizens to post comments on their page and then reacting to those comments. Comments are:

> embedded in a social contract of reciprocity. Comments are not left on politicians' profiles simply to be consumed by the aide who controls the profile: they are crafted to provoke a response by the politician or by anyone visiting the politician's page. (boyd, 2008b, p. 93)

It is then about interaction, engagement, and reciprocity. Whereas the first use of social network sites is about two-way communication, the second is more of a unidirectional broadcasting. Many politicians tend to use these sites as political marketing tools to push or test particular messages. boyd (2008b) argues that there is no willingness by politicians to engage citizens, but rather a desire to use the tool as a means of political marketing.

Canadian Members of Parliament and social network sites

In order to explore the ways in which Canadian MPs use social network sites to engage citizens in a public sphere where citizens can contribute to and inform MPs through communicative action, it is essential to look at Facebook in light of the previously discussed theoretical foundation—that is, to operationalize the theoretical concepts based on the specific features and technical terms of Facebook. The process of *political engagement through social network sites* can be theoretically conceptualized in light of the concepts of: *accessibility, universality,* and *rational/critical debate.* The key features and technical terms of Facebook are: *friending, supporters, wall, status, newsfeed, profile vs. page, discussion boards, page reviews, information tab, photo and video sections, tagging, events, notes, links, comments,* and *Facebook page insights.*

In doing so, this study utilized a qualitative research design through which two qualitative data collection methods were employed: in-depth interviews and content analysis. Complementing each other, the in-depth interviews measured the perceptions, views, and choice rationale of CMPs as they used Facebook while the content analysis examined the feature choices they made in establishing a framework for their page or profile. Guided by a non-probability sampling strategy—specifically a purposive sampling technique—CMPs interviewed were identified through Facebook searches and newspaper articles about politicians who use the site. Ethics approval by the University of Ottawa's Research Ethics Board was secured for this research study in order to protect the human subjects' privacy and other human rights. In-depth interviews were conducted in 2009 with four CMPs: three prominent CMPs who caucused with the Liberal Party of Canada and one prominent CMP with the New Democratic Party of Canada. Using a semi-structured interview guideline,[4] a series of questions asked of each CMP ranged from "How do you use Facebook?" to "Do you think Facebook should follow the tenets of the public forum doctrine?"[5] in order to explore their views and attitudes towards the social network site. Upon completion of the in-depth interviews, a qualitative content analysis was conducted on the Facebook pages and profiles of the CMPs who were interviewed to catalogue the features they chose to use. The content analysis catalogued the way in which CMPs used Facebook by observing the parameters they established through feature choices as well as how they interacted with users. Data were collected using a qualitative content analysis form with coding criteria that determined the style of Facebook page, friending openness, settings related to engagement features such as the wall, photos, discussion boards, comment features, information posted by the CMP, and level of reciprocity.

Political engagement

The features and technical terms of Facebook can help operationalize the process of *political engagement through social network sites,* which is theoretically conceptualized here in light of the concepts of: *accessibility, universality,* and *rational/critical debate.* Social network sites are defined as:

> Web-based services that allow individuals to 1) construct a public or semi-public profile within a bounded system, 2) articulate a list of other users with whom they share a connection, and 3) view and traverse their list of connections and those made by others within the system. (boyd & Ellison, 2007, *Social Network Sites,* para. 1)

These sites allow users to interact with one another, share photos and videos, post notes, blog, and engage others through instant messaging. They may also allow for creative ways of expression and interaction. All of this could lead to a form of engagement between users and public policymakers if they choose to use them as such.

Drawing on Habermas' (1981; 1962/1989) and boyd's (2008b) arguments, *political engagement through social network sites* can be broadly defined as an inclusive and accessible communication space based on a contract of reciprocity between politicians and other users that encourages rational and critical discussion. This is about creating an environment that reflects Habermas' public sphere and determining whether or not the uses of social network sites by Canadian politicians foster any of the public sphere's elements—*accessibility, universality,* and *rational/critical debate.*

Accessibility means having a discursive culture where everyone is allowed to participate. As a result, the issues that are debated are relevant because they reflect a grassroots and populist democracy. Topics that average citizens want to discuss are discussed, and are considered important as a result. This is a question of who is allowed in and who is not (Bennett et al., 2004). On Facebook, this happens through friending options and policies that allow for the exclusion or inclusion of users. CMPs then have the choice of whether or not to make their pages or profiles accessible to anyone who wants to join their networks of friends and supporters.

This study shows that accessibility is limited in several ways. First, it is restricted by the structural limitations of Facebook. For those who choose to use the Facebook profile style of site, there is a maximum number of friends one can accept in a network. Once users reach 5,000 friends, they are no longer able to add friends until they delete others. Those who use the Facebook page style are permitted to have an unlimited number of fans, but even those CMPs with pages (vs. profiles) managed their networks by periodically removing individuals from their list for a variety of subjective reasons. Interviewed CMPs deleted friends who posted comments that they found offensive or "crossed the line." For those who were able to gain access, there was a perception by CMPs that Facebook is a tool to communicate externally in a way that updates "followers" about the subjects and issues CMPs cared about. There was no sense that the users within their network were given much space to discuss the things that they wanted to discuss. They were neither given this space nor was everyone allowed to participate by joining the network freely. A number of limits were imposed by CMPs resulting in diminished capabilities.

In light of *universality*, the public sphere is a place where status is completely disregarded, and individuals are viewed as members of a common humanity. This raised the question of whether CMPs viewed those who posted comments on their Facebook walls, for example, as members of a common humanity who are equally able to contribute to public debate or discussion. Universality is about exploring the degree to which the Facebook pages or profiles of CMPs are user driven. Recognition, as the mediated public sphere's universality equivalent (Bennett et al., 2004), is about providing a space for discourse to those who are granted access. Recognition calls for Facebook to be used in a way that is participatory and allows the opinions of the grassroots to be considered as important as those of the CMPs.

Interviewed CMPs in this study neither allowed nor wanted their Facebook pages or profiles to be user driven. They all argued that Facebook was a communication tool that helped them communicate *to* constituents, Canadians, and other individuals. Again, this was about users following the CMP rather than the CMPs engaging users. None of the CMPs left room for the notion that users may have joined the network to contribute to a larger discussion and gain access to someone who is a government decision-maker sitting in the House of Commons. Instead, users joined *their* networks, and little weight was given to the idea that they were creating a public space with the potential for building community. With respect to Facebook features, none of the CMPs enabled the wall posting function for users of the network. The wall feature is the easiest and most popular way to communicate and share ideas with others in a public manner. Not enabling this feature greatly limits the way users within a network can interact with the CMPs as well as other users in that same network. The most popular features enabled by the interviewed CMPs included one-way communication functions such as the information tab, the status update function, notes, links, and limited photo sharing. It is the CMPs who framed discussions through posts, with users left only to react to their content. Even the type of content the CMPs contributed included such things as press releases, which were expressions of their views on various topics, material that promoted them, or items that supported their party or leader. Other non-controversial items such as music videos or jokes were also posted and were not meant to be political.

Creating an environment ripe for *rational and critical debate*, for Habermas (1962/1989), is crucial for his theory of communicative action and society's ability to advance what is right and what is wrong. Creating an environment for rational and critical debate is about how CMPs interact with constituents, Canadians, and other users who have joined their networks. This is also discussed in terms of responsiveness (Bennett et al., 2004), whether there is dialogue or mutual responsiveness in a mediated public sphere, and reciprocity (boyd, 2008b), whether CMPs respond to users' comments on their Facebook walls and engage in rational and critical debate.

Most interviewed CMPs responded to private messages through Facebook, but none of them responded to comments left on their Facebook pages or profiles. They were not interested in engaging in any dialogue or debate, let alone rational or critical debate. CMPs argued that they did not believe that they should constitute public forums. Instead, they referred to Facebook as "my space" and "my platform" and "my microphone." It was like their "office building," and they did not feel as though they "[had] to share it if [they] do not want to." CMPs placed many limits on what could be expressed and deleted many comments for various reasons.

Conclusion

The CMPs interviewed were not using Facebook in a way that engaged users. Instead, they used the site to broadcast information and market themselves while limiting user interaction. They posted marketing communication pieces such as press releases, speeches, and content that reinforced their leaders' messages. Their use of the social network site was individualistic and utilitarian, as CMPs were interested in advancing their agendas and pursuing their own goals. They viewed the site as a tool for themselves rather than a tool for society to build community, and the result may be at the expense of public trust in public institutions.

CMPs were not using Facebook to engage or empower users in a public sphere. Instead, they created an environment in which not everyone was welcome. For those who used the Facebook profile style, they limited accessibility by rejecting friend requests for a variety of subjective reasons. Each of the interviewed CMPs admitted to actively deleting users who signed up to their networks for reasons that ranged from wanting to make space in the network for other users to displeasure with the content of a user's contribution. Arbitrarily removing users from the network does not constitute accessibility. In addition to subjective choice, the structural limitations of Facebook do not allow for pages to be accessible because, for those who use the Facebook profile style, their network is limited to a maximum of 5,000 friends. This means that even if CMPs wanted to make their profile more accessible, they would not be able to.

CMPs did not use the social network medium to break down barriers and create an environment that disregards status. In fact, many viewed their site as an extension of their office where constituents may contact them for constituency support. It was considered to be a place where *followers* could update themselves on what their CMPs were doing, but not as a place for dialogue. Even for those users who posted information or opinions, the interviewed CMPs indicated that they did not treat those comments in a way that had an impact on their decision-making process. CMPs did not provide adequate space for users to contribute content or drive the site in a way that allowed users the ability to bring forward issues that mattered to them. The discourse space was not given to users because their Facebook walls were disabled, discussion boards were disabled or limited, and limits were imposed on users wanting to tag the CMPs in photos or videos.

CMPs were very much aware that they were not creating a space for discussion. They disagreed that the public forum doctrine applies to Facebook, and they subjectively deleted comments that crossed the line. Without the ability to dissent or expose others to differing viewpoints, there can be no true rational or critical discussion. Additionally, most CMPs did not read the comments posted on their pages or profiles and were therefore absent from any conversation that may have taken place. CMPs viewed their pages or profiles as private spaces designed to help them communicate with others, not the other way around. Interviewed CMPs used words such as "my platform," "my microphone," and "my space" to discuss their Facebook account. They gave a sense that the page was just another communication tool to reach wide audiences instantaneously with little or no cost. There is benefit to the speed, given that one of the golden rules of political communication is to sell one's message early—before anyone else—to frame the debate in one's own terms. If a politician can send short messages instantaneously through social network sites to get ahead of an issue, they can try to frame a debate before their opposition or traditional news organizations do so.

There was no indication in this study that the use of Facebook changed the power dynamic between political decision-makers (CMPs) and citizens. The tool and the way it was being used mirrored traditional means of communication—one-on-one interaction through private inbox messages and one-to-many communication through broadcasts and postings. Facebook as a tool changed nothing in the way these two groups interacted with one another.

Not only were the CMPs engaging in political marketing, they were also very much aware of it. They knew that they used the site to promote themselves, broadcast messages about the issues they were working on, and build a network that would provide an advantage in future elections. They understood that the type of content they posted was material that reflected their own opinions or attempts to legitimize them. This perception was also made clear by the way they viewed members of their networks. They were referred to as "followers" rather than "constituents," and were provided with very limited discourse space in which they could initiate their own discussions.

Although the CMPs used Facebook primarily as a political marketing tool, they did not do so in a scientific or systematic way. Rather, their use reflected an informal political marketing. They understood that social network sites can be powerful marketing tools, but their main goal was to broadcast messages rather than collect data on constituents, narrowcast to small segments of the population, test messages, or mobilize and demobilize constituencies. Although some of this did happen, it was either unintentional or conducted as small experiments.

Findings revealed that CMPs were not using social media to bring people together, create common experience, and generate discussion about various topics. Rather, sites were used as a tool to advance one's own goals. Whether their use actually works to engage a new generation of voters is still to be seen.[6]

The social shaping of technology theory helps to understand how technology affects society. The mere introduction of a particular piece of technology does not shape society. Rather, it is society that shapes technology and the way that it is used. As public debate around the role of social network sites and its effect on politics continues, it is important that studies around technology, politics, and public relations look at the choices made and views held by users who have adopted the technology. Facebook will not revolutionize democracy if it is not used for such a purpose. Similarly, a public sphere will not be realized online through the Facebook pages or profiles of CMPs unless they choose to use the site for such purposes.

Social networks sites are about how they are used. CMPs' use is important because they are the ones who make decisions in Canada's democracy. Constitutionally, CMPs are mandated with the task of approving or rejecting government legislation. From there, they have created larger roles for themselves as ombudsmen, facilitators, and constituent megaphones. They are elected by their constituents and have a responsibility to represent them. Understanding how they use social network sites is important because they are an integral decision making institution in Canada's democracy. Observing how they choose to use these sites helps to explain how their use will affect those institutions. This chapter looked at the choices, perceptions, and views of CMPs as they used the social network site, and it opens the door for future studies on effect—that is, what effect their choices have on the public sphere and political engagement. Future studies should look into effect by examining election campaigns where social network sites play dominant roles, as well as how they affect the relationship between citizens and politicians. A future study should also look at the wider political class from nomination candidates to election candidates, among others, as each has different motives than those who have already been elected, and it would be interesting to see how they use the sites as opposed to elected CMPs.

ENDNOTES

1   See www.davidakin.com for a list of politicians who use various social network sites.

2   danah boyd does not use capital letters in her name. See http://www.danah.org/name.html

3   Facebook was adopted as early as 2006 by Canadian politicians, and most Canadian MPs have a Facebook page or profile.

4   Probing questions were asked when needed to gain a deeper understanding of the issues being discussed.

5   The public forum doctrine, as outlined by Sunstein (2001), serves three important functions. It ensures that those who want to be heard have access to an audience, it allows speakers to have access to specific groups and institutions who may be the direct target of a speaker's opinion, and it increases the likelihood that people will be exposed to a wide variety of views.

6   As of this writing, CMP Facebook use has not yet made significant progress.

Demystifying Social Media
Use and Public Relations
Practice in China

Yi Luo

Much research has examined how social media are transforming relationship management in public relations practice. Most existing studies on this topic have focused on countries like the US or the UK where social media use is relatively unrestricted but not on a country like China where social media use is curbed by government regulations. This chapter fills this void and examines how public relations practitioners in China use social media for relationship management in China. The chapter opens by explaining the history of social media in China and the country's social media landscape as well as some unique characteristics of its social media users. This is followed by an explanation of how practitioners use social media for relationship management and how government regulations impact this.

Social media and public relations in China

China poses an interesting challenge for those wanting to understand how public relations practitioners use social media for relationship management. Organizations require quality relationships with their publics and social media have been shown to be useful tools for maintaining these relationships (Sha, 2007). The existing knowledge on how to use social media to manage organization-public relationships is primarily based on studies done outside China and on social media platforms unavailable in China. China's social media and public relations landscapes differ from the rest of the world. Before the 1980s, there were no public relations firms in China (Chen, 2003). Today, the country's public relations industry is the fastest growing in the world. China's social media's landscape is also unique. Due to government regulations, Facebook and other social media sites popular in the rest of the world are inaccessible in China. Despite these restrictions, China has a vibrant social media scene. Millions of Chinese spend hours on local social media sites like YouKu (similar to YouTube), Weibo (similar to Twitter), and dozens of others (Reisinger, 2010).

Various questions remain unanswered regarding social media, public relations, and relationship management in China. For example, how do practitioners in China engage publics on some Facebook-like sites? How do Chinese social media sites and their users differ from their counterparts in other countries? How does government regulation of the Internet affect public relations practitioners' efforts to manage relationships using social media?

This chapter thus provides insight on how public relations practitioners utilize social media for relationship management in China. It is organized as follows. First, it briefly outlines existing knowledge regarding social media's role in relationship management. Second, it describes how research was conducted to understand the state of social media, relationship management, and public relations in China. The third part explains the history and development of social media in China. Fourth, it discusses the unique characteristics of social media users in China. Fifth, it explains how public relations practitioners in China use social media for relationship management. Lastly, the chapter examines how government regulations influence the use of social media for relationship management.

Current research on relationship management and social media

A key to organizational success is dialogue that helps build quality organization-public relationships (OPRs) (Kent & Taylor, 2002; Porter & Sallot, 2005). When mutually beneficial relationships are present, an organization is more likely to achieve its various goals (Ledingham, 2006). Two factors that have been shown to promote dialogue are interaction (information exchange between an organization and publics) and debate (a process of providing statement and counterstatement) (Bruning, Dials, & Shirka, 2008). Social media have been shown to be valuable for relationship management because they facilitate interaction and debate (Pettigrew & Reber, 2010). Various studies (Smith, 2010b; Taylor, Kent & White, 2001; Waters, Burnett, Lamm, & Lucas, 2009) have demonstrated that social media can help distribute useful information, stimulate dialogue, and promote positive feelings (mutual control, commitment, and trust) that are crucial to effective relationship management.

Most existing studies regarding the Web, public relations, and China have focused on two issues. The first is how activist groups use social media to challenge organizations. Han and Zhang (2009), for example, examined how activist groups used social media to protest Starbucks' decision to build a branch near the Forbidden City. The second issue is how the websites of Chinese organizations are structured differently than websites for organizations outside China. Pan and Xu (2009) found that the websites of Chinese corporations offered more opportunities for user interaction than U.S.-based sites. Conversely, Yang and Taylor (2010) found that the websites of Chinese environment non-governmental organizations primarily focused on providing information and offered few resources for readers to mobilize. These existing studies, albeit interesting, do not shed light on how public relations practitioners are using social media for relationship management tasks. This chapter fills that gap.

Method

To understand how social media are used in public relations practice in China, a group of China-based experts and practitioners on social media and public relations were interviewed. These experts included the founder of Jiepang.com (China's version of Foursquare), a public relations director at Motorola China, a managing director at Rudder Finn (a public relations firm operating in five countries), a managing director of Weber Shandwick (the world's largest public relations firm), a strategy director of Resonance China (one of the first social media consulting firms in China), a chief executive at Wolf Group Asia (a public relations firm specializing in China), a senior vice president at Bite Communications (a communications consultancy operating in nine countries), a digital media specialist at Ogilvy, a research analyst and a manager at Corporate Intelligence Center (a research firm on social media in China), and a manager at SPRG (a Beijing-based public relations firm). In addition, journals, books, white papers, and other material on social media and public relations in China were reviewed.

History and development of social media in China

According to the Corporate Intelligence Center (CIC, 2011), a research firm that specializes in social media in China, the history of social media in China has three phases. During the first stage, incubation (1993–2003), the primary online social networks were discussion forums (e.g., Tianya, Xici) and instant messaging services (e.g., QQ) (Barboza, 2007; Hu, 2010). During the second phase, cultivation (2004–2006), blogs (e.g., Sina.com) and video sharing sites (e.g., Tudou) gained prominence in China. In this stage, Chinese social media users also moved from just consuming content to becoming active creators of content. During the third stage, proliferation (2007–2011), social networks, micro-blogging, location-based services, review sites, picture sharing services, social bookmarking, social commerce (e.g., coupon sites), and various other types of sites have flourished. Organizations in China have also taken a more active interest in engaging publics on social media sites by hiring staff or agencies to monitor social media and carry out campaigns aimed at social media publics.

China's social media landscape

Diverse players

China's social media landscape differs from the rest of the world because the key players in the world (Facebook, Twitter, and YouTube) are inaccessible in China. Not a single social media category is dominated by one company in China. For example, the social networking sector is split among multiple Facebook-like sites like Renren, Kaixin, Douban, Qzone, Pengyou. Qzone has garnered about 190 million users, followed by Renren with 95 million users, and Pengyou with 80 million users (Lukoff, 2011). In micro-blogging, the sector is split among several. For example, Sina Weibo has nearly 100 million users, and Tencent Weibo has about 160 million users (Kan, 2011). In location-based services, the sector is split between several players such as Jiepang, Qieke, Vld.sina, and K.ai. In social video, Tudou, Youkou, Sougua, and Qiyi share viewers. Similar patterns can be seen in review/ratings, music, picture sharing, social bookmarking, question/answer, and social commerce.

Segmentation among users

This multitude of players in each sector also means that China's social media landscape exhibits a characteristic not significantly evident in the US: demographic segmentation among users of top social media sites. In the US, the major social media websites have a multitude of users. Facebook and YouTube, for example, have users from various demographic sectors (e.g., young, old, middle-class, lower middle class). In China, however, social media websites tend to focus on a particular demographic group. For example, among social networking sites, Qzone is popular among teenagers and college students, young white collar professionals prefer Renren, rural dwellers prefer 51.com, and mothers/young children mainly use Taomee (Lukoff, 2011).

Characteristics of social media users in China

In addition to a difference in players in the landscape, how Chinese Internet users utilize social media differs from how Internet users outside (especially in the US and Europe) use social media. Various academic and industry studies (Forrester Research, 2010; Liao, Pan, Zhou, & Ma, 2010; Mandl, 2009; Zhao & Jiang, 2011), along with experts interviewed in this chapter, have shown that Chinese users differ from users from outside China in two primary ways: creativity and preference for bulletin board services (BBS) communities.

Creativity

Mandl (2009) found that Chinese blog programs offered users more ways to modify the appearance of their blogs, and Chinese bloggers tended to create more graphically oriented blogs in a comparison of blogs in China and Germany. Zhao and Jiang's (2011) comparison of social networks in the US and China found that on social networks, Chinese users were more likely to customize their profile images than U.S. users. Forrester Research (2010) surveyed 60,138 social media users in the US, China, and Western Europe. They found that Chinese users (about 44%) fell in the category called creators: They were more likely to generate or share content on blogs, wiki sites, discussion forums, and consumer review sities. In contrast, in the US, only 24% were creators, in the United Kingdom 15% were creators, and in Germany 9% were creators. Experts interviewed for this chapter explained that the popularity of creating content can be attributed to the fact that Chinese netizens find it fun to create funny and innovative online content.

An example that illustrates how content generation is an integral part of Chinese social media life is a campaign by Vancl, a Chinese clothing company. In 2010, the company published advertisements on its website. Chinese social media users were intrigued by the advertisements and began modifying Vancl's advertisements to create new ones that appeared to mock contemporary Chinese society. Thousands on social media sites posted redesigned advertisements and held competitions to judge the ads. Within two months, a relatively unknown company became a household name, and phrases from the mock campaigns became part of everyday Chinese language.

Preference for BBS communities

Another unique characteristic of Chinese social media users lies in their preference of engaging in conversations on bulletin board services (BBS) communities. BBS communities are the original social networks of the Internet. In most parts of the world, users have moved away from BBS communities as Internet access speeds have increased. Internet users in China, however, continue to show a preference for BBS communities. Tianya.cn, a BBS established in 1999, averages about 13 million visitors a day (Wolfram Alpha, 2011). BBS communities have significant influence in shaping how issues are discussed in China. There have been several cases where companies have been forced to make concessions after criticism from BBS communities. SK-II, a cosmetics multinational, was forced to withdraw its products after BBS users complained about harmful chemicals in its products. The KFC restaurant chain also faced significant BBS backlash after its commercials appeared to mock the Chinese educational system (Corporate Intelligence Center, 2011). Experts interviewed for this chapter stated that BBS communities allow Internet users in China to quickly form influential collective bargaining power, which they can use to confront corporations, government agencies, and other powerful entities.

Relationship management and social media in China

The experts interviewed for this chapter identified four primary approaches to relationship management for which public relations practitioners in China are using social media. These approaches can be labeled as (1) listening and analysis, (2) producing relevant content, and (3) community building, (4) engaging opinion leaders.

Listening and analysis

Social media have opened up new avenues for Internet users in China to express themselves. In the past, ordinary citizens had limited access to get their voices heard on radio, television, or print media. In the era of social media, opportunities for expression have increased. Government authorities have also shown some leniency toward online speech that is critical of corporations (foreign and local) and government agencies at the local level (e.g., town, village, city authorities). In this environment, social media users are often not afraid to express their displeasure at products, services, or policies they view as inferior. Faced with millions of users utilizing social media to post their opinions, organizations are increasingly investing in resources to listen to what these social media users are saying.

Listening to social media is coupled with analysis to gauge meaning. To conduct analyses, organizations in China use three approaches. The first is primarily software based. Computer programs are used to scan the Web for mentions of an organization. When these programs encounter a social media site that mentions a particular term (e.g., the name of a product a company uses), they automatically analyze and produce a report. This report is often a sentiment analysis that rates whether the social media mention was negative, positive, or neutral. A report is then delivered to a public relations practitioner to decide how to respond. This automated approach is the fastest and most cost-efficient because social media users in China create a large amount of content each hour. Unfortunately, automated approaches are not effective in detecting the layers or meaning embedded in conversations. The complex nature of the Chinese language as well as the constant evolving status of social media platforms means that computer programs can have problems with accuracy and interpretation.

A second approach is human based. In this approach, corporations hire staff or firms to monitor social media posts, analyze them, and recommend a course of action. This approach is expensive and tedious. It also requires employing multiple analysts to prevent critical issues from being missed and incorrectly analyzed.

A third approach is computer-aided human analysis. In this approach, computer programs surf the Web looking for relevant content. When relevant content is found, it is forwarded in raw form to a human analyst. The analyst makes an interpretation and decides on a course of action. This approach avoids the pitfalls of the computer-only approach that can miss critical social media posts and the human-based approach that requires tedious work.

Relevant content generation

A second approach practitioners use to manage online relationships is through creating relevant content. Experts interviewed explained that their organizations are always striving to produce content that can foster dialogue and interaction with social media publics. When this content is posted, practitioners examine how social media publics comment on, repost, recreate, and interact in other ways with the content. Content is created to spur interaction that helps meet the organization's goal. Experts gave several notable examples of successful content generation and several examples of failed content generation. One popular case cited by the interviewees involved a "Battle of the Bands" campaign by Pepsi that was aimed at teens. In July 2009, Pepsi used Chinese social media to spread content that encouraged Chinese teens in rock bands to submit videos that could be judged. The campaign was a significant failure. The campaign's content did not resonate with Chinese teens and failed to consider that rock bands (and in particular, alternative rock bands) are a relatively premature phenomenon in the country, and many teens lack the resources to produce the required content. In a country with millions of teens, only eight entries were received (Han, 2009). Experts interviewed in this study pointed to this case as an example of how relevant content is critical.

Community building

A third method used to maintain relationships is creating communities on social media platforms. The experts interviewed explained that organizations in China recognize that publics on social media are engaged in detailed and often emotional interactions. The goal for many of these organizations is to carve out a space on social media where publics can talk about their services or products.

Two approaches are used to accomplish this. The first approach is to use an existing third-party social network to create a community. This is equivalent of a fan page on Facebook. This approach is very popular, and many organizations in China have secured spaces on social networking, social video sites, and micro-blogging sites. An example of this type of community building can be seen in a campaign that was carried out by the China National Cereals, Oils, and Foodstuffs Corporation. In 2009, the company created a game and placed it on Renren (a social network site similar to Facebook). In the game, users could virtually grow plants, produce juices, and send virtual gifts to friends. Each week, the company awarded some players packets of real juice. The campaign was a success and managed to get Renren users to become engaged with the product virtually and in regular life (Han, 2010).

Using third-party networks poses challenges. Social networks like Renren have limitations on how you can engage publics on their platforms. For example, most third-party networks place strict restrictions on how organizations with fan pages can run contests, obtain private information from users, and remove disruptive users. This practice aims to ensure that organizations with fan pages do not anger the millions of ordinary Renren users. On the other hand, such restrictions can frustrate organizations seeking to communicate with publics in innovative ways. This is not unique to China. Facebook also has restrictions on how to engage publics. In addition, third-party sites in China expect organizations to pay, which some organizations frown upon because they view social media as a cost-effective alternative to traditional public relations and advertising.

To avoid the challenges posed by third party networks, a second approach to community building that organizations are using in China is to create custom built social networks. The advantage of these networks is their cost-effectiveness and increased control. An example of this type of community building was carried out by Lancome. In 2006, Lancome created Rosebeauty, a custom social network where users could obtain and exchange information. A pivotal moment for the network occurred in 2007 when a member of the community posted a moving poem explaining how Lancome's products had "enriched her life and made it more beautiful" (Lamy, 2010, para. 7). The poem received many accolades within the network (and outside) and prompted the company to launch a beauty contest in 2009 that was very popular. Users could participate by creating their own profile pages in this community, recommending a friend to participate, or by voting for the winner. Rosebeauty was a significant success. In addition to increasing discussion about the Lancome across the Chinese Web sphere, Rosebeauty has become one of the leading places where people buy Lancome products.

Engaging opinion leaders

A fourth approach practitioners in China use to manage relationships is by engaging online opinion leaders. In China, online opinion leaders are educated, urban, white collar professionals who spend about 55 hours a week online. These opinion leaders are viewed by regular Internet users as more reliable than news organizations and civil authorities (Corporate Intelligence Center, 2011). Organizations devote significant efforts identifying these influencers and engaging them. Methods of engagement vary from the basic (e.g., giving them early access to information) to the elaborate (e.g., inviting them on paid tours). The effort and attention given to these influencers matches the attention given to traditional journalists.

An example of opinion leader engagement can be seen in the public relations campaign carried out to promote the 2010 Shanghai Expo. Rudder Finn, a global public relations firm, was hired to engage online opinion leaders and inform them about the United States pavilion at the fair. By researching social media sites, the firm was able to identify influential investors, entrepreneurs, bloggers, and others in China and the US. These individuals were invited to an "East Meets West" event where they were encouraged to interact with other netizens through real-time platforms such as microblogging, social network sites, and location-based services. The event was a success and resulted in the influencers promoting the fair using various social media outlets.

Government influence

The Chinese government has made no secret of its desire to regulate social media websites. Access to Twitter, YouTube, and Facebook from within mainland China (policies are more lenient in Hong Kong) is often blocked by Web monitoring systems that also restrict access to pornography and other content deemed unacceptable. In addition, during events that the authorities deem controversial (e.g., civil unrest in other countries), mentions of these events is often discouraged either through written warnings or technological measures that block people from posting.

Little impact on social media campaigns

A great deal has been written about government regulation of the Internet in China. Despite the multitude of publications, surprisingly very little has been written on how the regulations affect how public relations practitioners manage relationships using social media. Experts interviewed for this chapter provided useful insight on this issue and revealed that the situation is very complicated. Government regulation of the Internet in China is often portrayed (by writers from outside China) as a clear cut picture: on one side are regulators working hard to meticulously police the Internet, and on the other side, are regular Internet users who grudgingly accept the controls. The reality, however, is more complicated. The experts highlighted two interesting characteristics of social media in China that are often ignored in discussions of government regulations of the Internet in China.

The first interesting characteristic is that rather than dampening social media, government regulations have actually encouraged the growth of social media platforms in China. As mentioned earlier, an effect of the restrictions of Twitter, Facebook, and YouTube is that they have allowed local companies to establish their own social media platforms without having to face competition from better-funded U.S. firms. In a market without outside competition, the Chinese social media websites have been able to develop a loyal following and construct their platforms in a manner that reflects local tastes. In addition, they also have signed contracts with local companies (e.g., film, music, books publishers) so that they provide the content that local people want. The reality, the experts cautioned, is that even if China were to open access to Facebook, Twitter, and YouTube, the foreign platforms would have a tough time in China because their Chinese counterparts have had years to learn how to cater to the Chinese social media user.

A second characteristic is that Chinese authorities are selective in what type of content they restrict on social media. Regulations of the Internet in China do not mean all critical content on social media is restricted. In the area of business speech, which is the realm most public relations practitioners operate in, a significant leeway is allowed. Social media postings that are critical of companies are tolerated and in some cases encouraged as a way of curtailing malfeasance by both foreign-based and local companies. Senior government officials, eager to curb administrative abuses at the village, town, or city level, appear to be willing to overlook social media content that is critical of mid- and low-level administrators whose actions harm citizen morale and serve to discredit the country's leaders as a whole.

Grassroots social media activism

The interesting result of this *laissez faire* attitude toward criticism of businesses and mid- and low-level bureaucrats is that it has spurred grassroots social media activists who aim to confront corporate and administrative misbehavior. These activists use blogs, bulletin boards, micro-blogs, and video-sharing sites to criticize what they view as problematic behavior by chief executives, business owners, mayors, government agency heads, and others. Examples of this type of activism are numerous. Following a devastating earthquake in 2008, Wang Shi, the CEO of Vanke (a large residential real estate developer) announced that his employees should donate no more than 10 RMB (US$ 1.50) each to earthquake relief fund (Zhong, 2008). Social media activists seized on this and organized a boycott of the company. The result was that company's stock price plummeted, the CEO apologized, and the company increased its donations to earthquake victims. Another example was when the son of a government official was driving drunk and killed a young woman. When confronted by the police, he yelled in Chinese "My father is Li Gang" (Wines, 2010, para. 2) as a warning that the police would face official censure if they arrested him. Social media activists transformed the phrase into a popular saying as they demanded justice for the crash victim. Eventually, the young man was arrested and prosecuted. Social media, rather than being completely restricted, are used by online publics (with tacit encouragement from senior officials) as a tool to serve as a check on the power of officials and businesses.

Avoidance of controversial issues

Another characteristic that is not discussed in regard to censorship is the behavior of Chinese social media users. Experts interviewed explained that from their observations, the majority of social media users utilize social media for mundane tasks (e.g., connecting with friends, searching for deals, engaging in playful banter, consuming entertainment content). Politics and other controversial policies are not discussed. And when they are discussed, it is during special contexts like the scenarios previously mentioned. Whether this avoidance of controversial topics is through self-policing, apathy, or a lack of knowledge is unclear. What is clear, however, is that when on the social media sites of organizations (e.g., corporations), most users steer clear of problematic discussions, and when the discussion gets heated, it is about controversies that are acceptable to get heated about. The experts did acknowledge there are cases when some individuals do break this status quo and bring forth touchy subjects. In these cases, the individuals are either ignored or, in some cases, external regulators in charge of monitoring Internet content do step in and remove the problematic content. These three factors (a vibrant social media sector, a seemingly *laissez faire* attitude toward speech that is critical of businesses and mid/low level bureaucrats, and avoidance of controversial topics by users) means that in day to day practices, public relations practitioners are minimally affected by government regulations of the Internet. Experts interviewed explained they are able to design and implement social media campaigns without worrying about government censorship. When creating content, they steer clear of sensitive topics, but because they are in the business of promoting corporations, avoiding touchy matters is part of effective public relations in China.

Conclusion

China has a vibrant social media landscape characterized by competition among top players and demographic segmentation among users. Users of social media in China demonstrate a preference for creating original content and utilizing formats (e.g., BBS) that have lost popularity in other countries. An emphasis in research suggests that the surge in social media communities has pushed public relations practitioners in China toward a more strategic approach. To use social media for relationship management, public relations practitioners primarily engage in listening/analysis, production of relevant content, community building, and opinion leader engagement. Government regulations of Internet use, rather than stifling social media, have encouraged the creation of a dynamic, social media scene occupied by local players who have thrived in the absence of foreign competition. Government regulators are lenient about curbing critical business speech, and users of social media generally do not use the social media pages of organizations carrying out regular public relations campaigns as a platform to confront government control. This chapter reveals several important implications for public relations practitioners seeking to use social media for relationship management in China. First, practitioners have to understand the complex, dynamic nature of the Chinese social media landscape and choose appropriate venues to reach specific demographics. Second, practitioners have to understand how Chinese users utilize social media and the venues at which they prefer to interact. Third, strategies for listening/analysis, content generation, creating communities, and engaging opinion leadersshould be carefully developed to manage relationships with social media publics. Fourth, the issue of governmental control of the Internet should be understood as complex, and that the realm of business speech is one in which official control is minor.

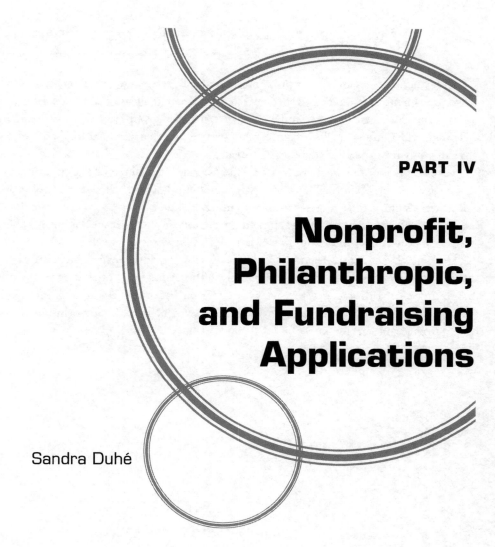

PART IV

Nonprofit, Philanthropic, and Fundraising Applications

Sandra Duhé

Overview

In recent years, the nonprofit sector has grown in large part because of a shifting in responsibility (and funding) for many social goods from the public to the private sector. This growth has enhanced competition within the nonprofit marketplace, with numerous charitable, environmental, educational, human service, and religious organizations vying for the attention of members, donors, and volunteers. Public relations is essential in these endeavors, and this section examines some successes and setbacks in that regard.

Sarah Merritt, Lauren Lawson, Dale Mackey, and Richard Waters examined the dialogic features of nonprofit blogs and found that the features were infrequently used. Too often, they argue, nonprofits are using blogs as extensions of their websites for one-way provision of information. The upside, however, is that the authors discovered that those organizations with dialogic blogs had greater connectivity on the Web.

Kati Tusinski Berg and Sarah Bonewits Feldner propose that "cause networks" result from the intersection of technology, philanthropy, and corporate social responsibility. They give us an overview of the captivating Girl Effect movement that is based on the idea that investment in girls enhances global economic development. Kati and Sarah provide a tour of Girl Effect's multi-media communication and networking initiatives and likewise raise some interesting points about Nike's sponsorship of those efforts.

Through the perspective of proximity, Denise Bortree and Xue Dou studied how the Sierra Club advocated environmental causes through Twitter. Their analysis of posts consistently revealed that while local groups focused on getting audiences to act, national groups fulfilled more of an information sharing role for a wider audience. This complementarity of roles between national and local groups is a fruitful area of study for other social media platforms.

Brooke Weberling, Richard Waters, and Natalie Tindall tested the situational theory of publics in the context of the "Text for Haiti" campaign and found donors who were in closer physical proximity to Haiti were significantly more aware, more involved, and perceived fewer constraints in the fundraising effort. To date, few studies have examined the effectiveness of text messaging in public relations or tested the situational theory of publics in fundraising. Their study did both.

Dedria Givens-Carroll completes this section by providing insight into how various faiths are establishing a presence on Facebook and connecting with "friends" who otherwise may not be part of a religious community. In addition to an overview of faiths on Facebook, she shares findings from her interview with the medical doctor who started Jesus Daily, which, at the time of this writing, is one of the most popular pages on Facebook.

If You Blog It, They Will Come

Examining the Role of Dialogue and Connectivity in the Nonprofit Blogosphere

Sarah Merritt
Lauren Lawson
Dale Mackey
Richard D. Waters

One of the principal characteristics of social media is connectivity, the state of connecting multiple entities together. The blogosphere is an excellent example of connectivity as organizations and individuals create original posts, link to others' posts on their own websites and social media accounts, share posts via email and text messaging, and ultimately communicate with one another through dialogue. This chapter examines the interrelatedness of Web connectivity and the role of dialogue in the blogosphere. Specifically, the research measured how well the blogs from the Philanthropy 400 nonprofit organizations incorporated the five dialogic principles and then examined whether those organizations with higher proportions of dialogic principles received more global Web traffic and whether they were connected to other websites and social media outlets more than the nonprofit blogs that used the dialogic principles with less frequency.

Introduction

Relationship cultivation management is a necessary component of nonprofit organization management, particularly in marketing and public relations. As dialogue is not a series of steps, but a product of ongoing communication and relationship, the dialogic principles in social media are key to building authentic relationships in the online environment, as part of a nonprofit's overall relationship cultivation management strategy.

Mattson and Barnes (2007) found that charitable organizations were ahead of the private sector in their use of social media in regards to relationship cultivation. More than 75% of charitable organizations were reaching out to stakeholders virtually using some form of social media including blogs, podcasts, message boards, social networking, and wikis. More than one-third of nonprofit organizations were blogging, and 46% reported social media as very important to strategic management of volunteerism, fundraising, and carrying out the nonprofits' programs and services. At that time, nonprofit organizations were blogging at a rate that doubled that of *Fortune 500* companies. Jin and Liu (2010) concluded that only 16% of *Fortune 500* companies had a blog for the general public.

Despite the use of social media by nonprofit organizations, studies have shown that their results from these efforts have been minimal and have mostly been focused on the dissemination of one-way messages (Waters, Burnett, Lamm, & Lucas, 2009). This study sought to help determine whether the largest nonprofit organizations in the United States were strategically using their blogs to facilitate dialogue with their stakeholders and whether participation in on-going conversations in the blogosphere helped the blog's traffic grow organically along with the number of other websites and social media profiles that linked others back to the blog.

Literature review

Dialogic communication theory

Buber (1965) explained dialogue as an intersubjective process based on an interhuman force or reality created between two people in a relationship. Buber's conceptualization of genuine dialogue hinges on ethics, where "communication should not be a means to an end, but rather...communication should be an end in itself" (Kent & Taylor, 1998, p. 324).

Kent and Taylor (1998) used this foundation to develop their principles of dialogic communication to gauge how technology is used by public relations practitioners and how it influences organization-public relationships. They expanded the definition of dialogic communication to cover organizational communication and concluded that dialogue was "any negotiated exchange of ideas and opinions, denoting a communicative give and take" (p. 325). Their original research described how this give-and-take dialogue could effectively be carried out over institutional websites. However, this approach has subsequently been applied to blogs (Seltzer & Mitrook, 2007), Facebook sites (Bortree & Seltzer, 2009), Twitter (Rybalko & Seltzer, 2010), and wikis (Hickerson & Thompson, 2009).

The principles of dialogue reflect the relational approach of public relations and the symmetrical communication concepts advocated for by Grunig (Grunig, 1989c; Grunig, 2001). Indeed, Kent and Taylor (1998) argued that the best way to cultivate relationships between organizations and publics is through strategic communication guided by a framework founded in dialogic communication theory. They proposed five strategies that comprise a dialogical framework, enabling organizations to create and change relationships with publics. They argued that dialogic communication as a framework outlined by these five principles can be utilized in online strategies of organizations to build relationships with publics.

The first of Kent and Taylor's (1998) five principles involves the creation of a dialogic loop. This feedback loop allows for organizational responses to the expression of questions and concerns on issues through public queries. Information provided by the organization should have quality, value, and significance to the needs and interests of publics. A qualified and trained organizational contact must provide feedback through a willingness and commitment to address public concerns. As with communication through traditional media, online public relations requires dialogue reflecting the same levels of professional standards and communication skill (Kent & Taylor, 1998). In the blogosphere, this loop can be obtained through allowing readers to post comments and replies to an author's entries in the public forum as long as the author reads and responds to them. This can also be done more privately by emailing the author directly.

The second principle focuses on the provision of useful information to the visitors to the site or blog. Generalized information of value to all publics must be available, even if targeted information is included for specific publics. Organization background and history is of ongoing value across publics, as long as the information is useful and trustworthy. Site hierarchy and structure contribute to audience ease in finding information, where genuine dialogue stems from questions and concerns being addressed through information. An organization can meet the consumer through mailing lists and discussion groups that provide information consumers need without their having to always go to the website. This reciprocity leads to goals of both parties being addressed and met.

Ultimately, the provision of useful information should help generate future visitors to the site, which is the third principle. Organizations can make sites attractive through updating information, including issues and commentaries, as well as the inclusion of special forums, such as online question and answer sessions between visitors and experts. Although updates of valuable information are the easiest way to appear credible and responsible, information updates still represent one-way communication. Interactivity is an effective method for providing information outlets that should be neither difficult nor time consuming to understand and use.

Visitors to websites should be able to intuitively navigate the site's content and understand where information is stored. This ease of navigation constitutes the fourth principle. Due to longer download times, text should outweigh graphics, especially because graphics often provide little additional information. Text-only information is more efficient and quick for information seekers than text and graphics combined. Online sites should be designed for accessibility by all current software, not just the latest versions, so that users with older software do not feel inadequate or intimidated.

The focus and priority of site building should be on valuable, informative content related to how the organization and product are presented to publics. Furthermore, overreliance on graphics can serve as a propaganda, marketing, or advertising tool that distracts from the emphasis of dialogic public relations and the priority of creating long-term, quality relationships with publics. Visual fluff outweighs text and taints organizational image.

The final principle focuses on keeping visitors on the website or blog on the website for their information seeking endeavors rather than having them go to other sites for information. It is important to avoid adding links to outside sites that could lead visitors astray because once users click away from a website, they may not return. As the goal of relational communication is to create relationships with users, not to entertain them, organizations should use essential links with clear paths for how to return. Also, avoiding sponsored advertising, or at least utilizing strategic placement of sponsored advertising, helps eliminate user distraction from valuable information, following the argument of Buber (1965) that dialogic communication should be the interactional goal and not a means to an end, such as marketing or advertising.

These five dialogic principles were operationalized to measure the use of dialogic communication strategies on activist websites. Taylor, Kent, and White's (2001) findings supported dialogue as "an important framework as public relations moves toward a relational approach (p. 264). Their findings suggested that it is not enough for organizations simply to have the required technical and design aspects for websites to fully engage publics in two-way communication. This study found that activist organization websites were not utilizing the dialogic capacity offered by the Internet. Interactive features and encouragement of visitor returns were both found lacking in terms of fostering relationship building.

Seltzer and Mitrook (2007) extended website-based research on dialogic principle use to an investigative analysis of blogs as an alternative online platform with equal or greater capacity for dialogue. They argued that blog structure and the interactive features of blogs foster ongoing discourse between organizations and publics, making them more dialogic in nature than traditional websites. The primary implication for public relations practitioners of Seltzer and Mitrook's work was that weblogs were effective for creating and cultivating online relationships. However, their analysis was based on activist organizations and failed to consider more mainstream nonprofit organizations.

Based on the rise and influence of blogging with particular regard to nonprofit organizations, this study replicated Seltzer and Mitrook's (2007) measurement of the dialogic principles in the blogosphere as they applied to a purposive sample of nonprofit organizations from the *Philanthropy 400* list. Specifically, this study was guided by one main research question:

RQ1: To what extent do nonprofit blogs incorporate the principles of dialogue into their design and management?

However, two additional research questions stemmed from a growing amount of literature on the topic of Internet connectivity as it pertains to the blogosphere. Practitioners advocate that one of the most important antecedents to developing a relationship with stakeholders online is getting them to participate in social media outlets. Because participation in a blog community requires more effort than merely becoming a fan of an organization on Facebook and following its status updates, participants were likely to be more committed to the organization if they actively connected to the blog. Corcoran, Marsden, Zorbach, and Röthlingshöfe (2006) cautioned that organizations may be drawn toward social media sites like Facebook that provide a larger, captive audience. However, organizational blogs were more effective in creating conversations that can result in stronger public relationships as well as inexpensive market research. Additionally, Burton (2005) concluded that individuals and organizations connecting through cyberspace ultimately resulted in stronger relationships that would reap social, political, and economic benefits for everyone.

Two measures of connectivity are a site's traffic score and the number of quality sites (e.g., non-spam) directly linking to the site. Several studies have concluded that websites that allow for dialogue opportunities whether through virtual chat rooms, feedback forms, or provision of email addresses generate higher Web traffic (Terlien & Graham-Cumming, 1997; Tierney, 2000). Therefore, it seems logical that nonprofit blogs that use Kent and Taylor's (1998) dialogic principles the most should generate more Web traffic than those that fail to include the principles. Additionally, studies have shown that increased dialogue opportunities results in more hyperlinks back to your website or blog from others (Fieseler, Fleck, & Meckel, 2010; Woodly, 2008). Kelleher and Miller (2006) concluded that blogs were a useful strategy for developing relationships with an organization's constituency because of their conversational nature and that active participation in the blog could result in increased Web connectivity. This study tested this notion with the following two research questions:

RQ2: Do blogs that have higher traffic scores incorporate more of the principles?

RQ3: Is there a relationship between the number of sites linking in to a blog and the utilization of the five principles?

Method

This project took a census of all blogs authored by nonprofit organizations on the 2010 *Philanthropy 400* List, which is compiled by the *Chronicle of Philanthropy*. The *Philanthropy 400* annually ranks United States nonprofit organizations based on their total annual revenues. Of the 400 organizations on the list, 126 organizations had institutional blogs. For this project, 124 blogs were coded since two of the blogs no longer existed even though it was on the organization's homepage.

Three coders went through a two-hour training session during which time they reviewed the variables that constituted the five dialogic principles by examining a codebook with detailed descriptions of each measure while comparing these definitions against *Fortune 500* blogs. The codebook contained 32 items from the Seltzer and Mitrook (2007) study that measured the five dialogic principles. The specific measurements for the five principles are shown in Table 1.

Additionally, data were collected from Alexa.com, an Amazon.com company that provides Web traffic metrics and analysis of websites and blogs to determine the traffic ranking and number of links connecting to each of the 124 nonprofit blogs.

After a two-hour training session, the researchers coded 25 *Fortune 500* blogs for their incorporation of the dialogic principles. The coders' intercoder reliability was calculated using Cohen's kappa and found to range from a low of $\kappa = .82$ for useful information to a high of $\kappa = .89$ for ease of navigation.

Results

From the *Philanthropy 400* list of nonprofit organizations, the 124 blogs represented a variety of nonprofit subsectors. Nine blogs were from educational institutions, 10 blogs represented arts and humanities nonprofits, 23 were managed by human service nonprofits, 23 blogs were from religious institutions, 26 discussed healthcare issues, and 33 represented public/society benefit nonprofits. Examining the site characteristics of the 124 blogs, 108 (87.1%) provided links to the host organization's home website, and 85 (68.5%) of the blogs were hosted within the organizational company's website rather than an outside, or independent, blog URL. The blogs were most often authored by non-communication staff members (91). However, the blogs were also written by communication staffers (17), executive directors (11), and volunteers (5).

The first research question asked to what extent nonprofit blogs incorporate the principles of dialogue into their design and management. The results indicated that the majority of nonprofit blogs only modestly incorporated the principles of dialogue, as only the conservation of visitors had a mean score ($M = 80.65$, $SD = 17.98$) that reflected significant inclusion of the variables by the sample. Nonprofit blogs failed to incorporate the remaining four variables in significant proportions. Ease of navigation was the second principle that was found most often in nonprofit blogs ($M = 62.23$, $SD = 23.44$), followed by usefulness to media ($M = 59.79$, $SD = 19.62$), and generation of return visits ($M = 52.23$, $SD = 22.40$). Two measures—creation of a dialogic loop ($M = 50.16$, $SD = 31.35$) and usefulness to publics ($M = 50.0$, $SD = 26.01$)—each were only implemented halfway by the nation's largest nonprofits' blogs. Although the dialogic loop had a slightly larger mean score than the public usefulness variable, it also had a much larger standard deviation, which means that publics were less likely to find a dialogic loop component if they randomly visited the blogs in the sample.

Table 1 presents the comprehensive breakdown of results for each principle using the 32-item[1] code sheet. For conservation of visitors, a loading time of less than four seconds was most used by the blogs

Table 1. Occurrence of Dialogic Features by *Philanthropy 400* classification.

DIALOGIC FEATURES	PHILANTHROPY 400 NONPROFITS' CLASSIFICATION						
	Total (n = 124)	Arts/ Humanities (n = 10)	Education (n = 9)	Healthcare (n = 26)	Human Services (n = 23)	Public/ Society Benefit (n = 33)	Religion (n = 23)
Ease of interface	62.23 / 23.44	60.00 / 36.51	55.56 / 24.00	58.65 / 23.99	69.57 / 23.28	61.36 / 27.03	64.13 / 23.91
Archive of posts	80 (64.5%)	8	6	14	18	23	11
Main links to rest of blog	90 (72.6%)	4	5	21	19	25	16
Search Engine	103 (83.1%)	10	7	19	19	26	22
Low reliance on graphics	36 (29%)	2	2	7	8	7	10
Usefulness to Media	59.79 / 19.62	52.5 / 22.52	58.33 / 15.43	61.06 / 21.51	61.96 / 27.47	59.85 / 19.76	59.78 / 24.68
Press releases	65 (52.4%)	3	5	14	15	18	10
Speeches	45 (36.3%)	3	7	8	8	13	6
Downloadable graphics	48 (38.7%)	7	5	10	6	13	7
Multimedia downloads	82 (66.1%)	8	5	17	17	21	14
Clearly stated positions*	97 (78.2%)	4	4	24	21	24	20
About Us section	101 (81.5%)	6	7	20	20	29	19
Identifies intended audience	53 (42.7%)	3	3	13	7	13	14
Logo prominently displayed	102 (82.3%)	8	6	21	20	27	20
Usefulness to Publics	50.0 / 26.01	48.57 / 27.95	39.68 / 26.34	47.25 / 23.99	57.76 / 38.29	48.92 / 24.21	51.55 / 25.03
Statement of philosophy*	99 (79.8%)	7	3	21	22	27	19
How to become involved	86 (69.4%)	8	6	15	20	22	15
How to donate	84 (67.7%)	7	4	17	20	21	15
Contact info for org leaders	80 (64.5%)	5	6	15	19	19	16

table con't on next page

Links to org podcast/webcast	36 (29.0%)	4	5	7	6	8	6
Ads for non-org products*	24 (19.4%)	0	0	6	2	8	8
Ads for org products	25 (20.2%)	3	1	5	4	8	4
Return Visits	**52.23 / 22.40**	**56.36 / 20.14**	**51.52 / 21.81**	**51.05 / 23.46**	**57.71 / 27.64**	**52.62 / 23.17**	**46.64 / 24.37**
Explicit invitation to return	11 (8.9%)	1	0	3	1	4	2
Presence of link to regular news	74 (59.7%)	5	5	17	17	18	12
FAQ	46 (37.1%)	4	3	6	13	14	6
Bookmark the blog	63 (50.8%)	5	4	17	13	15	9
Link to external web sites	105 (84.7%)	7	7	21	21	31	18
Calendar of events	40 (32.3%)	6	4	7	5	11	7
Downloadable org info*	46 (37.1%)	7	6	8	9	12	4
Request info from blog	89 (71.8%)	8	5	17	19	22	18
News items within last 30 days	62 (50%)	5	5	14	11	15	12
RSS feeds	93 (75%)	6	7	18	19	26	17
Encourage others to share	85 (68.5%)	8	5	18	18	23	13
Dialogic Loop	**50.16 / 31.35**	**40.00 / 23.45**	**46.67 / 28.76**	**47.69 / 31.90**	**48.69 / 31.11**	**60.00 / 34.71**	**47.83 / 33.68**
Opportunity to reply to blogs	101 (81.5%)	6	7	20	17	32	19
Author replies to comments							
Links to social media accounts	88 (70.9%)	6	5	16	18	26	17
Opportunity to send private msg or email	36 (29.0%)	2	2	8	7	12	5
Opportunity to vote on issues	8 (6.5%)	1	1	0	1	4	1

table con't on next page

Email subscription for updates	78 (62.9%)	4	5	18	13	25	13
Conservation of Visitors	80.65 / 17.98	94.00 / 8.94	75.56 / 21.37	73.85 / 25.43	82.61 / 18.70	**83.64 / 15.98**	**78.26 / 16.84**
Important info on front page	98 (79.0%)	10	7	19	20	25	17
Blog post within last 2 weeks?	71 (57.3%)	8	4	10	14	21	14
Short loading time	124 (100%)	10	9	26	23	33	23
Use tags to categorize posts	87 (70.2%)	9	6	16	15	26	15
Indication of last update	120 (96.8%)	10	8	25	23	33	21

Note: As in the Kent, Taylor, and White (2003) study, composite data for each dialogic principle index are provided, followed by frequencies for individual items present on the Philanthropy 400 blogs. The mean and standard deviation (M/SD) are provided for each of the five dialogic features.

* p<.05 Confidence level

(100%), and main blog postings made within two weeks of coding was least used (57.3%). For ease of interface, the presence of a search engine box was used most by nonprofit blogs (83.1%), and low reliance of graphics was the least used (29%). For useful information, a prominently displayed organizational logo was used most (82.3%), while advertisements for products and services was least used (20.2%). For generation of return visitors, links to external websites was most used (84.7%), and an explicit invitation to return was least used (8.9%). For dialogic loop, the opportunity to post replies to blog topics was most used (81.5%), and the opportunity to vote on issues was least used (6.5%).

The second and third research questions explored how the blogs' traffic and connectivity impacted their inclusion of the dialogic principles. Of the blogs sampled, 122 blogs had information on Alexa.com. The blogs averaged 5,968.14 sites linking to them (SD = 37,474.74) though the range of links coming into the blog varied considerably (minimum = 1, maximum = 408,226). Similarly, the range of website traffic varied considerably as the best performing blog was the 37th most visited website, and the blog that drew the lowest amount of traffic of the 124 blogs had a ranking of 21,207,545th. The average website traffic ranking was 2,703,508 (SD = 4,871,003.04).

The second research question asked whether blogs with higher traffic scores incorporate more of the principles. Organizations were organized into three roughly equal tiers, based on their traffic score ranking according to Alexa.com. Organizations were ranked as either in the top tier (the best rankings), middle, or lowest ranking tier (the worst rankings). A Chi-square test was conducted to assess whether blogs in the top third tier of traffic scores incorporated more applications of the dialogic principles. The organizations ranked in the top tier were more likely to include the particular dialogic principle behaviors.

The Chi-square test results indicated significance for the presence of 12 of the 41 principle items coded in the survey. The greatest significance for the top tier of blogs was the provision of updates within the last 30 days ($X2 = 13.14$, df = 2, p = <.001), and the least significant behavior was presence of a prominently displayed logo ($X2 = 5.98$, df = 2, p = <.05). The other 10 statistically significant behaviors were:

main links to rest of blog ($X2 = 6.53$, df = 2, p = <.05); press releases ($X2 = 7.81$, df = 2, p = <.05); statement of philosophy ($X2 = 7.38$, df = 2, p = <.05); how to become involved ($X2 = 7.86$, df = 2, p = <.05); contact information to organizational leaders($X2 = 8.14$, df = 2, p = <.05); important information on front page ($X2 = 6.64$, df = 2, p = <.05); blog post made within last two weeks ($X2 = 6.06$, df = 2, p = <.05); presence of and/or link to regular news ($X2 = 6.71$, df = 2, p = <.05); FAQ ($X2 = 9.17$, df = 2, p = <.01); and downloadable organizational information ($X2 = 6.03$, df = 2, p = <.05)

    The third research question asked whether there is a relationship between the number of sites linking to a blog and the inclusion of the five principles. Similar to the previous research question, organizations were ranked either in the top tier (the highest number of links), middle, or bottom third (the lowest number of links). The Chi square results for this research question were significant for 22 of the 41 principle items coded in the survey. The dialogic principle behaviors utilized among the top, middle, and bottom third behaviors were statistically significant across the five principles in the following way. Under ease of interface: archive of posts ($X2 = 5.76$, df = 2, p = <.05); main links to rest of blog ($X2 = 19.36$, df = 2, p = <.001); and search engine ($X2 = 6.01$, df = 2, p = <.05). Under usefulness of media: press releases ($X2 = 22.87$, df = 2, p = <.001); about us section ($X2 = 8.62$, df = 2, p = <.01); and logo prominently displayed ($X2 = 16.36$, df = 2, p = <.001). Under usefulness to publics: how to become involved ($X2 = 7.86$, df = 62, p = <.05); how to donate ($X2 = 14.83$, df = 2, p = <.001); contact information for organizational leaders ($X2 = 19.72$, df = 2, p = <.001); links to organization podcast/webcast ($X2 = 6.64$, df = 2, p = <.05); and runs ads for organizational products ($X2 = 6.26$, df = 2, p = <.05). Under dialogic loop: opportunity to

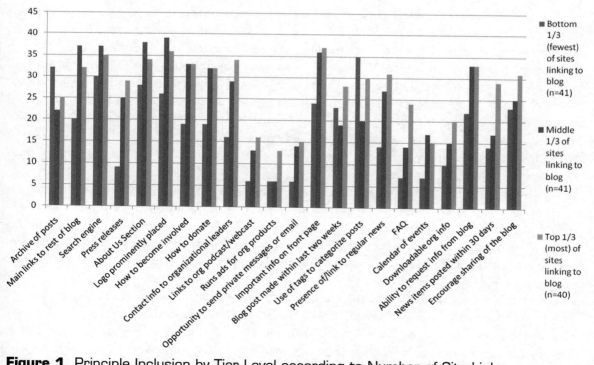

Figure 1. Principle Inclusion by Tier Level according to Number of Site Links

(Note: Alexa.com did not have data for two of the blogs in the study. Therefore, the number of blogs in this analysis totals 122 rather than the original 124.)

send private messages or email ($X_2 = 6.22$, df = 2, p = <.05). Under conservation of visitors: important information on front page ($X_2 = 16.89$, df = 2, p = <.001); blog post made within last 2 weeks ($X_2 = 6.05$, df = 2, p = <.05); and use of tags to categorize posts ($X_2 = 13.78$, df = 2, p = <.01). Finally within maintaining return visits: presence of/link to regular news ($X_2 = 16.92$, df = 2, p = <.001); FAQ ($X_2 = 16.22$, df = 2, p = <.001); calendar of events ($X_2 = 6.44$, df = 2, p = <.05); downloadable organizational information ($X_2 = 6.69$, df = 2, p = <.05); ability to request information from blog or blog author ($X_2 = 10.48$, df = 2, p = <.01), news items within last 30 days ($X_2 = 22.90$, df = 2, p = <.001), and encouragement to share blog with others ($X_2 = 7.4$, df = 2, p = <.05). Figure 1 displays the frequency breakdown of these items among the three tiers.

Discussion

The results of this study echo those of Taylor, Kent, and White (2001) as well as Seltzer and Mitrook (2007). Unfortunately, the principles of dialogue were only modestly incorporated into the organizational blogging efforts of the *Philanthropy 400* nonprofits. The most often used principle, conservation of visitors, showed that organizations understood that they did not want to lose website or blog traffic to their competitors. In fact, the three variables most often incorporated into their blogs were short loading time, indication of last update, and reduced reliance on external links. In particular, the presence of reduced reliance on external links as a top three variable indicated that these organizations realized the value of keeping visitors on their site. However, the blogs were not taking advantage of visitor retention because they were not incorporating the remaining four principles with significant effort.

The analysis of the principles indicated that most of the nonprofits in the sample seemed to use their blog as an extension of their website. Rather than taking advantage of the conversational nature of blogs, nonprofits more often used them to distribute organizational news and topical updates. When looking at the most commonly incorporated elements of the dialogic principles, it seems as if those behind the blogs were relying on others to help spread their information rather than being actively involved in the process. Nonprofit blogs were very likely to provide a search engine so that visitors to the blog could find information (83.1%), discuss their philosophy or mission statement (79.8%), provide clearly stated positions and news updates for the audience (74.2%), and brand themselves visually with their logo (82.3%) All of these strategies help build a strong blogosphere presence for the nonprofit, but it is merely a presence, not interactivity and dialogue.

The blogs promoted public conversation, as nearly 82% of the blogs offered the opportunity to post replies to the authors' posts. However, the ability to have a conversation by sending a private message or email directly to the author was used by less than one-third of the organizations (29%). Although this finding reiterated the results of earlier dialogic principles studies, it was somewhat disappointing to learn that the approach to dialogic communication online still has not changed despite an increased knowledge base about its impact and effectiveness.

However, the second and third research question demonstrate that there may be some signs of change, as the organizations that attracted the most Web traffic and had the most links back to the blog were generally the blogs that used the dialogic principles in greater proportion. It is important for public relations practitioners to understand the role of dialogue and how it can positively impact an organization's Web presence.

Conclusion

As Kelleher and Miller (2006) indicated, blogs have the potential to help organizations develop lasting relationships with their stakeholders. However, the nation's largest nonprofits are primarily using their blogs to replicate traditional public relations rather than embrace the interactivity brought about by new media. Using blogs as a one-way distribution channel echoes Grunig's (2001) summation that although two-way symmetry is the form of communication that practitioners should be moving toward, the industry has not yet fully embraced conversation and dialogue. In this sense, the findings of this study demonstrate that although blogs may be a part of the social media revolution, they have not significantly changed how public relations practitioners communicate.

However, there are signs that things may be shifting in the blogosphere given that the blogs that performed the best in terms of generating Web traffic and having sites and social media profiles link back to them were more likely to use dialogue. It seems that some practitioners recognize the benefits of virtual engagement with their stakeholders.

Although managing an organizational blog can be time consuming, it can be truly beneficial for nonprofit organizations. As Mattson and Barnes (2007) found, nonprofits have embraced social media in their fundraising and volunteer recruitment efforts. One of the earliest success stories in the nonprofit blogosphere focused on Oceana, an environmental nonprofit in Washington state. Shortly after the launch of its blog, Oceana reported a donation of several thousand dollars from a donor who was impressed with the organization's use of technology to educate the public. It has continued to update the blog several times per week to update its readers about environmental issues. These updates have been helpful in recruiting volunteers and activists, as it is able to communicate with other participants before the organization's events, thereby increasing excitement about the activities (Oceana, 2010).

Nonprofit organizations have been willing to experiment with social media largely because of the low cost of entering this virtual environment. Using free or inexpensive blogging platforms and creating free profiles on social networking sites have given nonprofits access to a socially active community that is willing to talk about the causes and concerns that they are passionate about. As evidenced by the study's findings that nonprofits were more likely to encourage others to share information about their blogs than proactively market the blog to audiences, nonprofits are trying to get others involved and help spread their word on their behalf.

By tapping into social media's opinion leaders and having them talk about the nonprofit and its causes, nonprofit communicators are using a tested principle of public relations practice, the third-party endorsement. Having others endorse and promote the nonprofit rather than spending employees' time and organizational resources on these efforts bring added credibility to the organization as well as a larger social mediated audience. Additionally, by tapping into a social network, the nonprofit organization does not have to use its own resources for publicity and communication outreach, and funds that would have been used for these purposes can be steered back into the delivery of the nonprofit's programs and services, which boost its financial and social accountability to donors, volunteers, and communities. With this strategy in mind, it is easier to understand why nonprofits may still retain some elements of a traditional, one-way communication style with their Web presence as they are relying more on their social network to help spread information about their work. However, as Oceana found, fully embracing the interactivity of the Internet, and specifically blogs, can reap huge benefits for the organization. For nonprofits, this could mean more donations being made as well as increased volunteer support.

Public relations and marketing consultants and agencies continue to push the adoption of new media tactics by organizations. However, not all organizations are prepared to dive into the interactive social media landscape. Organizations, such as nonprofits, that have traditionally carried out public awareness and educational campaigns may find it more difficult to transition to the symmetrical approach of public relations. However, they must be willing to experiment and explore new media to determine what approach works best for them. It may be that research is needed to understand what platforms an organization's audience is using so that a strategic decision is made about social media engagement.

Certainly, nonprofits can then make information available on their websites, blogs, and social media profiles so that others can share. However, as Kent and Taylor (1998) and a stream of scholarship has discovered, organizations that actively incorporate the five principles of dialogue into their Web presence will see positive results from their new media involvement. It takes time to develop a social media following, and expectations of an overnight return of investment for the efforts are unrealistic. However, as this study found, when carefully cultivated over time, the five principles of dialogue result in higher Web traffic, greater connectivity on the Web, and ultimately enhanced relationships with organizational stakeholders.

ENDNOTE

1   Table 1 includes 41 items. The coding sheet was numbered in a way that some of the additional items were sub-items and didn't have their own number. Each coded item is presented in the table.

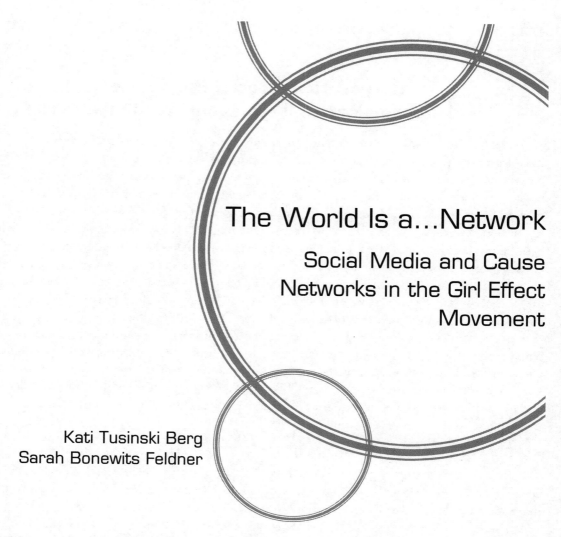

The World Is a…Network

Social Media and Cause Networks in the Girl Effect Movement

Kati Tusinski Berg
Sarah Bonewits Feldner

What if there were an unexpected solution to the problems of global poverty and disease? The Girl Effect, an online movement that began in 2008, offers an unexpected solution—a girl. This movement exists at the intersection of corporate social responsibility, corporate philanthropy, and the integration of social media in public relations practice. In this case analysis, we argue that the Girl Effect campaign is at the center of a "cause network" created through the use of social media.

In 2008, a provocative online video stating, "The world is a mess" began spreading virally (see http://girleffect.org/).[1] So began "The Girl Effect," an initiative created by the Nike Foundation to encourage philanthropic and government investments in girls. The campaign focuses on the value of providing opportunities for women and girls to make simple but meaningful changes that can impact their own lives, as well as their communities.

The Girl Effect is notable because it has expanded beyond a "campaign" to become a movement—entirely through the innovative use of technology. We argue that the Girl Effect represents a new era of networking and coalition building in public relations theory and practice. Ultimately, we suggest the concept of *cause networking* as a means of understanding the ways in which technology shapes public relations practice as it relates to philanthropy and corporate social responsibility. In this chapter, we examine the Girl Effect campaign to explore network creation in an online environment and its impacts on corporate foundation and philanthropic work.

Corporations, social enterprise and communication in an age of social networking

As a case, the Girl Effect is both intriguing and enigmatic. Because the movement exists primarily online, it is difficult to characterize the nature of the organization itself. Further, although the cause itself is clear (i.e., the need to invest in girls for global economic development), what is less clear to the casual observer is who is backing the initiative and who (if anyone) is seeking to benefit from the movement. A visitor to the Girl Effect website would not immediately be aware of the origins of the movement. Indeed, it took the authors several visits to the website to trace the funding back to corporate giant, Nike, Inc. The line of funding is indirect and masked. Nike, Inc. mentions its Nike Foundation in its annual report as a part of its corporate social responsibility efforts. Yet, the Girl Effect is not noted on the Nike, Inc. website (www.nike.com) itself. However, a visit to the Nike Foundation website (http://www.nikefoundation.org/) suggests that the primary program of the Nike Foundation is, indeed, the Girl Effect.

Clearly, a relationship exists between Nike, Inc. and the Girl Effect, which suggests the Girl Effect movement might best be considered in the context of corporate social responsibility (CSR) and corporate philanthropy. And yet, the movement does not seem to clearly fit within this framework as the links to the founding organization are masked on the primary online interface for the Girl Effect. We contend that the Girl Effect movement exists in the space of CSR and corporate philanthropy, but through its leveraging of the capabilities of social networking, it presents a new type of organizational form, the *cause network*.

Corporate roles in social enterprise

The Girl Effect is likely gaining traction because it fits within broader societal beliefs about the need for social innovations that solve social problems. Both scholars and practitioners are recognizing that publics have come to expect corporations to establish social value that exists beyond profits (Caroll, 1979; Kotler & Lee, 2005; Porter & Kramer, 2011; Werther & Chandler, 2011). At the same time, publics generally recognize that there are social problems that the civil sector cannot solve (Trivedi & Stokols, 2011). These two trends create a space for corporations to establish themselves as a part of the solution to social problems through CSR initiatives and corporate foundations.

The Girl Effect can serve as an example of social enterprise. Social enterprises are geared toward "reversing an imbalance in the social, structural and political system by producing and sustaining positive social change" (Trivedi & Stokols, 2011, p. 4). All of the messaging tied to the movement speaks to the goal of creating systemic change through "the girl effect." Social enterprise, while often considered in the context of NGOs, also can be tied to corporate efforts. Indeed, several major outlets (e.g., *Harvard Business Review*, the Page Society, and *The Wall Street Journal*) that advise and speak to corporate issues suggest that not only *can* corporations be involved in social enterprise but that they *should* be involved.

Within the corporate context, social enterprise might best be recognized as corporate social responsibility.

Corporate social responsibility is defined as a company, firm, or brand's "commitment to improve societal well being through discretionary business practices and contributions of corporate resources" (Kotler & Lee, 2005, p. 3). As such, CSR encompasses the "economic, legal, ethical and discretionary expectations that society has of organizations at a given point in time" (Caroll, 1979, p. 500). Corporate philanthropy is considered by many to be one particular aspect of CSR (Porter & Kramer, 2002). Corporate philanthropy can take many forms, but a primary vehicle for achieving philanthropic goals is the corporate foundation. These foundations are established by corporations with the aim of helping corporations fulfill the role of good corporate citizenship (Westhues & Einwiller, 2006). Corporate foundations often operate as separate entities but retain the goal of enhancing the reputation of the corporate founder. Given these defining attributes, one might imagine that the Girl Effect represents corporate philanthropy and foundation work. Yet, this case suggests that the Girl Effect's online presence and constantly evolving structure demand a more nuanced analysis.

Social networking and the networked organization

A primary challenge for classifying the Girl Effect is the nontraditional structure of the organization itself. The Girl Effect represents what many organizational communication scholars refer to as the "network organization" (see Cheney, Christiansen, Zorn, & Ganesh, 2011), which is made possible through its innovative use of social media and new technology. Social media, in this case, represent both the strategies used by movement organizers and the structure that holds the movement together.

Because social media allow for organizations to connect with multiple stakeholders, public relations practitioners and scholars are turning their attention to the role of social networking in professional communication with increasing frequency. Though there are multiple definitions of social media available, social media are characterized by the use of technology to allow for user participation and user supplied content (Waters, Burnett, Lamm, & Lucas, 2009). Kent (2010) argued that social networks are best characterized by real-time interaction, reduced anonymity, propinquity, short response time, and the ability to time shift.

To date, most research on social networking/media focuses on the ways in which various organization types utilize social networking and/or how users feel about their use of social networking (e.g., Curtis et al., 2010; Rybalko & Seltzer, 2010; Seo, Kim & Yang, 2009; Waters et al., 2009; Wright & Hinson, 2009b). These are certainly important factors to be considered. However, as Kent (2010) argued, there is more to be said about the role of social media in public relations practice. In particular, Kent asserted that public relations research should move beyond the *outcomes* of social media use to consider the *strategies* that will best inform practice. Further, scholars should seek to provide more theoretical and critical perspectives on the emerging role of social media in public relations. In this vein, we argue that although the Girl Effect certainly employs technology in innovative ways, the integration of social media in the Girl Effect creates a new kind of public relations initiative rooted in a particular type of organizational structure.

Advances in technology have led not only to changes in how we think about the interaction between organizations and individuals, but also changes in how we think about organizations themselves. The Girl Effect is referenced as a movement, but at the same time it represents a new organizational form that is structured by a network of relationships between several organizations. The network organization is "comprised of two or more organizational units from different organizations involved in a long-term, and more or less formalized relationship. The relationships of network organizations are often global in scope and reach" (Cheney et al., 2011, p. 167). The Girl Effect is an interorganizational network comprising funding organizations, philanthropic efforts, and government organizations. The website, coupled with social media, provides a vehicle for tying these organizations together into a single network.

Interorganizational relationships exist for several reasons. Shumate and O'Connor (2010) suggested three primary theories for explaining interorganizational alliances: transaction cost economics, stakeholder management theory, and collaboration theory. The Girl Effect best represents collaboration theory, which suggests that organizations partner when one organization alone cannot solve a problem. In this way, interorganizational relationships allow for a combining of resources (Cheney et al., 2011). The issue of global economic development and the struggles of girls within this context are formidable. By creating an online interorganizational network, the Girl Effect leverages the strengths of each organization involved, thus providing greater potential for large scale impact.

In the end, the Girl Effect is a campaign that exists at the intersection of CSR, corporate philanthropy, and new communication technology. The movement utilizes technology both to create and promote partnerships between organizations and to draw numerous users into the network through social networking sites (i.e., Facebook and Twitter). The movement represents a sort of hybrid organizational form that is emerging as both the capacity of technology and our understanding of its power expand. The Girl Effect may be best understood as cause networking. By this, we suggest that the communication efforts create a virtual network that is geared toward addressing a particular cause and that cause is then advanced through network creation.

Building a cause network: The case of the Girl Effect

The Girl Effect initiative was developed and funded by the Nike Foundation along with significant financial and intellectual input from the NoVo Foundation and Nike, Inc. (Elliott, 2010; Nike Foundation, 2008; Roberts, 2010). The Girl Effect is a self-described movement that seeks to invest in girls with the aim of raising the standard of living globally. The movement has developed via social networking and the creation of partnerships with other organizations that are tackling the same issues. The hallmark of the movement is the short online videos, which not only serve as the front page for the Girl Effect website, but also appear on YouTube and Facebook. These videos have been viewed more than 2 million times, and they serve as a launching point to draw stakeholders into the movement. Beyond the videos, the Girl Effect movement largely exists in the context of the website (http://girleffect.org/). The website includes information on how others can join the movement, a press kit for media, and several embedded links that connect site visitors to other partnering organizations such as the Coalition for Adolescent Girls, the Population Council, and the United Nations Foundation. The aims of the movement are to create awareness about the issues, educate multiple publics, and support others who wish to become involved either through grant opportunities or by assisting them in planning their own events/organizations.

Why girls? Framing a rationale for the Girl Effect

Public relations campaigns cannot be created in a vacuum, and the Girl Effect is no different. The cause of investing in adolescent girls was developed as the result of careful research and planning (Elliott, 2010; Roberts, 2010). The movement is grounded in extensive research on the global status of adolescent girls. The research was possible because Nike, Inc. allocated resources to address these initiatives.

As a brand, Nike is known for its innovative product design, edgy advertising, and ability to push boundaries. Nike has also been a long-time advocate for girls and young women. In 1995, Nike's advertising campaign "If you let me play" drew considerable attention and praise because it advocated the benefits of girls and young women participating in sports (Elliott, 2010; Nike Foundation, 2008). In 2004, Nike's focus shifted from sports to education when the corporation established the Nike Foundation. According to Nike's corporate website:

> The Foundation leverages the brand's drive for innovation and positive change, and its ability to inspire both. We believe that when girls receive support and realize opportunity for their futures, they can become an unexpected and powerful force in transforming their families, communities and the world. ("Nike Foundation," 2011, para. 2)

Subsequently, in 2008, the Nike Foundation began the Girl Effect.

The foundation provides a rationale for the campaign by suggesting that girls are the unexpected answer to global poverty. "Nike believes in the power of human potential to accomplish anything: on the field, on the court, in life. We're applying that belief to poverty in the developing world, an issue that impacts everyone's future" (Nike Foundation, 2008, para. 1). In many parts of the world, girls as young as 12 years old are forced to quit school, get married, and have children. During this time, they are also highly likely to contract HIV/AIDS. In a 2010 *New York Times* article, Leslie Lane, vice president and managing director of the Nike Foundation, explained that Maria Eitel, president and chief executive of the foundation, and Phil Knight, chairman of Nike, decided their "best investment" was an effort to "break the cycle of intergenerational poverty" by focusing on "the future mother of every child born into poverty" (Elliott, 2010, para. 15).

The Girl Effect website acknowledges that little research has been done to understand how investments in girls impact economic growth and the health and well-being of communities. However, research from a variety of sources, including but not limited to the United Nations Population Fund, Human Rights Watch, Global Coalition on Women and Aids, and the International Center for Research and Women, on issues such as population trends, educational gaps, child marriage and early childbirth, and health indicate a positive rippling effect when young women are empowered with resources (http://www.girleffect.org/learn/the-big-picture). For example, Fortson (2003) reported that when women and girls in developing countries earn income, they reinvest 90% of it into their families, compared to only 30% to 40% for men. Psacharopoulos and Patrinos (2002) found that an extra year of primary school boosts girls' eventual wages by 10% to 20%, and an extra year of secondary school increases wages 15% to 25%. A variety of other reports is available on the Girl Effect website (http://www.girleffect.org/learn/more-resources) that indicates the potential impact and rippling effect of investing in girls.

Benett, Gobhai, O'Reilly, and Welch (2009) described Nike as a brand that "extols the incredible potential of the human body—and the human spirit" (p. 14). So even though the Girl Effect does not include any swooshes on its website or mention of Nike, Inc. in its narrative, a movement focused on systemic change that begins with empowering individuals strategically fits with who Nike is as a corporation.

Impacting global poverty one girl at a time

Although the specific, measurable objectives of the campaign are unknown, it is clear that the Girl Effect seeks to raise awareness about the importance of empowering girls and women through three key components: Learn. Give. Mobilize.

Learn. As a movement, the Girl Effect wants to educate people about the importance of investing in girls and women around the world. Additionally, the Girl Effect encourages people to join the conversation, talk it up, and spread the world. According to Eitel (as cited in Roberts, 2010), "The first step has been to get the world to realize the power of the Girl Effect. When you improve a girl's life, everyone benefits: her brothers, sisters, parents, future children and grandchildren" (para. 6).

Give. In order to create holistic solutions, the movement needs funds to provide opportunities for women and girls to make simple but meaningful changes that can impact their own lives, as well as their communities. Thus, the Girl Effect also encourages visitors to donate money via online or text messaging. The Girl Effect website includes the following text on the *Give* page (http://www.girleffect.org/give, *How Can I Give?*): "Send a girl to school. Help fight her legal case. Give her a microloan. Start making a difference. Start the Girl Effect."

Mobilize. Since the Girl Effect is a movement driven by girl champions around the globe, the website (http://www.girleffect.org/mobilize/connect) encourages people to become agents of change by joining "the most important conversation on the planet" because "when the power and potential of girls is raised into the global consciousness then the Girl Effect really begins and change starts to happen" (http://www.girleffect.org/give, *Get Involved*). Furthermore, people are invited to connect with the Girl Effect networks via Facebook, YouTube, and Twitter. Followers are encouraged to share the videos, emails, tweets, and messages with others. Finally, site pages suggest that followers should wear the Girl Effect on their sleeves, on their computers, and in their offices by downloading the Girl Effect tool kit.

Ultimately, the Girl Effect is an educational awareness campaign that seeks to use strategic, innovative partnerships coupled with technology and social networking to spread the word about the importance of investing in girls and women around the world.

Championing girls: The Girl Effect's network of solutions

In part, any success of the Girl Effect can be attributed to the stylized nature of the campaign with its catchy videos and provocative messaging strategy. The Girl Effect website is the hub for information, networking connections, and motivation. However, in order to raise awareness and educate its audience, the Girl Effect relies on innovative partnerships and social networking to create a movement focused on changing the world by saving a girl.

As described on its website, "The Girl Effect is a movement driven by girl champions around the globe" (http://www.girleffect.org/about-us, *Girl Champions*), and these champions include a range of organizations including multi-national NGOs and government agencies, school teachers, mothers, and community leaders. In addition to the NoVo Foundation and Nike, Inc., other key partners include the United Nations Foundation and the Coalition for Adolescent Girls. The website also lists other girl champions including the International Center for Research on Women, the Population Council, CARE, the White Ribbon Alliance for Safe Motherhood, the Center for Global Development, Plan, and the Girl Hub. Additionally, BRAC (a development organization committed to fighting poverty) is included as a Girl Effect pioneer. And lastly, the Girl Effect thanks all those who work for girls every day, in schools, villages, cities, offices, agencies, and governments. Clearly, the Girl Effect is supported and endorsed by active partnerships with significant philanthropic organizations and leading non-governmental organizations. These partnerships have brought credibility, legitimacy, and international attention to the Girl Effect. The relationship between the Nike Foundation and other girl champions underscores our classification of the Girl Effect as cause networking because the organizations promote and support each other in order to solve a larger social issue.

In addition to its use of strategic partnerships, the Girl Effect relies on technology to educate, inform, share, connect, and continue the movement. In an online environment, the conventional role of the public relations agency shifts to the user who is expected to spread the message via the networking options. The campaign encourages people who visit the website to share, connect, and wear the message of the Girl Effect. In addition to the website, the campaign actively implements a social media strategy that includes YouTube videos, a Facebook page, and a Twitter account.

The Girl Effect utilizes social media to share facts and information, inspiring stories, and networking opportunities. In other words, the Girl Effect becomes the nucleus of the cause network. One way in which they do so is by posting status updates on Facebook that link to other organizations and causes. One example of a Facebook post is: "Fact: Pregnancy is the leading cause of death worldwide for women under 19. Disheartening? Yes. But, thanks to NextAid's 5Alive Campaign, here are five things you can do

RIGHT NOW to change this statistic: http://www.nextaid.org/women/." Even though NextAid is not a funding partner with the Nike Foundation, the Girl Effect is most concerned about sharing significant information with a captivated audience who wants to be the difference in the world. By using Twitter, the Girl Effect can respond to specific stories and opinions from other organizations such as when they tweeted, "@MicrofinanceWWB. We love this—so exciting! Thanks for sharing with us. Teaching girls to save: http://inv.lv/m5Sapk #banking #education." The Girl Effect also uses social media to praise its girl champions as they did with this tweet: "3 girl champs are completing 5 Ironman triathlons in 5 days 2 raise $ for the #girleffect—but they need ur help! http://inv.lv/hdqYwD." Social media also allow the Girl Effect to participate in dialogues about the various social issues facing girls and women around the world. Thus, the ability to share and connect via technology inevitably expands the cause network.

Has the Girl Effect changed the future?

The initial video suggests that investing in a girl will impact the future of humanity. Thus, one may ask if the Girl Effect has had such an impact. Since the Girl Effect is an ongoing campaign, we can only provide a programmatic evaluation of the campaign. If the number of videos viewed, Facebook likes, and Twitter followers is any indication of awareness, then the Girl Effect is making progress. As of this writing, the website has been visited more than 2.3 million times and shared more than 100,000 times. The videos have been viewed more than 2.8 million times. More than 247,000 people like the Girl Effect on Facebook, and the Girl Effect has more than 12,350 followers on Twitter.

Obviously, great strides in awareness are being made, but more substantial changes are still in the works. Jennifer Buffett (as cited in Roberts, 2010), president and co-chair of the NoVo Foundation explained:

> There has been a tremendous amount of awareness-raising about the importance of empowering girls and women. However, are we moving systems toward meaningful action in terms of modifying attitudes and patterns of behavior *and* moving larger dollars and resources for more equitable distribution? I think we are just scratching the surface. (para. 7)

Additionally, Eitel (as cited in Roberts) noted, "Together we have touched the lives of more than two million girls, and through the Girl Effect, their families, communities and nations" (para. 13). Because limited information is available about the specific programs and opportunities created through investments from NoVo and the Nike Foundation, it is difficult to evaluate the impact of the initiatives at this time.

Enacting global stewardship

The Girl Effect has been operating since 2008 and by all appearances the movement continues to grow. Facebook updates and tweets continue to suggest that more individuals and organizations are joining the network. In this way, the network created by the Girl Effect has contributed to a larger public dialogue about the role of adolescent girls in the global economy. Only time will tell the lasting effect that the movement will have. There is certainly room for growth, and yet, if we allow the programs and individual efforts to speak for themselves, the current impact appears substantial. The Girl Effect has advanced social change by creating a large online network that joins organizations and their stakeholders and leverages the resources of this network to advocate for change.

Conclusion

The world is a mess. All girls are valuable. Invest in a girl and she will do the rest. It's not a big deal. Just the future of humanity.

The messages of the Girl Effect movement are powerful indeed. However, to have the impact promised, the movement will need to be more than powerful messages. Although a comprehensive analysis of the success of the campaign cannot yet be offered, there are a number of key issues and questions that an examination of the Girl Effect raises.

First, the case continues to illustrate the power of social networking and online communication for distributing information to broad and dispersed audiences. The issue of girls and global poverty is not a new issue. Yet, relatively few people ostensibly know about the connection between investment in adolescent girls and global economic development. The Girl Effect movement helps to draw attention to the issue in important and effective ways. Certainly, the Girl Effect can be a model for other causes that seek to deliver messages in powerful ways. However, we argue that the lessons of the Girl Effect are far greater than viewing social media as a new and efficient way to communicate a message. The Girl Effect demonstrates the ways communication technology allows for the creation of emergent organizational structures that facilitate powerful partnerships and strategic networking. The website coupled with the Girl Effect's presence on social networking sites allows the movement to gather followers and collaborators, thus strengthening the movement. The ultimate effect is the creation of a network of organizations collaborating to address a cause or what we call here a cause network. In this way, the Girl Effect stands as a model for other organizations and individuals seeking to advance a particular cause or engage in social enterprise.

At the same time that the Girl Effect provides an example for network creation via social networking, the case also suggests some ways in which the use of social networking can improve. One of the cited advantages of social media is the opportunity for stakeholder participation and stakeholder created content. Yet, the Girl Effect seems to be largely focused on pushing out information and serving as a conduit for success stories rather than engaging stakeholders and creating new meanings for the Girl Effect. The movement could explore ways in which it might embrace technology to embrace a more dialogic stance (see Feldner & Meisenbach, 2007; Kent, 2010). Adopting a more dialogic stance might allow for better *enactment* of the empowering messages that the movement embodies, thus creating a more significant and substantial impact. The opening video suggests that individual viewers can be a part of the solution. However, it is unclear how the current strategy truly creates a sense of agency[2] in visitors to the website.

Although the Girl Effect is provocative and suggests new ways that public relations practitioners could use technology in the areas of corporate philanthropy and corporate social responsibility, we suggest a certain degree of caution in this regard. First and foremost, we contend that it must not be forgotten that the Girl Effect is tied in significant ways to a larger corporate initiative. In the end, its existence online allows Nike, Inc. to be noted in the movement's messaging without being highlighted. The connection between Nike, Inc. and the Girl Effect is not hidden, as it is noted on the About Us section of the Girl Effect website, but it is certainly not prominent. This necessitates a consideration of the ethics surrounding this strategy and the aims of this philanthropic effort.

From an ethical standpoint, Nike, Inc. has put itself in a position where skeptics might perceive there to be a lack of transparency and full disclosure. Although a bit obscure, Nike's sponsorship affiliation is stated on the Girl Effect website, in Nike, Inc. press releases, and in multiple news stories. However, since the Girl Effect is not a branded cause marketing campaign like the Nike Livestrong Collection (from which all profits from the sales of certain products are donated to Lance Armstrong's Livestrong Foundation), stakeholders might question Nike, Inc.'s involvement and intent.

Others might question whether Nike Inc.'s interest in girls and women in developing countries is a reactive approach to offset issues, such as labor and working conditions that have troubled Nike, Inc. in the past, rather than a proactive initiative to solve a social problem. In the end, The Nike Foundation denies that any commercial interests are tied to the Girl Effect (Elliott, 2010), but it does not put to rest whether this relationship is one to be modeled by other corporate foundations. In addition, we are left to wonder what the relationship between Nike, Inc., the Nike Foundation, and the Girl Effect means for each organization. That is, in what ways does the relationship enhance the reputation of each of the participating organizations? The Nike Foundation claims that the Girl Effect is purposely not branded as Nike so that the effort can more directly address issues of public policy without seeming to be directed by Nike's business needs (Elliott, 2010). Any response that we articulate at this point is largely speculative. However, we think the question is a significant one. The technology in some ways allows Nike to be connected—but only through a few hyperlinks. And, the connection and links are unobtrusive and subtle.

We will continue to monitor the Girl Effect, and we suggest that other public relations scholars and practitioners do the same. The campaign offers key insights into how we can think of technology as not simply a medium for public relations messaging but rather as a means of creating alternate organizational structures for public relations efforts. The creation of cause networks provides a simple, yet powerful solution to the challenge of making changes that address big social problems. It's not a big deal. It is just the future of public relations and corporate philanthropy.

ENDNOTES

1   At the time of this writing, the video was accessible through the home page, and the campaign was fully active.
2   The belief that one can act in ways that will actually lead to change.

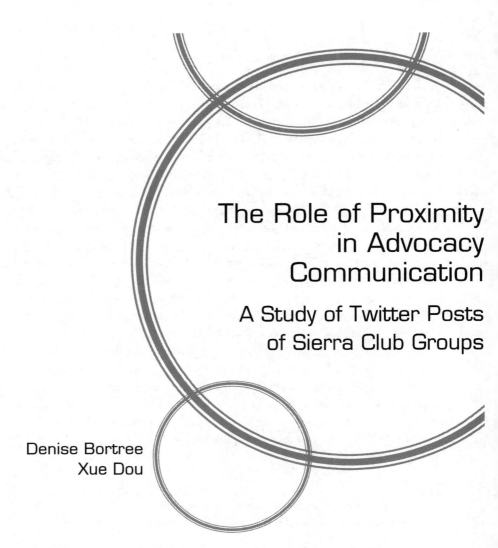

The Role of Proximity in Advocacy Communication

A Study of Twitter Posts of Sierra Club Groups

Denise Bortree
Xue Dou

This study sheds light on the way environmental advocacy groups currently communicate with key audiences through Twitter. A content analysis of 3,098 Twitter posts from national, statewide, and local Sierra Club groups found that proximity plays a role in the nature of communication, with local groups focusing more on motivating action and national groups sharing information to educate their audiences. The chapter offers an explanation for this pattern based on a number of communication theories. In addition, proximity appears to play a role in the types of communication strategies used in Twitter posts and the behaviors that groups attempt to motivate. The implications of these findings for relationship management are discussed.

Introduction

Local advocacy groups have begun to incorporate social media into communication plans as a way of engaging with key stakeholders: donors, volunteers, members, local government, and community members. Social media offer nonprofit organizations the power to connect with audiences in one-way information dissemination and two-way dialogue. However, the degree to which social media are used and the ways that they are used by advocacy groups, specifically environmental groups, are still understudied. This

* *This study was funded by the Arthur W. Page Center at Penn State University.*

study examined the content of Twitter posts of 30 Sierra Club groups over a six month period to track their use of the channel to engage with key stakeholders. As environmental groups struggle to gain (and maintain) a voice in the crowded sphere of "green" communication, offering regular, meaningful opportunities for members of key publics to connect with the organization is critical for sustained relationships. Using theory of hierarchy of effects and situational theory of publics, this quantitative content analysis examined how groups at different levels of the organization (national, state, and local) communicated with audiences through their Twitter streams. The study examined whether organizations may communicate for different effects based on proximity to the audience, with organizations that are physically closer to their audience (local advocacy groups) working to motivate behaviors more often than organizations that are farther in proximity from their audiences (national advocacy groups). The study categorized different target outcomes of messages, including information, attitude, and behavior, and it identified strategies used to build relationships with key publics through this channel. Specific suggestions are made for employing Twitter as a communication channel for environmental advocacy organizations.

Literature review

Public relations literature has begun to examine the role of social media in communication plans. Research has suggested that social media are useful for involving publics in dialogic communications (Bortree & Seltzer, 2009; Briones, Kuch, Liu, & Jin, 2011; Seltzer & Mitrook, 2007), allowing companies to implement relational maintenance strategies (Kelleher & Miller, 2006; Sweetser & Metzgar, 2007), and helping companies to connect with journalists (Waters, Tindall, & Morton, 2010). Social media are especially useful for nonprofit organizations given their ability to reach large numbers of publics with minimal cost. In fact, studies have reported that most nonprofit organizations use some form of social media (Curtis et al., 2010). However, many nonprofit organizations are not yet fully utilizing social media, and as a result miss opportunities to connect with publics, raise funds, and deepen relationships with key audiences (Bortree & Seltzer, 2009; Waters, Burnett, Lamm, & Lucas, 2009). Nonprofits appear to be uncertain of the value of social media and struggle in handling these new communication channels (Association of Fundraising Professionals, 2009). In particular, advocacy groups communicate with a variety of audiences using the same channels, and tailoring messages to reach different publics can be a challenge. Thus, more studies are needed to investigate how social media have been adopted to communicate key information and motivate audiences toward organizational goals.

Twitter

Among the various types of social media, the popularity of Twitter is especially growing. Twitter, a micro-blogging service, allows users to post short messages (tweets) that are 140 characters or less. Users have their own profile pages on which they can create posts and view others from Twitter streams that they follow. Users can follow others in the Twitter community, and once users become followers of others, their tweets show up on the users' profile pages in reverse chronological order (Rybalko & Seltzer, 2010). Twitter users often use signs (or markup language) in their tweets. For example, RT (retweet) represents that the post was reposted by others, and a sign of "@" followed by a username signifies that a tweet mentions or replies to a user. Further, "#" (hashtag) is another frequently used sign, which categorizes tweets. By using hashtags, users can spread information on Twitter, while allowing other users to easily search for and locate information that they want. These signs assist users in conveying more information within a strict limit of 140 characters per Tweet (Kwak, Lee, Park, & Moon, 2010).

Twitter differs from other social media platforms, such as Facebook and Myspace, in many ways. First, Twitter is relatively open compared to other major social networking sites. For example, people can see tweets without logging into the Twitter system, whereas Facebook and Myspace require users to login to view information. In addition, a Twitter user can follow other users without the need to get approval from them. In contrast, becoming a friend requires mutual confirmation (i.e., send friend request and get confirmed) for Facebook and Myspace. Second, Twitter is designed for sending short and concise messages, and thus it offers only limited functionality for activities such as photo and video posting. Most often, Twitter users upload multimedia content to third-party websites, such as TwitPic. Links to the websites are posted on Twitter in order to direct others to the multimedia content. Third, the markup cultures previously mentioned are (as of this writing) unique to Twitter. Users can use these signs to find information they need in a more effective manner compared to Facebook and Myspace.

Organizations' Twitter use

Given its unique characteristics, Twitter offers many opportunities for organizations to communicate with their publics in a prompt, direct, and open manner (Church, 2008). Studies have reported that Twitter has been used in many ways for supporting organizations' communication with their publics, such as answering publics' inquiries, gathering information from publics, and crisis communication (e.g., Briones, Kuch, Liu, & Jin, 2011; McCorkindale, 2010b). In addition, public relations practitioners are increasingly using Twitter in their daily practices. As of 2009, nearly 65% of practitioners used Twitter, and the number had doubled from 2008 (Hinson & Wright, 2010).

Several scholarly studies have been conducted for understanding Twitter use in public relations practices. For example, Rybalko and Seltzer (2010) investigated Fortune 500 companies' use of Twitter to facilitate dialogic communication. Their results suggested that responsive companies (i.e., companies that reply to tweets in a prompt manner) use Twitter for stimulating dialogue with publics, rather than as a place for simply disseminating information. In the same vein, McCorkindale (2010b) found that organizations that specify the name of individuals who manage the Twitter account are more likely to engage in dialogue with publics. Further, in their interviews with American Red Cross employees, Briones et al. (2011) found that the nonprofit uses Twitter, along with Facebook, for building relationships with publics by quickly addressing issues raised from publics and creating relationships with media by following reporters and being followed by them.

To explore the ways that advocacy groups engage with key publics through their Twitter streams, the following research question was asked:

> RQ1. To what degree do Sierra Club Twitter streams encourage engagement with publics through signs and functionality (RT, @, #, links)?

Whereas the previously discussed studies provided a general picture of organizations' Twitter use, we need to pay close attention to other factors that could possibly increase the effectiveness of Twitter as a communication channel. One such factor is message type. Because the Twitter post is limited to 140 characters, communications via Twitter need to be concise. Therefore, what information should be included or cut is an important issue to be considered. An example can be found in a recent study by Schultz, Utz, and Göritz (2011). They examined Twitter use for crisis communication through an experiment. Interestingly, their results indicated that participants' intention to share crisis information with others or react to the information differed depending on the message type. Specifically, messages of apology and sympathy were more likely to be shared than informational messages. These results signify the importance of message type in informing and motivating audiences.

In this chapter, we propose that organizations should use different types of messages depending on the nature of the organization and its psychological proximity to its publics. As proximity changes, the relationship between organizations and audiences takes on a different meaning and that has implications for communication.

Situational theory of publics

The situational theory of publics can be used to segment audiences of an organization into four categories—nonpublics (which experience no consequences from issues), latent publics (who experience consequences but are not aware of it), aware publics (who are aware of issues and consequences but have not become active on the issues), and active publics (who are aware of issues and actively seek information about the issues) (Grunig, 1989b). Audiences of a Twitter feed likely fall within the aware and active audiences for an issue/organization because individuals in these groups would share some concern about the issue an organization is addressing, enough to motive them to follow the organization on Twitter. In the case of a large national or international environmental advocacy organization that reaches out to audiences through many groups throughout a country or region, the situational theory of publics would be helpful in understanding who would follow the Twitter streams of groups at different levels of the organization. For example, it is likely that individuals who are connected to local branches of an advocacy organization would be more actively engaged in an issue and would see the relevance of the issue for their own community. Those who are associated with an environmental organization at the national level may be interested in the issue but may not make the same level of personal investment in it. As such, those who are involved on the local level may be more likely to be an active audience, and those who are involved at the national level may be more likely to be only aware audiences.

Further support for this reasoning comes from the theory of psychological distance, which argues that one's construal of information is influenced by four dimensions of psychological distance: temporal, social, spatial, and hypothetical. Specifically, people see psychologically distant events, objects, or things (in this case, an organization) with a higher level of mental construal, which is characterized as an abstract, simple, and decontextualized interpretation. In contrast, when an event, object, or thing is psychologically close to a person, he/she interprets it in a detailed and contextualized manner (Liberman, Trope, & Stephan, 2007). For example, we see events that occur overseas in an abstract manner (i.e., a natural disaster that happens on another continent), but interpret local events in a more detailed manner (i.e., a category two storm occurs in my town).

Because people adopt psychological distance to process information on a daily basis, it can be argued that they also develop certain expectations of information to be acquired based on their psychological distance from a particular event or information provider. For example, students would usually expect their teacher to provide a more detailed description about their assignment as the due date becomes closer. Similarly, people expect to get detailed suggestions from their families and friends (socially close) rather than strangers (socially distant). In addition, messages are found to be more persuasive when the message frames fit one's natural construal level (Castano, Sujan, Kacker, & Sujan, 2006; Thomas, Chandran, & Trope, 2006). For instance, Kim, Rao, and Lee (2009) found that abstract "why"-laden appeals are more persuasive when a voter's decision is temporally distant, whereas concrete "how"-laden appeals are more persuasive when a voter's decision is temporally close.

Based on these rationales and findings, we can argue that as the scale of organizations becomes bigger and as the activities in which organizations engage are physically farther from the individual, people will feel more socially and spatially distant from them. As a result, they would expect the national organizations to provide more general information, while expecting the regional organizations to provide more detailed information, such as how to become involved in an organization and what kind of actions should be taken. In addition, compared to national level organizations, local level organizations are more suitable to inform publics about taking specific action since it matches with publics' construal level of processing the information.

Theory of hierarchy of effects

The theory of hierarchy of effects proposes that individuals can be moved along a series of steps in the persuasion process, first building and strengthening information and awareness, and then changing or reinforcing attitudes, and finally encouraging behavior (Lavidge & Steiner, 1961). As individuals become more knowledgeable and begin to form opinions about an issue, they are more likely to take action on the issue. This theory has implications for organizational communication. Organizations communicate with a variety of audiences in an attempt to inform them, change their opinions, and motivate their behavior. Considering the audiences of advocacy groups, those who are more aware of an issue would be more likely to have formed attitudes about environmental issues. They would also be more likely to take some action to help further environmental causes (as suggested by the Situational Theory of Publics). Therefore, we argue that audiences of a local branch of an environmental organization, who are more likely to be active audiences, would be more easily motivated to action, while audiences of a national branch of an organization would be information processors but would be less likely to take action. Organizations that wish to use their Twitter streams most effectively to connect with audiences at these different levels should consider the needs of the audiences. At the national level, more information should be provided to audiences who are information processors. At the local level, more calls for action and opportunities to behave should be presented through Twitter streams.

To examine whether environmental advocacy organizations are using their Twitter streams as proposed in this study, the following hypotheses are offered:

H1: Organizations on the national level will be more likely to focus on information through their tweets than will state and local level organizations.

H2: Organizations on the national level will be less likely to try to motivate behavior than will state and local level organizations.

Environmental behaviors

Through social media communication, environmental advocacy groups have the opportunity to engage with audiences and motivate their behaviors toward environmental issues. These behaviors can fall into two primary categories—public-sphere behaviors or private-sphere behaviors (Lee, 2008; Stern, 2000). Public-sphere behaviors include group activities (e.g., campaigns, rallies, pollution cleanup) and individual activities (e.g., writing to congressman). Private-sphere behaviors are personal choices that can have direct effects on the environment (e.g., recycling, reduction of water use, green purchasing behaviors). To effectively address environmental issues, both types of activities should be promoted and maintained.

This study parsed the promotion of public-sphere behaviors into two categories: those that benefit only the issue/cause and those that benefit the organization (presumably in addition to the issue/cause). The purpose of this distinction was to assess the degree to which the organization focused its effort on generating behaviors toward the issue versus toward its own initiatives (i.e., generating donations, volunteers, or event attendance). In addition, the study examined differences in the behaviors promoted at different levels of the organization. To explore this, the following research question was investigated:

RQ2: Do organizations at national, state, and local levels attempt to motivate different behaviors through Twitter posts?

Methods

To identify trends in the messages disseminated by environmental advocacy groups, a content analysis was conducted on the Twitter stream of 30 Sierra Club groups. The streams were located through a thorough search of the Twitter site, and each stream was classified as national, regional (more than one state), statewide, or local, depending on its reach. Six months of tweets (January through June 2010) were captured from the Twitter site and pasted into a Microsoft Excel file for further analysis. In total, 3,098 tweets were analyzed in this study.

Five coders were trained in three sessions, and then intercoder reliability was established using a sample of 300 tweets. Reliability was calculated using Fleiss' Kappa, a statistic that is appropriate for multiple coders and closely related to Scott's Pi and Cohen's Kappa (Fleiss, 1971). An acceptable level of agreement using Fleiss' Kappa is .41, which indicates a moderate level of agreement. All of the items coded in the study met or exceeded this level except for the presence of volunteerism in the posts. That variable reached only .38. However, the score falls within the "fair" level of agreement and was accepted for the study.

The final codebook included 13 items reported in this study: group level, month, length, link (κ=.99), retweet (κ=.98), mention (κ=.50), hashtag (κ=.86), issue (κ=.46), effect (κ=.46), behavior type (κ=.60), donation (κ=.50), volunteer (κ=.38), and event (κ=.65). For group level, one of three levels was coded based on the focus of the group: national, regional (more than one state), state (pertaining to an entire state), or local (area smaller than a state). The length of the tweet was reported in number of characters. A number of the items were coded for simple presence or absence including link, retweet, mention, hashtag, and mention of donation, volunteering, or an event. For issue, coders were asked to indicate the prominent environmental issue addressed in the post. Issues included: pollution, greenhouse gas reduction, solid waste problems, species/habitat protection, energy efficiency, energy independence, and general state of the environment. Coders also indicated the type of effect that posts appeared to be trying to achieve. The effects included cognitive (information), affect (attitude or opinion), and behavior. And finally, the type of behavior was coded including personal, public, and organization-related.

Results

A total of 3,098 tweets were analyzed by the five coders. The average length of a Twitter post was 113 characters (SD=27). Approximately 82% of posts included links, 14% were retweets (they were passing along material from another Twitter stream), and 13% included mentions (@) of others. The number of posts per month steadily increased, beginning with 376 posts in January and ending with 528 posts in June. Of the 30 Twitter streams analyzed in this study, half (15) were maintained by state-level Sierra Club groups (1,708 posts), for example, Hawaii Sierra Club or Virginia Sierra Club. Ten of the Twitter streams were maintained by local clubs (338 posts), including groups such as San Francisco Bay area Sierra Club or university-based clubs such as Frostburg State University Sierra Club group. Four streams were national level accounts (926 posts), for example, U.S. Sierra Club or Canadian Sierra Club, and only one stream was categorized as regional (125 posts). That stream was collapsed into the national level data.

Of the 3,098 posts reviewed in this study, 2,110 specifically addressed an environmental issue. The most popular issues were pollution (37%), general state of the environment (25%), greenhouse gas reduction (22%), and species/habitat preservation (12%). The least popular environmental issues were energy independence (1%) and solid waste problems (1%).

To explore the first research question, a series of chi-square tests were run to see how groups at the three levels—local, statewide, and national—used Twitter signs and markups including retweets (RT), mentions (@), hashtags (#), and links to websites. The results can be found in Table 1.

The results support the idea that proximity affects communication for advocacy groups. At the national level, groups appear to be sharing more information through hashtags and links in their posts, while local groups appear to be focused more on retweeting information and mentioning others than are national level groups. Hashtags and links would be indicative of information sharing, while retweets and mentions would suggest relationship building. The greater focus on relationships at the local level could be due in part to the personal connections that develop from local advocacy work. All four chi-square tests were significant. It should be noted that mentions were only a slightly higher percentage among local advocacy groups than national groups. Also, hashtags dropped off dramatically at the local level, suggesting that local groups have not adopted this communication strategy.

H1 proposed that national groups would have a greater focus on providing information than state or local groups. A chi-square test found that the observed and expected frequencies were significantly

Table 1. Use of Twitter signs and markup language at three levels of the Sierra Club groups.

ORG LEVEL		NATIONAL LEVEL	STATE LEVEL	LOCAL LEVEL
Retweet (RT)*	Included RT	11%	14%	22%
	Not included RT	89%	86%	78%
Mentions (@)**	Included @	15%	12%	17%
	Not included @	85%	88%	83%
Hashtag***	Included #	34%	23%	5%
	Not included #	66%	77%	95%
Link****	Included link	84%	82%	73%
	Not included link	16%	18%	27%

*($X2$ (2, n = 3,095) = 26.93, p < .001)
**($X2$ (2, n = 3,094) = 11.47, p = .003)
***($X2$ (2, n = 3,092) = 118.64, p <.001)
****($X2$ (2, n = 3,097) = 21.43, p <.001)

Table 2. Chi-square of national and regional/state orgs use of information and behavioral posts

ORG LEVEL		INFORMATION	ATTITUDE	BEHAVIOR	TOTAL
National and Regional	Observed	444	317	287	1,048
	Expected	397	263	388	1,048
	% within type	42%	30%	27%	
State level	Observed	621	419	661	1,701
	Expected	644	427	629	1,701
	% within type	37%	25%	39%	
Local level	Observed	105	39	193	337
	Expected	128	85	125	337
	% within type	31%	12%	57%	
	Total	1,170	775	1,141	3,086

($X2$ (4, n = 3,086) = 111.43, p < .001)

different, and the trend was in the direction of both hypotheses (see Table 2). Posts on national level streams tended to focus most on information, followed by state and then local.

H2 proposed that national groups would attempt to motivate less behavior than state and local groups in their posts. Using the same chi-square reported above (Table 2), the observed and expected frequencies were examined, and the trend was in the direction hypothesized. Interestingly, the data suggest that at the national level, organizations tend to share information more than they attempt to motivate behavior. At the local level, the focus runs in the opposite direction with local groups focused much more on motivating behavior than sharing information. At the state level, the numbers of posts that share information are about equal to the number that attempt to motivate behavior.

RQ2 explored the differences in behavior type at various organizational levels. To examine this, a chi-square test was run for the organization types and the three behavior types (personal, public, and organizational). The results suggest that all three types of organizations focus more on organizational behaviors than other types of behaviors. At the local level, the focus on organizational behavior is greater than expected, and for the national and state level, the focus on organizational behavior is less than expected. Considering that organization-focused behaviors are considered public behaviors, the posts seem to be promoting public actions toward the environment much more strongly than personal actions (see Table 3).

To further explore the types of actions that organizations are promoting, analysis was run to see to what degree organizations are encouraging three popular behaviors—volunteering, donating, and attending events (see Table 4).

The results of the three chi-square tests were significant and suggested that the behaviors tend to appear in local advocacy groups' Twitter streams more often than national groups' streams. This was most pronounced in event promotion, with 56% of local groups' posts mentioning events and only 10% of national groups' posts referring to events. Overall, proximity appears to change the content and intended outcome of the posts for advocacy groups.

Conclusion

Table 3. Chi-square crosstab of three types of behaviors motivated by tweets from three organization types.

ORG LEVEL		PERSONAL BEHAVIOR	PUBLIC BEHAVIOR	ORGANIZATIONAL BEHAVIOR	TOTAL
National and Regional	Observed	59	38	190	287
	Expected	51	31	205	287
	% within type	21%	13%	66%	
State level	Observed	119	73	469	661
	Expected	117	71	473	661
	% within type	18%	11%	71%	
Local level	Observed	24	12	157	193
	Expected	34	21	138	193
	% within type	12%	6%	81%	
	Total	202	123	816	1141

$(X2\ (4, n = 1141) = 13.53, p = .009)$

The study found that the use of Twitter among one national environmental organization is on the rise. A closer look at the different levels of the organization found that in the Twitter streams of the different branches of this organization, national level branches were more focused on information sharing while state and regional branches were more focused on motivating behavior. The most commonly promoted behaviors were public behaviors that were organization sanctioned (donating, volunteering, attending organization events) rather than personal actions or other types of public behaviors toward the environment. Local groups were most strongly focused on organization-related behaviors. They also appeared to be more focused on relationship building through retweets and mentions of other users. At the national level, organizations appeared to be sharing information through links and helping users find information through the use of hashtags.

These findings suggest that at the national level, environmental advocacy organizations are aware of the fact that their audience is seeking information but may not be interested in taking actions toward the environment. At the local level, branches understand that their audiences are more invested in the issues and are more likely to take some kind of action. One would expect to see personal behaviors promoted at the national level and more public behaviors at the local/state level. Although there seemed to be some suggestion of this, overall, behaviors that were promoted were almost entirely organization-focused. As such, Twitter posts focused more on event attendance than fundraising or volunteer recruiting efforts. It is possible that activist groups use traditional channels of communication for their volunteer communication and donor relations/fundraising. They may feel more comfortable with those channels and may find them to be more effective for relationship building with volunteers and donors. Social media are still relatively new for activist groups, and it appears that, in this case, the Sierra Club views Twitter

Table 4. Inclusion of three types of behavior in advocacy groups' Twitter posts—donations, volunteering, and events.

ORG LEVEL		NATIONAL LEVEL	STATE LEVEL	LOCAL LEVEL
Donating*	Mentions donating	1%	2%	3%
	Not mention donating	99%	98%	97%
Volunteering**	Mentions volunteering	2%	2%	9%
	Not mention volunteering	98%	98%	91%
Attend event***	Mentions event	10%	26%	56%
	Not mention event	90%	74%	44%

*($X2$ (2, n = 3,095) = 11.89, p = .003)
**($X2$ (2, n = 3,094) = 60.97, p < .001)
***($X2$ (2, n = 3,095) = 304.24, p < .001)

as a way to promote its events and activities to a broader audience in hopes of attracting more attendees. Event attendance is a one-time behavior that is more easily motivated than volunteerism or donations. It is possible that through these events the organization conducts fundraising or recruits volunteers.

A primary goal of advocacy groups is to educate key publics and motivate action toward a cause or issue. The results of this study suggest that that use of Twitter among local advocacy groups focuses on behavior motivation toward that goal. Due to proximity, it may be difficult to move beyond education in Twitter streams. Local chapters were least likely to tweet and, on average, post less than the other groups. This may suggest that local groups have a tool that could be used to reach their goals of both educating and motivating behavior, but they need to increase usage.

This study sheds light on the way environmental advocacy groups currently communicate with key audiences through Twitter. It is important for organizations to share information about their work and also offer publics opportunities to connect with them. Twitter is a communication channel that many advocacy organizations are beginning to adopt, and this study examined messages that groups used at multiple levels of the organization to reach their audiences.

The degree to which Twitter can be used as a relationship management tool can be debated. In general, it appears to be useful as one of many strategies that advocacy organizations can use to build relationships. Through the use of Twitter functionality and signs such as retweets, mentions, hashtags and links, organizations can engage with audiences and share important information about the work that they are doing as well as invite audiences to participate with them. Organizations in this study were using these tools, but it appears that more could be done, especially at the national level, to encourage more relationship building.

At its core, relationship management is about the actions and communications that occur between an organization and its partners and/or publics. However, relationships occur at many levels, and it appears that the proximity between an organization and its audience may impact its communication strategy. In this study, it was clear that at the national level, the Sierra Club communicated differently than it did on the local level. This may be due in part to the manpower and expertise of communicators at different levels, but it was also clear that the nature of the relationship between the organization and its audiences influenced the content, with local branches, more so than national branches, focusing on events and volunteerism. This study raises the question about the nature of communication and relationship building when organizations are proximally distant from their audiences. As advocacy groups plan their communication strategies, they should bear in mind the needs of their audiences and the expectations for information and behavioral engagement with the organization.

The Role of Text Messaging in Public Relations

Testing the Situational Theory of Publics for Mobile-Giving Campaigns

Brooke Weberling
Richard D. Waters
Natalie T. J. Tindall

The purpose of this study was to determine how previous American Red Cross donors from two separate chapters—one in Florida and one in Illinois—reacted to the "Text for Haiti" campaign following the 2009 earthquake. Specifically, the study used the situational theory of publics to analyze awareness of and involvement in the text messaging campaign. Through a survey of 271 donors, the study found that text messaging may be a quick way to communicate with stakeholders, but it may not be the most effective for developing lasting relationships, as the situational theory demonstrated that active publics participated at significantly higher rates than their aware counterparts. Donors to the "Text for Haiti" campaign sought information about the earthquake and the America Red Cross' relief efforts and had fewer constraints to their involvement in various aspects of the campaign than non-donors, even those who had contributed to other disaster relief efforts.

Introduction

On January 12, 2010, a series of 6.5 to 7.3 magnitude earthquakes struck the island nation of Haiti, killing approximately 200,000 people, leaving another 1.2 million homeless and more than 3 million dealing with injuries, loss of loved ones, and other disastrous effects (American Red Cross, 2010a). As people around the world began hearing news about the disaster, nonprofit organizations began mobilizing employees, volunteers, and donors to help provide emergency relief assistance in the form of food, water, and medical supplies, along with services, shelter, and support for survivors. As part of the effort to help the people of Haiti, the American Red Cross turned to new technologies to raise money through an innovative fundraising campaign.

The "Text for Haiti" campaign allowed mobile phone users to make a $10 donation to the American Red Cross by typing and sending the word "Haiti" to the number "90999" via text. More than $32 million was raised in a record-breaking mobile-giving effort (American Red Cross, 2010a). These efforts were promoted through numerous traditional and social media outlets, including CNN, the *New York Times*, Facebook and Twitter, a live telethon called "Hope for Haiti Now" broadcast on most major television networks, and public service announcements featuring First Lady Michelle Obama (American Red Cross, 2010b; Philanthropy News Digest, 2010).

Although text messaging as a fundraising strategy had been pursued in the past, this national effort was the first time that it had succeeded in producing results that surpassed the $1 million amount. The United Way attempted the first national text messaging fundraising campaign during Super Bowl XLII. Their 10-second spot encouraged viewers to text "FIT" to "UNITED" (864833), which would result in a $5 donation to youth fitness campaigns. Though the broadcast was seen by an estimated 95.7 million viewers in the United States, the campaign resulted in less than stellar results. The United Way never released figures specific to the text messaging campaign. However, fundraising analysts cited the lack of post-campaign publicity as a key indicator that the first text messaging campaign failed to produce strong results (Jones, 2008). Yet the Mobile Giving Foundation (quoting the United Way) called the test "profoundly successful" (para. 6) on its own website and confirmed that roughly 2,000 donors contributed to the campaign (Mobile Giving Foundation, 2009).

Though this national rollout was somewhat unsuccessful, the Mobile Giving Foundation continued to develop a technological framework that would allow nonprofits to sign up for dedicated numbers to conduct their own text messaging campaigns to raise funds for their organizations. This foundation currently has more than 800 nonprofit clients with coordinated fundraising campaigns, and competitors, such as mGive Foundation and Mobile Active, have been created to provide nonprofits with additional hosting sources for text messaging campaigns. Since 2008, numerous nonprofits have conducted local text messaging campaigns with tremendous success. For example, "charity: water," a nonprofit organization seeking to bring clean and safe drinking water to people in developing nations, tapped into the power of local markets to raise more than $250,000 during its Twestival, when more than 200 international cities hosted special events and encouraged people to use Twitter to inform others about the events, the cause, and ultimately to text a donation to charity: water (charity: water, 2009).

The failure of the national text messaging campaign and the success of the smaller text messaging campaigns poses an interesting dilemma for fundraisers: Can text messaging be used as a viable fundraising channel for national organizations and national campaigns? This study sought to answer this question by applying the situational theory of publics to the "Text for Haiti" campaign. Specifically, the study employed an online survey to explore attitudes and behaviors related to Haiti disaster relief efforts. The findings of this study may help practitioners and scholars to better understand the attitudes and behaviors of various publics regarding donating in the wake of a disaster, and could help inform future communication campaigns involving text messaging.

Literature review

Text messaging as a public relations medium

Having evolved from memo pagers in the 1980s, today's short message service (SMS), or text messaging, allows users to exchange short messages between fixed Web-based lines or mobile phone devices. The length of standard text messages varies between 120 and 160 characters (Groves, 1998). The increase in the use of text messaging by the public is one of the fastest growing technological adoptions ever seen. The International Telecommunication Union estimated that since 2007, the number of text messages sent by individuals had tripled to reach more than 6.1 trillion messages in 2010. This figure translates into roughly 200,000 text messages being sent globally every second (International Telecommunication Union, 2010).

The vast majority of these messages were sent by individuals to other individuals and were personal in nature. However, Faulkner and Cutwin (2005) predicted that text messaging use by organizations would grow as the potential for revenue and stakeholder relationship growth demonstrates promise in the marketplace. Indeed, fast food restaurants, such as Papa John's Pizza, have begun taking delivery orders via text messaging, and banks, such as Chase, allow their customers to check balances and make payments with text messaging. Additionally, mobile marketing campaign companies have sprung up to help organizations send mass text messages to their stakeholders to mobilize them to action in consumer, political, and nonprofit settings.

To date, little research has been done on the role of text messaging in public relations though there has been considerable support given to the technology's ability to help organizations cultivate relationships with their publics. Grunig (2009) argued that text messaging and new media technologies may have a greater ability to lead to mutually beneficial relationships with publics than any other traditional public relations tactic because of the focus on interactivity and being able to communicate with publics anytime and anywhere. Text messaging and social media are viewed as critical forces in current public relations efforts because they allow practitioners to become part of ongoing conversations with stakeholders that have previously been difficult to communicate with (Neff & Hanson-Horn, 2010).

Pavlik (2007) agreed that text messaging is becoming one of the standard tools of public communication and interaction. However, he cautioned that practitioners must assess the medium critically. Even though text messaging is convenient, the impact of the technology—especially the time and resources involved to manage these ongoing exchanges in conjunction with other duties—must be considered before it is adopted by practitioners. Additionally, messages have to be developed and delivered in a creative and extremely brief manner, and they have to fit within the structure, culture, and management of the organizations that are seeking to use text messaging. But, perhaps most importantly, he suggested that the impact of text messaging adoption has to be considered in light of the relationships between the organization and its publics by asking whether text messaging will enhance or harm the relationship.

For instance, political communication scholars have suggested that text messaging campaigns appear to have helped political candidates connect with their constituents, though more research needs to be conducted to determine their causality (Kiousis & Stromback, 2011). The success of local text messaging fundraising campaigns by nonprofits echo this apparent relationship growth with the public. However, as James (2007) suggested, research must be conducted using theoretical frameworks to determine whether public relations can fully accommodate new media, including text messaging, given the struggles of practitioners to embrace and incorporate them into their public relations campaigns. This study sought to help fill that void by exploring text messaging in the context of the situational theory of publics.

Situational theory of publics

Organizations have many different stakeholders, which when organized into groups around common issues or characteristics, become important publics with which organizations communicate or otherwise interact. The situational theory of publics provides a framework for exploring the various factors involved in different publics' attitudes and behaviors toward an organization based on their perceptions of an issue or situation (Grunig, 1989a; Grunig, 1997; Grunig & Hunt, 1984; Hamilton, 1992).

According to the situational theory of publics, three independent variables—problem recognition, constraint recognition, and level of involvement with the issue—predict two dependent variables—information seeking and information processing. However, the dependent variables have changed in research over the years (see Kim & Grunig, 2011). Problem recognition is defined as the moment when people recognize that something should be done about an issue or situation, and stop and think about what to do. Constraint recognition happens when people perceive that there may be obstacles in the way of acting related to the problem, and level of involvement is the extent to which people connect with the issue or situation (Grunig, 1997; Grunig, 1989a; Hamilton, 1992).

The dependent variables, information seeking and processing, may reflect passive or active forms of communication. Passive or low levels of information seeking and processing may simply imply that an individual simply receives or consumes information that is presented to them. Active or higher levels of information seeking and processing, on the other hand, implies that individuals expend effort to locate or consume information about an issue or situation. As Grunig (1989b) stated, "people communicating actively develop more organized cognitions, are more likely to have attitudes about a situation, and more often engage in a behavior to do something about the situation" (p. 6). Based on these variables, individuals can be considered *latent, aware,* or *active* publics, a classification system that can help organizations determine information dissemination strategies and create communication campaigns (Aldoory & Sha, 2007). Additionally, those publics who are not impacted by an issue are labeled as nonpublics.

Subsequent studies since Grunig's work have found support for and advanced the situational theory of publics. For instance, Aldoory (2001) focused on *involvement* related to health communication targeting women, while Sha (2006) highlighted the importance of cultural identity among various publics. Research related to political and international issues has also been important in advancing the theory. Atwood and Major (1991) found that involvement with an international political issue was more highly correlated with interpersonal communication than mass media use. However, they suggested that information seeking may not take place until an issue reaches crisis proportions. Hamilton (1992) studied the theory related to media use and public perceptions of a governor's race, and suggested that drive and habit may be important variables in determining communication preferences and subsequent actions like voting. These findings are important when considering a medium like text messaging, an international disaster like the earthquake in Haiti, and taking action such as donating to a fundraising campaign.

Although there has been significant research on the situational theory of publics, applications to fundraising and international relief efforts seem lacking. The current study fills this gap by examining public attitudes, perceptions, and behaviors related to the earthquake in Haiti and subsequent fundraising and disaster relief efforts by the American Red Cross. More specifically, this study looked at donor response from two chapters of the American Red Cross—one in Florida and one in Illinois. Given the framework of the situational theory of publics, those uninvolved with the American Red Cross would be considered nonpublics, and latent publics would be those who are impacted by the organization's programming or relief efforts though they fail to actively recognize the impact. Specifically, this study focuses on the differences between active and aware publics. Aware publics were defined as those who had donated to the American Red Cross chapters in the past but did not contribute to the "Text for Haiti" campaign while the active publics donated to the campaign. Based on previous research on the situational theory of publics, this study sought to answer the following research questions:

RQ1: Do active and aware publics differ in their levels of problem and constraint recognition, involvement, and information seeking and processing of the American Red Cross "Text for Haiti" disaster relief efforts?

Although the first research question explores the descriptive nature of the situational theory of publics, the ultimate test of theoretical perspectives lies in their ability to predict whether described conditions are accurate. The situational theory of publics argues that problem recognition, constraint recognition, and level of involvement predict information seeking and processing, which should ultimately predict whether someone contributed to the fundraising campaign in the context of the current study.

To test the predictability of the situational theory of publics in the current context, the study's second research question was posed:

RQ2: Does the situational theory model accurately predict the interconnectivity of the six variables in the context of the American Red Cross' "Text for Haiti" campaign?

Method

Six months after the Haitian earthquake in 2010, data were collected from two American Red Cross chapters that gave the research team permission to survey a random sample of 500 donors from each chapter's donor database. An Internet-based survey was used to survey the donors. Of the 1,000 survey invitations sent via email, 271 useable surveys were returned for a response rate of 27.1%, which is comparable to previous online survey response rates (Sheehan, 2001).

The survey for this study used existing scales to collect data to answer the research questions. The five situational theory of publics variables—problem recognition, constraint recognition, level of involvement, information seeking, and information processing—were modified slightly from Grunig's (1997) measures to more accurately reflect the context of the "Text for Haiti" campaign. Each variable was measured using a modified seven-point Likert scale where higher values indicated level of agreement. Additionally, the study asked whether the previous donors to American Red Cross fundraising campaigns donated specifically to the "Text for Haiti" campaign through text messaging as opposed to more traditional crisis fundraising responses (e.g., Internet-based fundraising, 1-800 telephone hotlines, donating at church services, etc.). Demographic and geographic information was also collected to aid in the analysis.

Cronbach's alpha values were calculated for the five situational theory of publics variables. All items were deemed to be moderately reliable though three lacked higher levels of statistical significance (e.g., alpha values greater than 0.75). However, as Carmines and Zeller (1979) noted, existing scales that are modified to meet new situations may not produce reliability values that match those resulting from the study's original conditions. For this study, the highest alpha values were for information processing (α = .79) and information seeking (α = .76). The values for problem recognition (a = .63), constraint recognition (a = .61), and involvement (a = .54) indicated moderate reliability for these scales. The authors believe that these lower values were likely the result of using one question that focused exclusively on text messaging, one question that focused on the American Red Cross relief efforts, and one question that focused on the Haitian earthquake for each of the scales. The three differing focal points in each scale may have contributed to a less than desirable Cronbach alpha value.

Results

Females outnumbered male survey respondents considerably as they represented 82.3% of the participants in the study. Half of the study's respondents (50.9%) were Caucasian. Significant numbers of participants also came from African-American (19.5%), Hispanic/Latino (17.3%), and Asian (10.7%) populations. Less than two percent of the sample chose Middle Eastern, Native American, or "Other" as the race they self-identified with. The participants in the study were well educated, as 41% had a bachelor's degree and an additional 33.5% had a graduate or professional degree (Ph.D., J.D., M.D., Ed.D.). The average age of the participants was 30.8 years (SD = 6.47 years) though the ages ranged from 18 to 63.

In terms of their mobile phone usage, the participants indicated that they use their phone for communicating with others an average of 71.6 minutes per day (SD = 93.58 minutes). It should be noted that this time includes phone calls, sending emails, sending text messages, and any type of communication activity that the users' mobile phone allows. Additionally, the survey respondents sent an average of 17.4 text messages per day (SD = 22.89 messages). In terms of their mobile phone usage and the "Text for Haiti" campaign, 179 participants (66.1%) used their mobile phone to text a donation to the campaign.

The first research question sought to determine whether the situational theory of publics was able to describe the conditions surrounding the "Text for Haiti" campaign. Overall, the 271 previous American Red Cross donors were aware of the 2010 Haitian earthquake and the relief efforts designed to help the Haitian people (M = 5.24, SD = 1.19), and they had relatively few constraints preventing their involvement in the campaign (M = 3.67, SD = 0.74). However, their involvement was just slightly greater than the neutral point on the seven-point scale (M = 4.48, SD = 1.71). The participants indicated that they were neutral in terms of actively seeking out information about the Haitian situation (M = 4.11, SD = 1.23) as well as actively processing the information to which they were exposed (M = 4.31, SD = 1.39).

To further test the situational theory's ability to describe the current situation, one-way ANOVAs were conducted to compare the five variables along the lines of whether the survey participants donated to the "Text for Haiti" campaign. Following the descriptions of the public types in situational theory literature, those who did donate with their mobile phone were deemed active publics while those who did not donate but had previously donated to American Red Cross relief efforts were deemed aware publics.

Table 1. One-way ANOVA on Situational Theory of Public Variables by Public Type.

	AWARE PUBLICS (N = 92)	MS	F(1,269)	P-VALUE
Problem Recognition/ Awareness	4.67 (1.31)	44.79	35.65	.000
Constraint Recognition	3.12 (0.64)	41.46	106.75	.000
Involvement	3.22 (1.31)	225.42	107.07	.000
Information Seeking	3.59 (1.47)	36.11	26.11	.000
Information Processing	3.52 (1.05)	85.73	52.56	.000

As Table 1 shows, strong statistical differences emerged from the ANOVA tests as active publics were impacted more significantly by the situational theory variables. It should be noted that higher scores on the constraint recognition variable indicate a lower measure of constraint given the reversed-wording of the measures.

For the second research question, path analyses were run using AMOS 17.0 to determine whether the situational theory of publics model could accurately be used to describe the data measuring stakeholder involvement with the "Text for Haiti" campaign. The original model, shown in Figure 1, tested all possible paths between the situational theory variables. Using criteria outlined by Raykov and Marcoulides (2000), the current data fit the model, as all statistical measures were successfully met ($\chi 2 = 1.08$, CFI = 1.0, GFI = .99, NFI = .98, RMSEA = .02).

Table 2 highlights the resulting statistically significant paths from the tested model. The model revealed that problem recognition, constraint recognition, and involvement in the issue were all statistically significant predictors of information seeking and processing, which, ultimately, successfully predicted participation in the campaign by text messaging a donation to the American Red Cross.

Table 2. Path Analysis of Situational Theory Variables on the Participation in the "Text for Haiti" Campaign.

PATH	*STANDARDIZED COEFFICIENT*	*STANDARDIZED ERROR*
Problem recognition → Information seeking	.12*	.06
Problem recognition → Information processing	.17**	.07
Constraint recognition → Information seeking	.09*	.08
Constraint recognition → Information processing	.10*	.07
Issue involvement → Information seeking	.23***	.05
Issue involvement → Information processing	.28***	.04
Information seeking → Donating to campaign	.19***	.08
Information processing → Donating to campaign	.21***	.10

*$p<.05$, ** $p < .01$, *** $p < .001$

Conclusion

Although a relatively new medium in the public relations context, text messaging proved to be highly successful for the American Red Cross during the "Text for Haiti" campaign. This study found that the situational theory of publics accurately describes the stages of the situational theory of publics, and it was able to successfully predict campaign participation by following the model described in public relations literature. Given these findings, one might conclude that text messaging as advocated by many scholars should be explored as a viable public relations medium (Grunig, 2009; Hiebert, 2005; Neff & Hanson-Horn, 2010).

The overall analysis of the data supports this claim. However, it is worth returning to the first fundraising campaign conducted via text messaging to assess how far text message-based fundraising has come. The national text messaging campaign conducted by the United Way in 2008 was largely considered a failure given the low response by roughly 2,000 donors that contributed nearly $10,000 to the national organization. By basic measures, the American Red Cross' "Text for Haiti" campaign improved leaps and bounds over the United Way campaign, as $32 million was raised from an unspecified number of donors. At $10 per text, it is feasible that the number of donors was 3.2 million. However, people may have given multiple $10 donations through the text messaging efforts. The American Red Cross has never revealed nor estimated the number of individuals who donated to this campaign.

Little is actually known about who the donors to the "Text for Haiti" campaign actually are. Fundraising records are generally considered to be private contracts between a nonprofit and the individual, and the Internal Revenue Service does not require the American Red Cross to discuss the details of its fundraising efforts beyond totals of organizational expenses and revenues. But, given the increased use of text messaging from 2007 to 2010 as documented by the International Telecommunication Union and the documented successes of local text messaging initiatives that users of the Mobile Giving Foundation and mGive Foundation implemented, it is possible to hypothesize that text messaging may be growing in popularity and be most effective for local or regional campaigns.

This statement is further supported by post hoc analysis of the current data. The two American Red Cross chapters that participated in the study were in Florida and Illinois. The Florida chapter was less than 250 miles from Haiti compared to the roughly 1,800 miles separating Haiti from the Illinois chapter. Fundraising literature and practitioners argue that campaigns are most successful when donors have strong connections to the campaign. Locals are more likely to see the results of the fundraising efforts and are more likely to want to help their own community (Kelly, 1998; Waters, 2011a). In 2009, the U. S. Census Bureau estimated that there were roughly 1,000,000 Haitians living in the United States and that the vast majority of these individuals live in Florida. Haitian immigrants represent one of the largest demographic groups in Florida (Stepick & Stepick, 1990). According to the 2000 U. S. Census, 23 of the largest 35 Haitian communities in the United States were in Florida (U. S. Census Bureau, 2000).

Table 3 illustrates the impact of the situational theory's variables on the survey participants based on which chapter they were involved with (Florida or Illinois). Participants who had connections and previous involvement with the Florida chapter were significantly more aware of the Haitian earthquake, were more involved with the situation, and had fewer constraints to involvement than their Illinois counterparts. Donors to the Florida American Red Cross chapter also were more likely to seek information about the earthquake and more likely to process the information that they received.

This post hoc analysis provides significant insights into how text-message campaigns should be used by public relations practitioners. As James (2007) suggested, public relations scholars need to test existing

Table 3. One-way ANOVA on Situational Theory of Public Variables by American Red Cross chapter

	FLORIDA DONORS (N = 196)	V	MS	F(1,269)	P-VALUE
Problem Recognition/ Awareness	5.35 (1.16)	4.94 (1.23)	9.04	6.50	.011
Constraint Recognition	3.76 (0.72)	3.42 (0.73)	6.75	13.05	.000
Involvement	4.74 (1.68)	3.84 (1.65)	43.16	15.51	.000
Information Seeking	4.19 (1.18)	3.88 (1.34)	5.34	3.57	.060
Information Processing	4.42 (1.40)	4.01 (1.22)	8.89	4.64	.032

theory to determine whether new technologies can be described using the existing theoretical perspectives. This study found that the situational theory of publics can describe the "Text for Haiti" campaign, at least for active and aware publics. An individual's participation in a campaign can be predicted based on his or her awareness of and involvement in an issue, obstacles preventing further involvement, and desire to seek and process information about the issue. But, the distance from that issue appears to have a significant impact as well.

Although new media and text messaging campaigns have the ability to eliminate geographic boundaries for campaigns, practitioners may need to reconsider their desires to take a campaign to a national audience. As the United Way case demonstrated, mass-mediated messages to national and global audiences appear to produce results that may not be as effective as appealing to smaller, local or regional audiences. Just because text messaging allows practitioners to reach national audiences does not mean that they should forget one of the fundamental rules of public relations campaigning: segmenting the audience.

The situational theory of publics provides mental process variables that help describe the situation, but the differences between the Florida and Illinois donors are striking in how they experienced the "Text for Haiti" campaign. The Florida donors were significantly closer to the devastation caused by the 2010 earthquake, and they responded more favorably than the Illinois donors in terms of their donation to the relief efforts.

Scholars have argued that new media can aid in the growth of relationships with stakeholders regardless of their physical proximity to the organization. However, practitioners must realize that the most successful results often stem from targeting people in their own communities, especially for nonprofit organizations. There are ample case studies showing that small donations via text message campaigns are quite fruitful for nonprofit organizations when conducted at the local level. As charity: water's Twestival demonstrated, texting donations and spreading information about an organization's programs and events via mobile phone communication can reap huge benefits for organizations that recognize the impact of local communities.

Just because text messaging and new media have the power to reach across time and space does not mean that the pursuit of these endeavors will result in the most beneficial results for organizations. As with any other public relations endeavor, a significant portion of the campaign planning efforts has to be dedicated to researching the audience to determine who the appropriate publics are. When that research is done, organizations will be in a better position to determine whether their audience is national, regional, local, or even global in nature and how to best reach their audiences. However, without the research, the campaign may generate lackluster results like the first text message-based fundraising campaign.

Every study has limitations that must be taken into consideration when interpreting the findings, and this study has several limitations worth noting. Most importantly, this study did not attempt to measure nonpublics or latent publics outlined in the situational theory literature. No suitable framework for measuring the reactions of the general public were available to the research team to capture the full gamut of public types. Additionally, the adaptation of the situational theory variables did not produce indices that were as reliable as they have been in previous academic studies.

Finally, it is also worth noting that the American Red Cross, despite repeated attempts by the researchers to obtain some general demographic and geographic breakdown of the "Text for Haiti" campaign donors, did not share information about their donors. It is speculation based on the chapter comparison that leads the researchers to conclude with the cautionary warning about ignoring local boundaries for campaigns.

Gimme That Ole Tyme Religion in a New Fangled Way

Faiths Connect, Build Relationships via Facebook

Dedria Givens-Carroll

Faith-based organizations are increasingly taking their messages online via social media. Facebook leads the pack for religious groups who are trying to connect, communicate symmetrically, and build relationships with their stakeholders and target publics. Religious fan pages look toward creating, building, and maintaining long-term relationships with their followers via trust.

> Go therefore and make disciples of all the nations, baptizing them in the name of the Father and of the Son and of the Holy Spirit, teaching them to observe all things that I have commanded you; and lo, I am with you always even to the end of the age. Matthew 28:19–20, *The Holy Bible*, New King James Version, accessed from YouVersion.com

Religious organizations, including churches, mosques, temples, and synagogues, have long been networking centers—where congregants fellowship with others—but these organizations are now taking their messages online. Faith-based groups are building relationships outside their walls in the extroverted world of Facebook, blogs, email, Twitter, texting, Skype, and YouTube, or the religious version, GodTube (http://www.godtube.com/). More than ever, religious groups are using the Internet, and specifically social networks, as two-way symmetrical communication to parlay connections for prayer requests, group therapy, evangelism, or just simply sharing social cares and concerns in their communities. Evangelicals seem to lead the cyber pack, but other Christian faiths, along with Jews, Muslims, and Buddhists are connecting via social networks to regularly minister to friends' needs. The giant social network Facebook is the new community gathering spot, but does it allow users to feel connected to their faith?

One church garnered attention with its outdoor sign that read: "Does Jesus have a Facebook page?" and "One new friend request from Jesus—Confirm or Ignore?" The pastor (as cited in Longwell, 2009) responsible for the display was reported as coming to appreciate the "sign of the times" (p.1), admitting that he was unsure if visitors were drawn to the church because of the sign but said, "I am sure it makes people curious about what is going on inside" (p.1).

One idea is for certain: Religious organizations are using new media as a public relations tool to disseminate their various messages. Faith-based organizations are clamoring to connect with their target audiences via whatever medium seems to be appropriate for their message. As Marshall McLuhan (1967), born in 1911, wrote in his book of the same title, "the medium is the message."

Religious groups using public relations strategies and tactics trace the relationship to a linkage that has endured over time. According to Cutlip, Center, and Broom (2004), Greek theorists wrote about the importance of the "public will" and Romans "coined the expression vox populi, vox Dei—the voice of the people is the voice of God" (p. 102). Peter G. Osgood (as cited in Wilcox, 2005), former president of Carl Byoir & Associates, noted: "St. John the Baptist himself did superb advance work for Jesus of Nazareth" (p. 28). Other examples of the ties between religion and public relations include the Roman Catholic Church being among the first organizations to use the word "propaganda" with the establishment by Pope Gregory XV of the College of Propaganda "to supervise foreign missions and train priests to propagate the faith" (Wilcox, 2005, p. 28). Brown (2003) contended that the apostle Paul, a contemporary of Jesus, was one of the first public relations practitioners and used his "letters" as tactics to communicate with his stakeholders at the churches he helped establish. "Historians of early Christianity actually regard Paul, author and organizer, rather than Jesus himself, as the founder of Christianity" (Brown, 2003, p. 229). According to Wilcox and Cameron (2006), "Saint Paul, the New Testament's most prolific author, also qualifies for the public relations hall of fame" (p. 44). Grunig and Hunt (1984) concurred:

> It's also not stretching history too much to claim the success of the apostles in spreading Christianity through the known world in the first century A.D. as one of the great public relations accomplishments of history. The apostles Paul and Peter used speeches, letters, staged events, and similar public relations activities to attract attention, gain followers, and establish new churches. Similarly, the four gospels in the New Testament, which were written at least forty years after the death of Jesus, were public relations documents, written more to propagate the faith than to provide a historical account of Jesus' life. (p. 15)

Modern public relations tactics used in the religious realm cover the whole gauntlet of media. "We are online, like it or not," said one pastor (Daniel, 2009, p. 27), who admitted that:

> We are not the first generation of Christians to struggle with multimedia communication. In his second letter to the Corinthians, Paul wrote: "I do not want to seem as though I am trying to frighten you with my letters." There is no one right way to communicate. God has given us many ways to get our point across and to listen to others. The world that isn't online becomes invisible. (p. 27)

Many church leaders contend that new media have not only spread the gospel, but also help them connect with their congregants, for example, by "conducting weekly prayers via conference calls, posting sermons online and reaching members through social media platforms such as Twitter and Facebook... new media and technology may become the only way that some people still worship" (Horton, 2010, p. 20).

Religious organizations are employing a variety of media to get their messages across. After the Haiti earthquake, one missionary group used social networking, especially Twitter, to share news and photos, solicit donations, and ask for help in finding missing loved ones (Keen, 2010). The leader of a small church in Cape Coral, Florida, has a Twitter account, Facebook page, YouTube channel, and actively creates iTunes podcasts and blogs (Ruane, 2010). Even the Roman Catholic Church, which has been reluctant to change, has turned to new media to reach the digital youth culture:

> Pope Benedict XVI has tried to bring the Vatican into the Internet age by launching a YouTube channel earlier this year. Officials say he also e-mails and surfs the Web.... The Vatican's top communication official, Archbishop Claudio Maria Celli, has said a key priority of the Catholic Church is to be able to use new technologies to spread its message, particularly to the young. "Our dream in this global village created by new technologies is that the church and Jesus' disciples can have their tent—Jesus' tent—so that the attention of men and women who walk the streets of the world is turned toward it," he said recently on Vatican Radio. ("Facebook, Wikipedia Execs," 2009, paras. 4, 10–11)

There are even apps that can be downloaded via iPhones for users to connect to their faiths, like iConfess, iRepent, iAdmit, and Penance, which is a free app game that features public confessions (Falsani, 2011).

Other faiths are flocking to social media, including a group of Jewish rabbis who blogged about whether people should "fast from Facebook" (Ruane, 2010, p. 3A) during Passover. In reference to attempting to contact members of his faith community, one rabbi said: "It's amazing because you could call or e-mail them, and you'd never hear back, but post them a note on Facebook, and you hear back from them in a moment" (p. 3A). One study demonstrated how moderators of a Buddhist message forum, E-sangha (http://www.e-sangha.com/), "use religious community narratives to frame Web environments as sacred community spaces (spaces made suitable for religious activities)" (Busch, 2011, p. 58). A Sikh association in New York has a Facebook fan page (http://www.facebook.com/group.php?gid=4964128758) for followers (Freedman, 2009). Niche social networks cater to their followers. Xianz (http://www.xianz.com/) is a Christian social network that bans "cursing and derogatory words." Christianster (www.christianster.com) is a network that focuses on friendship, fellowship, and spiritual growth. Shmooze (www.shmooze.com), Yiddish for casual chat, is a site catering to the global Jewish community, and Naseeb (http://www.naseeb.com/), which in Arabic means destiny, is a social network for Muslims (Holahan, 2007).

Facebook has emerged as the social network giant because it has more than 500 million active users. The average user has 130 friends, and people spend over 700 billion minutes a month on the network ("Statistics," 2011). At the time of this writing, the popular site used the slogan, "Facebook helps you connect and share with the people in your life." The site claims that more than 250 million active users access the network via their mobile devices, and those users are twice as active as their counterparts. Users are obsessed with the network: One out of every 13 people on earth is an active user, half are logged in on any given day, and 48% of 18- to 34-year-olds check their page right when they wake up (O'Dell, 2011).

Roach (2011) proposed that "churches are turning increasingly to social networking tools as ministry aides and Facebook is by far the most popular tool" (para. 1). LifeWay Research Director Scott McConnell (as cited in Roach) added, "Churches are natural places of interaction....Congregations are rapidly adopting social neworking, not only to speed their own communication, but also to interact with people outside their church" (para. 5). McConnell's study found that 81% of congregations with 500 or more in average worship attendance use Facebook, compared to 27% of churches with 1 to 49 attendees. Accordingly, large-city and suburban churches were more likely to use the network to communicate with their audiences than small-city and rural churches. Roach commented on how social media are being used:

> Among churches that utilize social networking tools, 73 percent use them for interacting with the congregation, 70 percent for distributing news and information in an "outbound only" manner, 52 percent for fostering member-to-member interaction and 41 percent for managing the church's group ministry. (para. 8)

Curtis Simmons (as cited in Roach, 2011) of Fellowship Technologies highlighted the importance of churches being involved with social media for relationship building purposes:

> Social networking tools have become an integral part of most people's daily lives and relationships. If churches desire to connect with their congregation and community in meaningful ways, then they need to establish a strategy for

actively engaging in the social media conversation. Thousands of individuals are sharing support and encouragement through these tools. The church needs to be an active participant in these conversations and connections. (para. 10)

An October 2010 survey of 1,000 Protestant pastors revealed "that many pastors are personally using social media to interact with their congregations" (para. 11). Nearly half (46%) were found to use Facebook, 16% were bloggers, 6% used Twitter, and 84% sent emails to groups. By early 2011, nearly half of U.S. Protestant churches, including Baptist, Episcopal, Lutheran, Methodist, and Morman faiths, had a presence on Facebook (Singer, 2011).

Social network sites (SNSs) such as Facebook provide religious groups the opportunity to become "friends" and connect with people who they may not otherwise contact. After joining the network, users are prompted to identify others in the network with whom they have a relationship. Facebook allows users to set up a profile, which includes a religious status choice that has been expanded to include a description. Facebook users may select their individual religion choice on their profiles, but many may not be honest with their portrayal (Keating, 2010). A 2009 survey of Millennials (ages 18 to 29) noted that members of this demographic "consider themselves spiritual but not religious and feel that none of the traditional religious categories are a good fit for them" (Keating, 2010, p. 20). Jerry Rice (as cited in Bailey, 2010), who wrote *The Church of Facebook*, had a difficult time profiling his religion on Facebook. Once, he put "follower of Jesus" as his religious view. Because he was from San Francisco, he felt uncomfortable being categorized as right-wing and divisive, so he finally decided to "leave the category blank" (p. 14).

Age was once thought to be a factor amongst social networking and religiosity. A core finding of Pew's "Religion Among the Millennials" report showed that Americans in the 18–29 category may be "less religiously affiliated" than their elders and will be "browner, more comfortable with rapid change, higher tech, more upbeat and unworried by tattoos" (Stephen, 2010, p. 9a). The 35-plus demographic now represents more than 30% of the entire userbase on Facebook, although the 18–24 (college) demographic grew the fastest at 74% in one year (O'Dell, 2011). Religious millennials were also found to spend plenty of time on social networking sites. Social network users find various reasons to join, and religiosity becomes a mediating factor for some (Nyland & Near, 2007). The results from Nyland and Near's exploratory factor analysis indicated five main uses for social networking sites: meeting new people, entertainment, maintaining relationships, learning about social events, and sharing media. Furthermore, the authors found that "religious individuals are more likely to use social networking to maintain already existing relationships" (Nyland & Near, 2007, p. 2), which complements Ellison, Steinfield, and Lampe's (as cited in boyd & Ellison, 2007) finding that Facebook is "used to maintain existing offline relationships or solidify offline connections, as opposed to meeting new people" (*Bridging Online and Offline Social Networks*, para. 1). "People are [thinking of] Facebook as an evangelistic tool," said Ron Edmondson, a pastor at Grace Community Church in Clarksville, Tennessee. "That's not the point; the point is social media—building relationships in a community" (Bailey, 2010, p. 14).

Rice (2009) believed the social network encourages real relationships with others to better understand community. "It is simply my contention that the gospel (literally, 'good news') of Jesus is particularly well suited for helping us understand, adapt to, and even thrive among the challenges of living within a hyperconnected culture" (p. 114). He contended that the understanding the "shifting social tides" (p. 114) will help enrich relationships.

One of the most obvious uses for social networking in the religious realm is relationship building and nurturing, an important aspect of public relations. Waters, Burnett, Lamm, & Lucas (2009) studied nonprofit organizations' Facebook use and how they connect with stakeholders and foster growth in relationships. Nonprofits, which could include religious organizations, were found to "use social media to streamline their management functions, interact with volunteers and donors, and educate others about their programs and services....organizations seek to develop relationships with important publics" (p. 103). They found that "interactivity plays an important role in developing relationships online with stakeholders" (p. 103), but admitted that one of the limitations of their study was specifically "assessing the effectiveness of Facebook as a relationship building tool" (p. 106). They agreed with other studies that indicated "social networking sites can be an effective way to reach stakeholder group if organizations understand how their stakeholders use the sites" (p. 106). Campbell (2010) in *When Religion Meets New Media* focused on how new techology helps congregations build relationships with their audiences. She offered an alternative ideology of religious engagement with media technology and suggested that one "considers religious individuals and communities as active empowered users of new media who make distinctive choices about their relationship with technology in light of their faith, community, history, and contemporary way of life (p. 6)".

Hon and Grunig (1999) have spent years studying relationships and proposed guidelines for measuring their key components: "control mutuality, trust, satisfaction and commitment" (p. 3). In addition, they defined two types or relationships: a) exchange relationships, in which "one party gives benefits to the other only because the other has provided benefits in the past or is expected to do so in the future"; and b) communal relationships, in which "both parties provide benefits to the other because they are concerned for the welfare of the other—even when they get nothing in return (p. 3). They argued that "for most public relations activities, developing communal relationships with key constituencies is much more important to achieve than would be developing exchange relationships" (p. 3).

The key component of trust may be one of the most important factors for religious organizations. One pastor addressed the trust measure as vital to the religious experience:

> Religious vitality requires a balancing act between private contemplation and public conversation, and both practices require the establishment of trust. Without trust, there can be no deepening of a personal relationship with God nor can there be an enrichment of relationships with other people. This is true regardless of whether the connections are made through Facebook or through a congregational small group. (Brinton, 2010, para. 11)

In fact, trust is a cornerstone to any Christian's faith base: "And in his name shall the Gentiles trust" (Matthew 12:21). A Christian's faith requires the belief that Jesus Christ is trustworthy. In the Bible, Jesus teaches Peter about trust when he allows him to walk on water until the disciple is distracted by the wind begins not to trust Jesus' outstretched hand to hold him: "Lord, save me," (Matthew 14:30) he cried. "And immediately Jesus stretched forth his hand, and caught him, and said unto him, O thou of little faith, wherefore didst thou doubt" (Matthew 14:31). In this vein, Brinton (2010) asserted, "In order to continue to grow, Facebook and Google are going to have to show good faith to their members, and prove that they are trustworthy." (para. 14).

Grunig, Grunig, and Dozier's (2002) excellence model applies to trust and relationship building, particularly in the context of religious organizations, which should be practicing communication ethically. The authors contended that "two way symmetrical communication produces better long-term relationships with publics than do the other models of public relations" (p.15) and is innately more ethical and effective. Tilson and Venkateswaran (2004) recognized religion and faith traditions as dimensions of culture that have an impact on the practice of excellent public relations.

From a research perspective, analyses of Facebook pages could reveal fans' level of trust in the organizations they affiliate with online. Waters et al. (2009) suggested that "case studies should be conducted to help offer insights for other [nonprofit] organizations based on efforts that have both succeeded and failed" (p. 106). With so many religious fan pages to analyze, one approach could be to perform case study analyses of sites with the highest levels of engagement.

In May 2011, a website (www.allfacebook.com) that collects statistics about Facebook listed organizations that rank highest in user engagement. User engagement was measured by number of fans and interactions (i.e., commenting, posting, and liking the page). According to engagement statistics at the time, of the top 20 Facebook fan pages, 25% had a religious suasion, followed by sports, musical pop stars, and television. The top religious pages were: Jesus Daily, the Bible, Jesus Christ, Dios es Bueno!, which in Spanish means "God is Good!," and We Are Khaled Said, for Muslim followers.

So why is Jesus Daily so popular among its fans? Jesus Daily, according to its own Facebook fan page (www.facebook.com/JesusDaily), was established in 2009 by Aaron Tabor, MD, with the mission to "help 50,000 people accept Christ this year". The fan page is linked to Web pages www.everystudent. com/features and www.everyarabstudent.com, in Arabic, and maintained by Campus Crusade for Christ, a non-denominational Christian organization that reaches out to students. In addition to posting and commenting on posts, interactive clicks included (as of this writing) the ability to invite family and friends to join, purchase Christian art featured on the page, and participate in a prayer request comment thread. A free iPhone/iPad/iTouch Jesus Daily App was available for download on the page and at iTunes. Users were encouraged to be interactive with the page so that their friends will bring in "seekers." Likes and interests, which were all clickable, included: Jesus Diario (Jesus Daily in Spanish), TrueLife.org, Revival Soy Protein, GOD winks, Breast Cancer Medical News with Dr. Tabor, Join to Stop Diet Scams!, and Dr. Tabor's Diet. Looking at the fan page from a two-way symmetrical communication standpoint, because of its interactivity, would gauge Jesus Daily as leaning toward excellence, which according to Grunig, Grunig, and Dozier (2002) would produce better long-term relationships. Although highly interactive, Jesus Daily clearly advocates a certain philosophy and does not deviate from it. The fan page adheres to several tenants of symmetry, which includes listening to and responding to stakeholders and publics—a quality that apparently pushed the fan page to the number one spot for page engagement.

One characteristic that defines quality relationships is trust, which according to Hon and Grunig (1999) has several underlying dimensions, including integrity, dependability, and competence. The Jesus Daily fan page is honest and transparent in its objectives, which include encouraging users to "strive to follow Jesus daily by contemplating his sayings daily." The dependability dimension indicates that the organization will carry through with what it says it will do. The fan page has a dependable reputation in that at least once daily, if not much more, a status update, which contains words from Jesus, is posted on the wall. The updates usually include a photo, graphic, link, or video to attract target audiences, along with a scripture verse. Thousands of "likes" and hundreds of comments on each status update support symmetrical relationships are being developed that have the dimension of integrity, which is "the belief that the organization is fair and just" (Hon & Grunig, 1999, p. 3). The page's huge following satisfies the competence dimension of trust, or "the belief that an organization has the ability to do what it says it will do" (Hon & Grunig, 1999, p. 3), with the mission of reaching 50,000 people for Christ. Although there is no evaluative statistic available to know how many of the fans have experienced conversion or some other type of faith-based religious behavior change as a result of the page, there are still more than 5 million fans who follow Jesus Daily, and about half of those fans, 2.3 million, interact with the page.

Another influence on the credibility and thus the integrity of the page may be that the administrator is Aaron Tabor, MD, who studied at Johns Hopkins School of Medicine, and has his own fan following. After Dr. Tabor graduated from medical school, he became concerned with his own mother's struggle with menopause and in turn developed nutritional supplements to help. He also developed a soy-based weight loss program and other soy food products. He started Jesus Daily in 2009 while he was browsing through Facebook and thought if people could see a Bible verse every day it may make a good page. Tabor (as cited in Sun, 2011) said:

> We focus on bringing in seekers through the viral nature of Facebook by asking Jesus Daily members to like, share, and comment on the content we post. This spreads the message to their Friend's List and spreads further from there. (para. 24)
>
> People are hungry for a relationship with God, particularly during these rough economic times, wars, uncertainties, and the breakdown of a family unit. People are wounded emotionally and need Christ to heal their broken hearts. (para. 26)

As a physician, Tabor focuses on the need for physical healing, as well as spiritual recovery.

In an interview with the author, Dr. Tabor said that Jesus Daily fans were engaged because:

> They are passionate about their beliefs and want to help others find the fulfilled, abundant life we have in Jesus Christ. The fans' true passion is the only thing that can account for being the number one most engaged page the past three weeks in a row. We believe that mankind is separated from God through our sin, but by accepting Christ's free gift of salvation through his death, burial and resurrection, one can make peace with God. No works are required. (A. Tabor, personal communication, May 18, 2011)

Tabor conducted research prior to launching the page and said:

> The main religion sites were only quoting a Bible verse and I believed that people would be more passionate about a personal relationship with Christ, so Jesus Daily focuses on His words daily. We do include other content from time to time. Jesus Daily has turned fans into "fanatics" for Christ.

In regard to engagement, Tabor said that he builds relationships with fans by "trying to interact as much as possible with every fan that posts a comment or like. It is impossible to keep up now, but I am trying to hire additional staff to help." He also sought to reach a diverse audience of fans with "an unmet need." Millions of fans now follow the page from around the world.

"We have people in Nigeria posting prayer requests for cancer that are being prayed for by people in Indonesia. It is so encouraging," said Tabor.

Conclusion

Further longitudinal studies are needed to quantify relationship measures, in addition to the trust characteristic, to know if Facebook fan pages are actually creating, building, and maintaining long-term relationships. Other dimensions could be added to include credibility. Research from a practical standpoint could focus on what makes religion-based Facebook fan pages user friendly. Why are some organizations' pages more popular than others? After interviewing church leadership, Davis (2011) compiled a list of 10 Facebook tips to effectively use the medium for "in-reach" to stakeholders and "outreach" to publics: begin well, keep it short, add a graphic, post regularly, keep it positive, connect, develop a relationship with the reader, use video clips, add different groups, and wait a minute before posting (pp. 1–2).

Few studies have focused on impact of various beliefs by a variety of religions via the social media. A glimpse into fan pages on Facebook and how they are changing the face of religion reveals how new media are being used in an old-fashioned way to disseminate unchanging messages. Grunig (2009) has been critical of public relations practitioners for using "new media in the same ways they used the old—as a means of dumping messages on the general population rather than as a strategic means of interacting with publics and bringing information from the environment into organizational decision-making" (p. 1), noting a simple transfer of traditional media techniques to online environs is not sufficient to fully leverage the promising potential of social media for public relations. Grunig, in effect, made the argument that much of the public relations use of social media is from a press agentry standpoint, acknowledging: "these media have the potential to truly revolutionise public relations—but only if a paradigm shift in the thinking of many practitioners and scholars takes place" (p. 16). Digital media, he argued, provide the tools for the paradigm shift. However, a look at a greater variety of Facebook fan pages could provide the data to support or deny Grunig's claim that old media are being used in new ways. Certainly, Facebook fan pages for religious organizations are more strategically planned and provide more outlets for two-way communication than the old medium of Paul's letters to the churches he planted on his missionary journeys, but does that make the messages any less relevant to their respective target audiences?

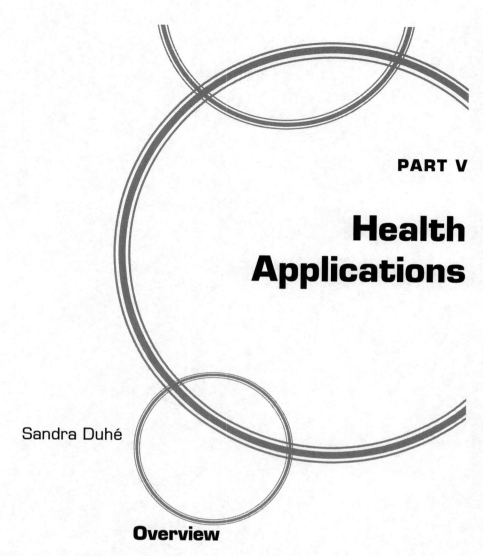

PART V

Health Applications

Sandra Duhé

Overview

Public relations offers a variety of rewarding environments in which to practice. Health-related public relations may be among the most impactful of careers, as practitioners have the opportunity to influence and witness life-enhancing behaviors among their publics. Health is of global concern at the individual, community, and state level, and this section offers a public and a private view of health communication in action.

Lucinda Austin studied how the U.S. Centers for Disease Control and Prevention framed messages and built relationships through Twitter and Facebook. She discusses differences in how the social platforms were used and how safety, prevention, individual responsibility, and other themes were conveyed. She also offers research-based recommendations for how the agency can improve its online presence.

Jeong-Nam Kim and Kelly Vibber examined the benefits of social networking from the personal health perspective of chronically ill patients. Using sociology and other theoretical foundations, they investigated how our networked society is related to "cybercoping." In an age where we often question the superficial nature of online exchanges, these authors propose that individuals who use blogs as part of their health-related problem solving can significantly improve their health-coping outcomes as a result. They offer a fascinating model that explains the process.

Framing Health Through Social Media

A Web Analysis of the U.S. Centers for Disease Control and Prevention's Use of Social Media

Lucinda L. Austin

Exploring the U.S. Centers for Disease Control and Prevention (CDC) and its use of social media, this study examined how health information is framed in CDC's social media, characteristics of CDC's use of social media, and how CDC builds relationships with publics via social media. Through a purposive sample of 622 posts over a 6-month period from among select CDC Facebook and Twitter pages, this study employed a qualitative, grounded theory approach to analysis. Findings revealed key differences between the use of Twitter and Facebook and themes in CDC's framing of health information. Despite the potential of these media for interactivity, dialogue, and tailored communication, CDC largely uses the sites to disseminate general information.

Social media have emerged as an innovative way for public health organizations to communicate with publics, and research indicates that social media are changing the practice of public relations (Wright & Hinson, 2009a). Studies have found that social media are being used increasingly by journalists as a source for news generation (Lariscy, Avery, Sweetser, & Howes, 2009a). Still, little is known about how public health organizations or government agencies use social media to build relationships with publics.

Government's use of social media and its social media policies have the potential to influence use of social media for public health initiatives on a larger scale (e.g., CDC, 2010a; U.S. HHS, 2010). As Aronowitz (2008) suggested, the framing of policy and communication initiatives can shape population health by affecting health and illness beliefs and behaviors, attitudes about health policies and practices, and socio-demographic dynamics.

Though many studies have focused on media frames represented in mass media channels such as television and newspapers (e.g., Liu, 2009), Zoch and Molleda (2006) have called for research to focus on frames produced by the source or the organization, in addition to frames from the receiver's standpoint. This research answers that call, as the purpose of this study is to understand how health information is represented to the public through government organizations' use of social media. Exploring one specific government health organization, the CDC, this study examines how health information is framed in CDC's social media, characteristics of CDC's use of social media, and how CDC uses social media to build relationships with publics. This study fills a gap in this emerging area of research by showing how health organizations use social media to communicate with publics.

Social media in public relations

"Social media" has been used as an umbrella term to describe Web-enabled, digital tools and applications that facilitate interactive communication and content exchange among and between audiences and organizations, involving user-generated content and comments (Lariscy et al., 2009a; Pew Internet & American Life Project, 2010b). These media may include text as well as audiovisual components (Wright & Hinson, 2009a). A range of social media is relevant to the study of public relations, such as blogs, micro-blogs (e.g., Twitter), forums, message boards, photo sharing, video sharing, Wikis, social bookmarking, and social networking (e.g., Facebook and LinkedIn).

Online sources can be ideal for generating timely communication (Taylor & Perry, 2005) and interactive, two-way conversations with publics (Kent & Taylor, 1998; Seltzer & Mitrook, 2007). The majority of public relations research on computer-mediated communication has focused on websites (e.g., Bucher, 2002) and blogs (e.g., Sweetser & Metzgar, 2007). However, research on use of other types of social media in public relations is beginning to grow (e.g., Wright & Hinson, 2009b).

Framing theory

Framing theory highlights how social media can portray health information in a way that may influence publics' perceptions of that information and their perceptions of an organization or issue (Entman, 2003; Hallahan, 1999; Scheufele & Tewksbury, 2007; Tankard, 2001). As Entman (1993) described, framing involves selecting aspects of a perceived reality to make more salient in communication. Framing theory acknowledges that framing of issues may influence receivers' perceptions of the reality of the issues, but that receivers also come to the table with their own frames and are situated in a culture that contains frames (Entman, 2003).

Zoch and Molleda (2006) stated that frames serve four functions for public relations practitioners, each related to specific problems: define the problem, determine the source, make judgments, and propose remedies. Most public relations framing research has focused on frames used in organizational communication to publics or frames used in media coverage (e.g., Murphree, Reber, & Blevens, 2009). For example, Zoch, Collins, Sisco, and Supa (2008) studied framing devices used on activist organizations' websites to see how these organizations used catchphrases, depictions, exemplars, metaphors, visual images, and promotion of the organization as a solution.

Many framing studies also examine framing effects in the influence of news coverage to understand how organizations' frames in press releases and organizational communication are utilized in media coverage of the organization (e.g., Danowski, 2008; Liu, 2009; Reber & Berger, 2005). Some public relations studies on intentional media framing have shown how frames can influence publics and audiences through the language that organizations choose to employ. For example, Darmon, Fitzpatrick, and Bronstein (2008) studied how Kraft intentionally framed messages regarding obesity through press releases and found that Kraft's messages were reflected in media coverage. They argued that organizations should seek to actively frame messages about important issues in their correspondence to media. Using framing theory as a guide, this study examines how health information is framed in CDC's social media as well as characteristics of social media sites and postings.

Research questions

RQ1: How is health information framed in CDC's social media?

RQ2: How does health information vary (or not) by type of social media site?

Method

This study employed a qualitative content analysis of CDC's social media sites to understand how health information is represented in these media and the relational messages and strategies this government agency employs. Qualitative analysis provides a richness of data that can further understanding in this exploratory study of how health information is framed (Denzin & Lincoln, 2003). Comparative methods of content analysis can help to reveal framing and clearly highlight differences (Entman, 1993).

Sampling

The CDC—one of the larger agencies within the U.S. Department of Health and Human Services (HHS)—is one of the most visible government health organizations. CDC has become active in its social media efforts and has developed standardized policies for use of social media. At the time of this research study, CDC's social media use included at least 10 Twitter accounts, two Facebook accounts, a Myspace account, a YouTube Channel, a Daily Strength support group site, a Flickr photo site, a Second Life site, and at least six different blogs from different divisions within CDC.

To address the research questions, a purposive sample of 622 posts from among two of CDC's most popular text-based social media applications—Facebook and Twitter—was analyzed qualitatively to understand how health information is framed. To limit the scope, posts over a 6-month period—November 2009 to April 2010—were analyzed. CDC platforms directed toward general or mixed audiences were analyzed, including CDC's main Facebook page with 191 posts during this 6-month time period and 55,712 individuals who "liked" the site (formerly referred to as "fans" of the Facebook page). Select CDC Twitter pages during this 6-month period were also analyzed including: "Act Early," "Bio Sense," "E-Health," "Emergency," "Flu," and "Hepatitis." These different Twitter sites ranged greatly from 604 to 1,244,200 followers and from 10 to 131 posts in the 6-month time period.

Analysis

Because there have been no published studies (as of the time of this writing) for this particular issue, open coding was used for this study via a grounded theory approach to analysis (Corbin & Strauss, 2008).

Postings were analyzed qualitatively through line-by-line open and axial coding. Emergent themes were explored to show how health information was framed and how this organization used social media to communicate with its publics. Using axial coding, initial categories were then grouped under broader similar themes.

Description of characteristics of social media use included information such as what content government organizations were posting, how often they were posting, whether or not they allowed for two-way communication, and who seemed to be their primary audiences. Another area for exploration was how health information varied by type of social media site. More research has been done about blogging than other types of social media, and little research has examined how their content varies.

Results

RQ1: How is health information framed in CDC's social media?

Safety as related to health. CDC listed on its website a tagline for "Safer... Healthier... People." This theme of safety and health together integrated throughout all of CDC's social media pages. Safety was inextricably tied to health through these posts. For example, on CDC's Facebook page, posts stated, "Be Healthy and Safe in the Garden. Stay safe and healthy while enjoying the benefits of gardening;" and "Safer healthier people depend on a healthy environment, so learn what you can do to go green!" An example from CDC's Emergency Twitter page stated, "No one can stop winter weather, but you can be ready for it when it comes. Learn how to stay healthy & safe."

Prevention versus response. Most content was framed as either prevention or response. Although most of the focus was on prevention—particularly on CDC's Facebook page—some of the Twitter sites, such as CDC Emergency, CDC Flu, and CDC Hepatitis also included response frames. For example, a statement on CDC's Facebook regarding prevention stated: "Essential public health services are key to preventing injuries and illnesses, enhancing public health preparedness, and reducing the risk from climate change." An example of a prevention frame from CDC's Flu Twitter page stated, "While traveling, wash hands often and cover coughs and sneezes to keep yourself and others well."

Mentions of response on CDC's Facebook page were mostly limited to the then recent earthquake in Haiti or to recent H1N1 flu events. For example, "CDC's Injury Center Responds to Haiti Earthquake. CDC's Injury Center is supporting an agency wide effort to work with partners and immediately address the public health needs of the Haiti earthquake survivors..." Response information on Twitter pages included information such as "CDC issues Health Alert Notice for travelers leaving Haiti" (CDC Emergency).

Individual responsibility. Health information was framed as the individual's responsibility for herself and her loved ones, which was stressed in posts geared toward the public and limited in posts directed at health professionals. Individuals were encouraged to be responsible for their own prevention and protection. For example, CDC's Facebook page stated, "Spring break means an escape from the daily grind. For high school and college students, it can be a rite of passage or an annual tradition. This is your time. It is all about you, and YOU are in charge of your health, safety, and well-being." This post capitalized "YOU" for extra emphasis on the individuals' role in health.

This focus on individuals' responsibility extended beyond the individual to the individual having responsibility for his or her loved ones as well. Posts on CDC's Facebook page made statements such as: "Show love for yourself and others this Valentine's Day by celebrating the safe and healthy way;" and

"Tips from CDC's Injury Center on motor vehicle safety can help you protect yourself, your passengers, and your family and friends." A post on the Act Early Twitter page stated, "Less than half of children w/ a delay receive treatment before starting school. Don't let this be your child!"

Collaboration and community. Although not as often as individual responsibility, posts also emphasized the importance of collaboration and community. For example, a Facebook post stated, "Community design directly affects your health. Choose to live in communities that encourage physical activity as part of your daily routine. Join with your neighbors to make your community as healthy as possible." Also, CDC's E-health Twitter page included statements such as, "April is STD Awareness Month. Get info & tools to mobilize your community."

Timely and event-specific. A predominant theme on both Twitter and Facebook was timely and event-specific information centered on events such as "National DNA Day," "Sexual Assault Awareness Month," "National Child Abuse Prevention Month," "National Native HIV/AIDS Awareness Day," and many others. For example, nearly every tweet on CDC's Act Early Twitter page was related to autism during the month of April, such as "It's Autism Awareness Month! Has your child been screened? The AAP recommends autism screening at 18 and 24 mo."

Information was also season-specific with news and information regarding "staying safe and healthy" during "Winter Weather Winter storms and cold temperatures" and "Unpredictable Spring Weather." Opportunistic information highlighted issues such as "Be[ing] Physically Active in the New Year," as this is a time that individuals typically think about New Year resolutions and weight loss.

Evidence for the burden of health problems. CDC's Facebook and Twitter pages used a variety of support and evidence to show the burden of health problems. Most of the evidence on Facebook was statistical in format; however, actual statistics were more limited on Twitter due to character limitations. Many tweets instead directed users to research studies or other postings where they could view statistics about the problem. Representative examples of CDC Facebook posts using statistics included, "In the United States, 1 in 6 women and 1 in 33 men report experiencing rape in their lifetime. April is Sexual Assault Awareness Month," and "Sickle Cell Disease (SCD) Visits to Emergency Departments (ED): Between 1999 and 2007, approximately 197,333 sickle cell disease ED visits occurred each year."

Tweets using statistics were shorter than Facebook posts, but formatted similarly. An example from CDC's Hepatitis Twitter page stated, "In the United States, over 4.4 million people are living with chronic viral hepatitis." Another example from CDC's Flu Twitter directed users to a site where they could find the statistics: "New summary of CDC's updated estimates on 2009 H1N1 cases, hospitalizations and deaths."

In addition to statistical evidence, CDC also used narrative evidence, such as individuals' stories. However, narrative evidence was employed much more limitedly than statistical evidence. For example, identical posts on CDC's Facebook and E-health Twitter pages included the following: "Ida was in disbelief when she learned she had HIV. Learn from Ida's Story," and "Think an STD couldn't happen to you? Watch Molly's story." Both of these posts linked to YouTube videos of individuals' stories.

The science of health. In addition to the focus on the burden of health issues, CDC also discussed the importance of evidence-based approaches to health issues and framed their efforts as "science." For example, CDC's Facebook page included posts such as "Scientists have made great advances in understanding important environmental causes of obesity as well as identifying several genes that might be implicated. Major efforts are now directed toward assessing the interactions of genes and environment in the obesity epidemic," and "Health Protection. Health Equity. Scientific Excellence. At the CDC, we are dedicated to protecting health and promoting quality of life through the prevention and control of disease, injury, and disability."

Expert, authority, and resource provider. Posts regularly discussed CDC and other "experts" that could give advice and recommendations on particular health topics. For example, posted on the CDC Facebook and E-health Twitter pages was the post, "What can you do about STDs? Experts talk about how to prevent them here." CDC also made specific recommendations and endorsed particular courses of action for individuals. For example, CDC posted on Facebook, "The meningococcal conjugate vaccine is recommended for all 11–18 year olds. Kids should get this vaccine at their 11–12 year old check-up with other preventive services…" On the Hepatitis Twitter page, an example read, "People with hepatitis should get the H1N1 vaccine as soon as it is available."

In this role, CDC also provided research and information on a variety of topics, working to increase awareness and knowledge. A post from CDC's Facebook page stated, "Did you know that reptiles and amphibians like turtles, lizards, and frogs can carry a harmful germ called Salmonella? If there are young children in your home, reptiles and amphibians might not be safe pets for your family." Another post from CDC's Hepatitis Twitter page stated, "A person can have viral hepatitis and not know it. But even though they do not have symptoms or feel sick, liver damage can still occur."

CDC also directed individuals to resources and served as a resource provider on topics for the general public and health professionals. For example, on CDC's E-health Twitter page CDC featured release of "a new FREE online training program: Health Literacy for Public Health Professionals" and a "new interactive state map widget to add to your website or blog to access state-by-state reports on tobacco control." CDC's Facebook page also highlighted resources as well as CDC publications such as "Preventing Chronic Disease (PCD)… a peer-reviewed electronic journal published by the CDC's National Center for Chronic Disease Prevention and Health Promotion."

Health as a serious issue. The tone of most posts conveyed that health was a serious issue. Information was stressed as important and often severe. For example, on CDC's Facebook page a post read, "Actor/musician Terrence Howard talks about his mother's death from colon cancer. He says, 'This is personal. Let my heartbreak be your wake-up call.'" Other posts discussed the burden of injury through statistics as mentioned above.

Although most sites portrayed health as a very serious issue, CDC's Act Early Twitter page was an outlier in that it also provided fun, interesting, and light-hearted facts to parents about their children. Many of these facts did not include a call to action, just a point of information. An example of this type of post read, "At 1, babies respond to the sound of their name. The most popular baby name in 09 was Ethan for boys & Emma for girls."

Participation. Two frames of participation occurred in CDC's social media. One frame occurred quite often, which was to encourage individuals to spread the word to others regarding specific health issues. For example, CDC's Facebook and E-health Twitter pages both promoted sexual health with "Spread[ing] the word about GYT! Get Yourself Talking. Get Yourself Tested." Individuals were encouraged to complete a variety of activities such as to "…Grab, share and promote GYT banner ads, logos and icons, and desktop and cell phone wallpaper. And connect to GYT on Facebook and Twitter" (Facebook); "…print & distribute fact sheets…" (Emergency Twitter); and, in regard to a specific disease, to "add new CDC Diabetes Risk Test widget to your Web page or blog" (E-health Twitter).

Another frame of participation, not occurring as often, was encouragement for individuals to engage in dialogue with CDC and other partners, inviting feedback and active participation in this manner. On

CDC's Facebook site, individuals were invited to "…discuss your views, concerns and ideas regarding public health and chemical exposures as part of the National Conversation on Public Health and Chemical Exposures," and to "comment on selected articles" within a CDC online magazine. On Twitter, individuals were asked questions and asked to comment. For example, on CDC's Act Early Twitter individuals were asked, "Two year olds can talk in two-word phrases like 'drink milk' or 'play car.' What's your baby's favorite phrase?" Individuals were also asked to provide survey feedback on Twitter pages such as: "Help us improve our Twitter profile! Click here to complete a short survey."

RQ2: How does health information vary (or not) by type of social media site?

General versus audience-specific. CDC's Facebook page contained information on a wide variety of topics for multiple audiences, as opposed to CDC's separate Twitter pages, which had particular foci and audiences. Some Twitter pages, like CDC's E-Health account, were general and more closely mirrored CDC's Facebook page.

Cross-posting. Despite the specific topics for some of CDC's Twitter pages, Facebook and Twitter were very similar in the information they displayed, often cross-posting the same information on multiple Twitter pages and between Facebook and Twitter pages. For example, CDC's Facebook page posted, "What can you do about STDs? Experts talk about how to prevent them here" an hour after the E-health Twitter page posted the same statement—both linked to a CDC Podcast. Facebook and Twitter pages also linked to CDC's other social media and Web pages. For example, two minutes apart both CDC's Facebook page and E-health Twitter page had a story posted about a new CDC blog about traumatic brain injury and concussion. On Twitter pages, multiple postings for the same message were often in the form of a "retweet" of announcements from other CDC sites and other government agencies' sites, such as the FDA, NIH, or HHS. For example, CDC's Emergency Twitter page retweeted emergency announcements from CDC Flu, such as "RT @CDCFlu Voluntary Non-Safety-Related Recall of Specific Lots of MedImmune Nasal Spray Vaccine for 2009 H1N1 Flu."

Re-posting. CDC's Facebook page had a tendency to repeat the same information several months apart for varying topics. For example, in December 2009 CDC posted: "There is no known safe amount of alcohol to drink while pregnant. There is also no safe time during pregnancy to drink and no safe kind of alcohol. CDC urges pregnant women not to drink alcohol any time during pregnancy" and, in April 2010, reposted the same message. CDC's Twitter pages were much less likely to repost the same information in the same wording within the time frame analyzed.

Personal versus organizational identity. CDC's Twitter pages featured direct comments to others on Twitter via tweets. Facebook included fewer responses from "CDC" to commenters, although there were quite a few individuals commenting on posts. Instead, it appeared as though other CDC staff and partners were replying to comments using their personal Facebook identities.

Stories and linking. Most likely due to the character limitations of Twitter, CDC's Twitter pages included fewer actual stories or "messages" and more links to other outside sources such as social media, CDC websites, RSS feeds, and blogs, than did the Facebook page. Both, however, linked to CDC podcasts and YouTube videos and other CDC pages quite frequently. The sites did not often link to any sites or stories outside of CDC resources, except for the occasional link to a partner organization's campaign site or information.

Guidelines for interaction. Due to the platform nature of Facebook that allows comments, CDC posted more guidelines for response on the Facebook page than the Twitter pages. For example, twice on the Facebook page in different months CDC posted, "CDC actively removes comments that violate our comment policy. Please remember that while CDC's goal is to share ideas and information with as many individuals as possible and our policy is to accept the majority of comments made to our profile, comments will be deleted if they contain hate speech, profanity, obscenity..." CDC's Facebook page also included the disclaimer, "Disclaimer: Posted comments and images do not necessarily represent the views of CDC." CDC's public-facing Twitter pages included in this analysis did not include a disclaimer of this kind.

Conclusion

Most of CDC's Facebook and Twitter posts directed participants to take some sort of action, whether this was a prevention or response action, or geared toward encouraging participants to increase awareness and knowledge of specific health issues through outside resources. Some posts focused instead on highlighting the burden of particular health issues and announcing new information. CDC's social media highlighted the link between safety and health, which was messaging from CDC's mission.

Most information was geared toward preventing health problems. Health was generally framed as a serious issue, and CDC was framed as an expert, authoritative source, and resource provider. Most posts framed the importance of individual responsibility, although some posts also focused on the importance of collaboration and community.

CDC's Twitter and Facebook posts closely mirrored each other in terms of content and somewhat in regard to the actual frames. Because Twitter is more limited in the amount of characters allowed per post, some frames were not highlighted as strongly or as often through Twitter. Also, Twitter pages often directed individuals to other sites for more information.

CDC took advantage of the timeliness of the social media platform (Taylor & Perry, 2005) in that posts reflected timely and often event-specific information. Posts were largely opportunistic, relating to a specific period of observance for a health issue, seasonal issues, emerging research, and in response to crises. The inclusion of event-specific and timely information reflects journalistic news values and is more likely to be engaging for audiences (Bivins, 2011).

Despite the interactive potential of social media, there appeared to be a lack of strategies to encourage dialogue, particularly on CDC's Facebook page. Previous research has found that perceived interactivity can enhance individuals' perceptions of their relationships with organizations and increase their intentions to share information (Jo & Kim, 2003; Yang & Kang, 2009). Since many of CDC's posts encouraged others to share information by reposting, adding buttons or gadgets to their Web pages, etc., developing more opportunities to actively engage in dialogue may aid in this goal. Also, as CDC did not respond directly to comments on the Facebook page, responses to commenters may also show that dialogue is encouraged and welcome and lead to perceptions of interactivity.

Both Twitter and Facebook posts were directed at very general audiences, perhaps because of their character limitations. Although this may have limited the content, the pages lacked focus on diverse audiences' needs and concerns. A few Facebook posts referenced diverse audiences (e.g., "Renowned Hollywood actor and Cherokee tribal member, Wes Studi, explains how American Indians and Alaska Natives are at risk for Seasonal Flu, and urges Native people to Take 3 as the best way to protect themselves, their families and their tribal communities from seasonal flu and the H1N1 virus.") Twitter posts were even more limited in their focus on diverse audiences, although CDC's E-health Twitter post did include an occasional post in Spanish. Social media have been touted for their ability to more effectively tailor information to diverse audiences with diverse needs (Harrington, 2008). To more effectively communicate with all audiences who may have unique risks and needs related to specific health issues, CDC should further explore ways to individualize messages for diverse audiences and publics.

Also, as CDC focuses on partnering with other professional organizations and encouraging community involvement, linking to partner organizations' sites and retweeting partners' information could help to develop and foster a greater sense of community and partnership. Linking to partners' sites and helping partners' promotion efforts could also create a sense of reciprocity, encouraging partners and affiliates to also promote CDC's sites and links; although, as a government agency, CDC may be hesitant to endorse other partner organizations. CDC already includes disclaimers on some social media sites that the information provided on the pages does not necessarily represent the views of CDC or imply an endorsement.

Most posts focused on prevention and responsibility of the individual. However, more focus on social determinants of health and encouragement of community collaboration could also aid in prevention efforts and the fostering of supportive communities and environments. Messages to individuals such as teachers, doctors, and others who could support and assist individuals in their prevention could increase prevention efforts beyond the role of the individual.

Future research should explore how audience members and publics use social media platforms to communicate with organizations and how they make meaning of this communication. Many resources and efforts have been placed into social media without knowing the effects or outcomes of these utilities on publics' perceptions of relationships. Although some research has explored this in an experimental setting (Jo & Kim, 2003; Yang & Kang, 2009), more research is needed in applied and naturalistic settings as well to explore this phenomenon.

Information in CDC's social media was framed as timely, important, and science- and evidence-based. CDC used statistical evidence to a greater extent and narrative evidence to a lesser extent. Information focused on prevention, individual responsibility, safety, and to a lesser extent, community collaboration. Cross-posting and re-posting of information was common between and within Twitter and Facebook sites. Twitter and Facebook sites were similar in terms of content, although the Twitter pages targeted more specific audiences and linked to more outside information. CDC's Facebook page included more statistical evidence than the Twitter pages, and individuals posting responses to publics' posts on Facebook were most often individuals affiliated with CDC, not CDC as the organization.

Although participation was encouraged in the form of sharing messages with others, active dialogue was not encouraged as often on the social media sites. Increasing the focus of messages beyond individual responsibility, extending more toward community efforts, may aid in prevention. Additionally, as these platforms targeted very general audiences, using these platforms more interactively and to better tailor health messages to diverse audiences can better enhance the unique benefits that these social media platforms can provide.

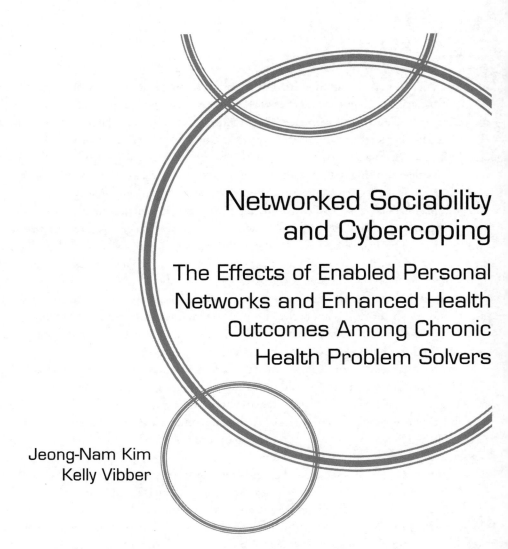

Networked Sociability and Cybercoping

The Effects of Enabled Personal Networks and Enhanced Health Outcomes Among Chronic Health Problem Solvers

Jeong-Nam Kim
Kelly Vibber

Digital communication technology has been deeply integrated into our routines, greatly influencing our interactions with others as well as the ways in which we create, maintain, and even terminate relationships. Our emergent daily experience of relating to and interacting with others through digitalized communication networks can be aptly described as networked sociability (Castells, 2004b).

However, this migration from offline, analog, and direct personal interactions to online, digital, indirect interactions comes with tradeoffs. For example, on one hand we are largely free from economic and geographic constraints so that we are quantitatively more social than ever before. Yet on the other hand, some think we may only be spreading ourselves thin but not thick in relational experiences. Although the new digitalized communication environment makes the scope and density of today's personal relationships incomparable to those of the pre-digitalized era, many still express concerns that much of online human relations are superficial and shallow, instantaneous, quick, and lacking depth, leaving us connected better in quantity but poorer in quality. Although these debates will undoubtedly continue, the goal of this chapter is to test the good that can come out of these digitalized, essentially 24-hour social networks. It examines whether the experiences with digitalized social relationships make positive contributions in dealing with chronic health problems. Specifically, we want 1) to observe whether the digitalized communication environment and enabled "networked sociability" help individuals with chronic health issues better cope with their problems (i.e., cybercoping) and 2) to explain, if such positive effects are found, what the key condition is (i.e., when it happens and when it does not) and the process for achieving enhanced coping. There is a relatively small amount research on the effects of enabled personal networks and enhanced health outcomes, especially from the patient's perspective. Thus, this chapter investigates the association between increases in relational interactions via digital communication use and perceived improvement of health coping outcomes. We will discuss the concept of cybercoping, health coping outcomes, and a mediating process between online characteristics and health coping outcomes using survey data drawn from online communication users with various chronic health problems.

Communicating, cybercoping, and networked sociability

In this chapter, we use the term *cybercoping* to refer to one's problem-solving efforts in cyberspace regarding various chronic health problems (Kim, Park, Yoo, & Shen, 2010). Kim and Ni (2010) coined this term based on findings that people with problems tend to use cyberspace as "a tool for amplifying their problem perceptions and as leverage for empowering through identifying and connecting with fellow problem solvers" (Kim & Ni, 2010, p. 46). Despite the increasing research attention dedicated to online digitalized communities such as blogs, the effects of Internet use and relational enhancement through blogs on health outcomes, especially among those patients with chronic health issues, needs further examination.

In both cybercoping and networked sociability, communication and communication behaviors are pivotal. Communication is something that individuals do when they encounter problematic life situations (Carter, 1965; Kim, Grunig, & Ni, 2010). It is epiphenomenal in that it is an instrumental tool but not a determining factor for successful problem solving (Kim & Grunig, 2011). Communicating problem holders, those engaging in information acquisition, information selection, and information transmission, will possess and activate a useful means for overcoming problematic consequences (Kim, Grunig, & Ni, 2010). These three communication behaviors can be performed in both active and passive manners as outlined in Kim, Grunig, and Ni's (2010) reconceptualization of communicative action of publics (CAPS). This sub-theory extended work on the situational theory of publics and allowed for closer examination and explanation of the information behaviors of publics. In the CAPS framework individuals can engage in these behaviors actively or passively. The three behaviors of interest (information acquisition, information selection, information transmission) can each be further delineated into their active and passive behaviors respectively, as follows: information seeking and information attending, information forefending and information permitting, information forwarding and information sharing (Kim, Grunig, & Ni, 2010). The use of these active and passive communicative behaviors will be examined within the sample for their impact on the cybercoping abilities of the participants.

Individuals with chronic illness who are problem solvers are motivated to engage in communicative actions in both offline and online communicative channels. Of particular interest in this chapter are digitalized communication environments that enable patients access to greater informational and relational resources. In addition, this chapter specifically examines online blogging, which has been one of the fastest growing online communication media. Kovic, Lulic, and Brumini (2008) studied the adoption of medical blogs in the healthcare industry and reported that blogs have become a notable platform for publishing and sharing information.

Patients now turn to the Internet as a primary resource to search for health information and advice (Pew Research Center, 2000). Using the Internet as a source of information regarding their life problems is commonplace because of its efficiency. One can easily collect a vast amount of directly relevant as well as related information while exerting low levels of effort (Murray et al., 2003). In this vein, many studies have examined online information behavior from a number of various perspectives. Some studies have focused on the quality of online information. Dutta-Bergman (2003) examined the influence of health information search motivation and information completeness on attitudes and behavioral intention regarding health issues. Other researchers investigated the cognitive process of judging credibility of information sources, specifically, the effect of one's content-related knowledge and perceived source expertise to the credibility judgment of online health information (Eastin, 2001). Still other studies examined the predictors of information seeking behavior on the Web (Cotten & Gupta, 2004; Rains, 2007). Finally, some studies also tested differential influences from sociodemographics to motivations of online health information seeking (Warner & Procaccino, 2007). These studies and the related body of health communication literature focused more on motivation and cognitive process of health information consumption for the purpose of reducing uncertainty caused by health problems and enhancing efficacy in dealing with health problems such as choice about and adoption of (preventive) medical treatments (Brashers, 2001; Griffin, Dunwoody, & Neuwirth, 1999) but relatively less on emotional and psychological aspects of health communication behaviors (Kim, Lee, & Guild, 2009).

These latter issues are of great importance because our communicative behaviors regarding health problems are also closely related to *relational aspects* of Internet use or networked sociability (Castells, 2004b). In other words, we not only seek out information about health problems, such as learning others' health knowledge, through the Internet but also give out our affective states or emotional residuals to others through our personal networks formed on the Internet (Kim & Grunig, 2011). In doing so, we build and maintain, consciously and unconsciously, personal relationships through both offline and online communication channels. Through our relational networks we not only learn and share "information" but also, simultaneously, we learn and share "emotion" of networked neighbors. Focusing on coping with health problems, the two processes—exchanges of information and exchanges of emotion related to health problems—will jointly influence coping outcomes, and these two processes are intertwined with one's building and managing online personal relationships with others.

As mentioned previously, we are particularly interested in the online social networks associated with digitalization, specifically, whether increased use of digital communication (Internet) and the enhanced online personal networks (blogging friends) will improve the manageability of one's chronic health problems. Importantly, we pay greater attention to the social embeddedness (density of online relationships) as a more critical predictor of improved coping outcomes. In a nutshell, we emphasize individuals as *social actors* surrounded by social relationships (Kim, Grunig, & Ni, 2010). This is a more critical factor than one's mere use of digitalized communication technologies even when examining communicating behaviors for individual life problems (e.g., chronic health problems). In addition, this contextualization of the individual and the individual's problem within a larger network of individuals who share that problem or information on that problem allows us examine this interaction as part of a group problem solving effort or a public.

Our study in this chapter will further delineate the conditions when digitalized communication can enable and enhance health problem solvers' coping capability. We postulate that digitalized communication will only enable and enhance coping with chronic illness when it increases relational density but not merely the use of the Internet. We contend the need for more understanding of individuals as *social, collective,* and *interactive in nature,* especially in their digitalized communication behaviors in managing their chronic health problems. In the following, we will posit hypotheses and research questions and specify a model delineating presumed effects on health coping outcomes from Internet use and online relational characteristics.

Coping outcomes and Internet use

People use the Internet as a major source of information for various problems from booking flights to comparing medical treatments. The convenience and cost of information seeking through the Internet is unrivaled by other communication and information media that have existed throughout human history. In addition to its informational use, the Internet is also employed for entertainment and relational purposes—looking for humorous information and interacting with people who are socially detached (e.g., commenting on pop stars' websites) and/or geographically distant (e.g., friends and family living distantly). Individuals tend to experience the Internet as a multifunctional tool for various purposes. Although Internet use could include some information behaviors that increase manageability of chronic health problems (e.g. information seeking, information forwarding), it may also include many irrelevant communicative behaviors. For this reason, the simple amount of Internet use among individuals with chronic health problems cannot be assumed to contribute to increasing one's manageability of health issues. Thus, Internet use among chronic health problems will not be a sufficient condition in itself for improvement of health outcomes.

To test the effects of Internet use and networked sociability, we also need to clarify the concept of coping outcomes. A coping outcome of chronic health problems such as diabetes, depression, or cancer should be construed at least in two dimensions. The first dimension of coping outcomes is *psychological coping*—how one with a chronic health problem can maintain affective stability and remain confident regarding their ability to manage health issues. The second dimension of coping outcomes is *physical coping*—how one can decrease negative health symptoms and impede development of other illness as well as increase physical resilience. Because these are both significant health coping outcomes, we examine the desirable effects of Internet use and the networked sociability one builds in digital communication networks in terms of both psychological and physical coping outcomes. However, given the previous statement about Internet use, we predicted the following effects of Internet use on health coping outcomes in Hypotheses 1a and 1b.

H1a. Increase of Internet use will NOT be associated with the enhancement in psychological coping outcomes.

H1b. Increase of Internet use will NOT be associated with the enhancement in physical coping outcomes.

Coping outcomes and relational density

Social psychologists such as Lazarus (1993) and Folkman (1997) distinguished coping as two distinct processes: problem-focused coping and emotion-focused coping. Problem-focused coping happens when a person makes endeavors to modify one's relationships with the environment, and emotion-focused coping happens when one makes efforts to modify the way one interprets what is happening, even when the actual person-environment relationship is intact.

Research shows a variety of support processes for health problem solving can enhance health outcomes. In addition, quite a few studies paid special attention to the dimensions of emotional (or relational) support, in the sense that online interactions and virtual communities could become "a source for emotional support and practical advice, a blend of comfort and counsel" (Whitney, 1999, p. 14). Indeed, many communication studies and articles discuss the positive physical and psychological effects of social support (e.g., Albrecht & Adelman, 1987; Burleson, Albrecht, Goldsmith, & Sarason, 1994). For example,

Eysenbach (2003) found that one's sense of community, through the interactions from virtual communities, could help cancer patients. Researchers also found that health-related online message boards and chat rooms provide advice and information, as well as emotional support among users (Macias, Lewis, & Smith, 2005). Mo and Coulson (2008) found through content analysis of messages on an HIV/AIDS support group that two types of common support are informational and emotional support among those users of online groups.

From the problem solving theory and information behaviors (Kim & Grunig, 2011), the role of information and emotion can be considered as resources that facilitate the problem-solving (coping) processes. Problem solving necessitates both decisions and resources (Kim, Grunig, & Ni, 2010; Ni & Kim, 2009). In general, in successful problem-solving situations, problem solvers identify several options as solutions and select the most adequate solution. The chosen solution then improves his or her overcoming of the problematic situation by changing the problem-causing context ("external inquiring" and "individual effectuating," Kim & Grunig, 2011). Effective information acquisition and selection contributes to one's problem-solving capability because it improves decision-making in the given problem situation. As problem solvers seek and share relevant information and experiences, the connected information users can create better access to the intellectual resources that help problem solvers improve decision-making and problem-solving approaches. Here, pooled information and created accessibility are considered a type of problem-solving resource.

Furthermore, health problem solving also necessitates emotional and psychological resources from support providers and fellow problem solvers. Successful problem solvers tend to identify and connect with other problem solvers with similar problem states. As one finds and connects with other problem solvers, they organize (possibly loosely) as a group, and can make collective efforts toward their common problem. In this way, problem solvers will better mobilize the required intellectual or material resources toward problem solving and will lower the costs in bringing the desired resources ("collective effectuating," Kim & Grunig, 2011). In addition, interactions within a (loosely) connected community tend to increase the number of acquaintances and thus the frequency and quality of interactions among members. Such emerging relationships among problem solvers can generate emotional support and could thus improve one's health coping efforts and outcomes. We postulated that the enhanced coping outcomes can be observed only when online relational ties and density of social network come together with one's increased Internet use. Here, pooled and emotional support through relational ties and frequent interactions are considered as another type of problem-solving resource. We thus predicted the effects of increase of relational ties and interactions on health coping outcomes in Hypotheses 2a and 2b.

H2a. Increase of online relational density (i.e., number of blogging acquaintances) will be associated with the enhancement in psychological coping outcomes.

H2b. Increase of online relational density will be associated with the enhancement in physical coping outcomes.

In testing the hypotheses of Internet use and relational density on health coping outcomes, it is also interesting to examine the effects of key sociodemographics such as education, income, and gender. We thus posited the following research question:

RQ1. What are the effects of sociodemographics—age, gender, income, and education—on the psychological and physiological coping outcomes?

A mediation process through relational enhancement

In addition to the first part of the study, we also take a closer look at underlying processes of how digitalized communicative behaviors and networked sociability would enable and enhance health coping outcomes. If we can conceptually specify and test all the identified variables and proposed hypotheses within a single conceptual model, then this will help us acquire an overall perspective, not fragmentary pictures of the health coping processes.

Given this proposition, we posit that the effect of Internet use and relational density, if any, on coping outcomes will be mediated through the enhanced relational quality and satisfaction. In other words, simply spending more time in digitalized networks and increasing the number of online acquaintances cannot be equated with relational enhancement. Further, we postulate that only when one experiences relational enhancement will the volume of digitalized communicative behaviors and relational density have effects on improved health coping outcomes for managing chronic illness. Therefore, we posit the following hypotheses:

H3. Increase of Internet use will _not_ increase relational enhancement.

H4. Increase of online relational density (i.e., number of blogging acquaintances) will increase relational enhancement.

H5. Increase of relational enhancement will enhance psychological coping outcomes.

H6. Increase of relational enhancement will enhance physical coping outcomes.

Similar to the first set of hypotheses, it is interesting to test the effects of key sociodemographics such as education, income, and gender. Thus, we posit the following research question to examine the effects of key hypotheses while controlling key sociodemographics:

RQ2. What are the effects of sociodemographics—age, gender, income, and education—on the psychological and physiological coping outcomes in the mediation model?

Method

Data collection and sample

A total of 254 individuals with a chronic illness participated in an online survey. The distribution of chronic illness in our sample was as follows: lupus 21.7%, diabetes 12.5%, cancer 8.2%, depression 7.0%, HIV/AIDS 6.3%, arthritis 6.3%, fibromyalgia 5.9%, Parkinson's disease 5.5%, and others 26.6%. In recruiting participants we only considered individuals whose age was eighteen or older.

In the survey we referred to the term "blog" for online activities where participants would read or exchange information in places such as individual Web logs, Internet forums, and discussion boards. We identified and contacted candidate participants through Google and Technorati's blog search engines and posted recruitment messages to those identified personal blogs. We also identified blogger communities such as Livejournal, Blogspot, and Typepad for recruitment. In addition, we further searched and sent invitations to members of various online health communities whose users and members have chronic illness conditions such as depression, HIV/AIDS, lupus, cancer, arthritis, and diabetes.

The final sample we tested consisted of 81.4% (n = 206) female and 18.6% (n = 47) male participants. The age groups represented were as follows: 15.0% 18–29 years (n = 38), 20.1% 30–39 years (n = 51), 30.3% 40–49 years (n =77), 22.4% 50–59 years (n = 57), and 12.2% 60 or more years (n = 31). Education was distributed with 26.4% having a high school education (n = 67), 19.7% with associate degree (n = 50), 35.3% with bachelor degree (n = 89), 14.2% with master's degree (n = 36), and 4.3% with doctoral degree (n = 11). Survey participants' daily Internet use was 13.0% at 1 hour or less (n = 33), 28.7% at 1–2 hours (n = 73), 18.1% at 2–3 hours (n = 46), 14.2% at 3–4 hours (n = 36), 11.4% at 4–5 hours (n = 29), and 14.6% at 5 hours or more (n = 37). In addition, participants' number of blogging acquaintances were 38.2% with 0–3 people (n = 97), 12.6% with 4–6 people (n = 32), 8.3% with 7–9 people (n = 21), 9.4% with 10–12 people (n = 24), 3.9% with 13–15 people (n = 10) and 22.4% with 19 or more people (n = 57).

Instrumentation

We measured the following key constructs to test the proposed effects of Internet use and relational density and the mediating processes on relational enhancement. Participants' psychological coping outcomes were measured with four measurement items that capture the emotional as well as relational aspects of coping (α = .94). Participants' physical coping outcomes were measured with six measurement items that focus on patients' health behaviors and treatments (α = .82). Internet use was measured by the daily use of Internet, and relational density was measured by the number of people the participant knew through online blogging. Relational enhancement through online social interactions was measured with six items that focus on the quality of relational experiences one has (α = .91). In all measurement, except the time of Internet use and the number of online blogging acquaintances, we used seven-point Likert-type scale ranging from 1 (*not at all*) to 7 (*very much*).

Analysis

To test the first set of hypotheses (H1a-b, H2a-b) and RQ1, we used multiple regression analysis. Internet use and relational density were entered in to separate regression equations as independent variables where we set two key coping outcomes as dependent variables—psychological coping outcome and physical coping outcome. To examine the possible effects of sociodemographics, we entered gender, age, education, and income as control variables in testing the hypotheses.

To test the second set of hypotheses (H3-H6) and RQ2, we specified a structural model for testing the hypotheses (Byrne, 1994; Kline, 1998). In the first step of the measurement phase, we analyzed and selected the best measurement items for three constructs with multiple measures (i.e., online relational enhancement, psychological coping outcome, and physical coping outcome) and then made a composite variable (item parceling) to test the specified model. To test RQ2, the effect of sociodemographics on key dependent variables, we also specified the sociodemographics from the first set of hypotheses (i.e., gender, age, education, and income) as predictors of psychological and physical coping. In testing models, we followed conventional model assessment criteria—Comparative Fit Index (CFI), Normed Fit Index (NFI), and Goodness-of-fit Index (GFI). In general, these three model test indices indicate the testing model as viable when the fit statistics turn out equal or higher than .90. Our structural model produced in this study reached viable data-model fits based on multiple-fit indices.

Results

Hypothesis testing and findings of research questions

Our first set of hypotheses (H1a-b and H2a-b) examined the effects of Internet use and online relational density through blogging activities on one's health coping outcomes of chronic illness. We used hierarchical Ordinary Least Square (OLS) regression and regressed each dependent variable of health coping outcomes on the two key independent variables of Internet use and relational density. To examine RQ1 and for control purposes, participants' demographics (gender, age, education, income) were entered into the regression models first. Two independent variables were then entered into the regression model (see Table 1).

In H1a and H1b, we expected no significant effect from increase of Internet use on health coping outcomes. As predicted, findings from H1a and H1b tests showed no significant regression coefficients to coping outcomes (See Table 1). Internet use ($\beta = .037$) is not significantly associated with the dependent variable of psychological coping or with physical coping ($\beta = -.023$), indicating that merely an increase in one's time use of the Internet cannot enhance one's health outcomes.

In H2a and H2b, in contrast, we expected significant effects. Our predictions were that one's increased relational ties (density) through online blogging activities can enhance both psychological health coping (H2a) and physical health coping outcomes (H2b) among chronic health problem solvers. As predicted, findings from H2a and H2b tests showed significant regression coefficients for coping outcomes. H2a, increase of relational density, was positively associated with psychological coping ($\beta = .482, p < .001$). H2b, increase of relational density, was also positively associated with physical coping ($\beta = .317, p < .001$). The findings indicate that individuals' efforts to increase online relational ties and interactions—networked sociability—will reward them with increased manageability of health outcomes both psychologically and physically (see Appendix for the specific aspects of health coping outcomes). This supports our original

Table 1. Regressions Predicting Psychological Coping Outcomes and Physical Coping Outcomes (*N* = 254)

	PSYCHOLOGICAL COPING OUTCOMES	PHYSICAL COPING OUTCOMES
Control Variables		
Gender (Female)	.102	−.038
Age	−.175**	−.182**
Education	−.010	−.131*
Income	.027	.019
R2 (%)	7**	7**
Main IVs		
Internet Use	.037	−.023
Relational Density	.482***	.317***
Incr. R2 (%)	24	9
Total R2 (%)	31***	16***

Note: Entries are standardized regression coefficients (final betas).
*** $p < .001$, ** $p < .01$, * $p < .05$

predictions, that simply increased use of digital communication media could not be a lone predictor of enhanced coping among users with chronic health problems. Yet, as one's digital communication efforts result in higher relational ties and more within-network interactions, this could help them enhance manageability of chronic health problems.

In RQ1, we also checked how sociodemographics would moderate the effects of relational density on health outcomes. As summarized in Table 1, we found no significant effects of gender on either health coping outcome. Similarly, income has no substantial effect on coping outcomes. However, age has significant effects on both outcomes—as one's age increases, their psychological and physical coping outcomes decrease: $\beta = -.175$, $p < .01$ and $\beta = -.182$, $p < .01$ respectively. For education, there is no significant effect on psychological coping, but we found higher education will be negatively associated with physical coping, $\beta = -.131$, $p < .01$.

Our second set of hypotheses examined a meditational process of how Internet use and one's degree of relational ties improve health coping outcomes via relational enhancement. In testing we paid special attention to *simultaneous effects* of presumed predictors through increase of one's satisfaction with online social connectivity and interactions while controlling the sociodemographics.

According to conventional structural model fit criteria, our final structural model reached good model fit (i.e., higher than .90). In addition to the key structural paths in the initial model, we specified direct paths from gender, age, income, and education to both of the dependent variables, psychological

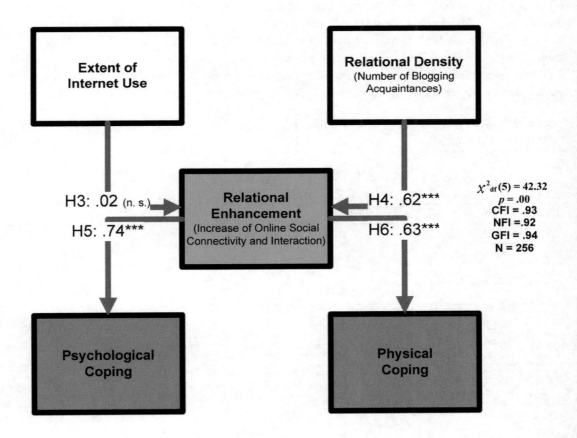

Figure 1. Result from hypotheses testing.

and physical coping. This model test was to examine RQ2, and results showed that no significant paths from any of sociodemographics to the two dependent variables at $p < .05$ level. The separate regression model tests suggest weak effects from age and education, but these effects in structural model testing would not be substantial enough to be significant. One possible reason is that as those key variables are simultaneously tested with new mediator variables, some spurious relationships that were not captured in a separate regression tests, have been clarified.[1]

Thus, we proceeded to construct a final model without specifying sociodemographics as control variables, and the final result is summarized in Figure 1. The final testing model in Figure 1 reached good model fit, and thus we interpreted the model parameter estimates to test the hypotheses. We posited four paths, the effects from Internet use to relational enhancement (H3), from relational density to relational enhancement (H4), and from relational enhancement to psychological coping outcome (H5) and physical coping outcome (H6).

In H3, we expected no significant effect because the mere increase of Internet usage will not necessary bring enhanced relational experience for users. As predicted, we found an insignificant effect (.02) of the estimated standardized path coefficient and thus conclude there is no effect of Internet usage on relational enhancement among chronic health problem solvers. In H4, we have a substantial effect from relational density to relational enhancement. H5 is important in particular because we can further break down the direct effect we observed from the first set of hypotheses (i.e., H2a–H2b). In other words, as we found the strong and significant effects through relational enhancement to both aspects of health coping outcomes among individuals with chronic health problems, we can now delineate the boundary conditions and specific theoretical process of how it works and to what extent they are substantiated through digitalized social relationships among health problem solvers.

The path of H4 was strong and significant, as the parameter estimated for the path was .62, p < .001. This indicates that the effect of increased relational ties in a digitalized communication environment will enhance one's online socializing and the level of rewards reaped from the formed online networks. Furthermore, the paths mediated via relational enhancement to psychological coping outcomes (H5) are also strong, .74, p < .001. Likewise, the path from relational enhancement to physical coping (H6) was also strong and significant, .63, p < .001. Overall, these paths suggest that increase of relational ties (blogging acquaintances) could help individuals with chronic health problems (e.g., depression, HIV/AIDS, arthritis, lupus) form and solidify personal relationships with friends and fellows with similar situations and allow them to feel empowered and earn some comfort from these interactions. The resulting enhanced relational interaction, that is networked sociability, can in turn increase one's coping capabilities in affective and physical dimensions.

In summary, from both sets of hypotheses and research questions, we found some evidence of the possibility and importance of collective effectuating of individualized health problems using digitalized communication technologies. Specifically, this will enhance their coping capabilities in dealing with their chronic illness when one's experiences of using the Internet or digitalized media can enable chronic health problem solvers to form and maintain personal relationships.

Conclusion

The purpose of this chapter was to investigate the phenomenon of cybercoping and to examine whether experiences of digitalized social relationships have positive consequences on managing chronic health problems. To do this, we conducted a survey of individuals with chronic health problems and tested the effects of Internet use and relational density on health coping outcomes. We examined whether enabled "networked sociability" helps individuals cope better with chronic health problems and then took a closer look at how and in what process cybercoping occurs.

A review of literature suggests little research on cybercoping and on the effects of digitally created social interaction—networked sociability—among chronic health problems. We took the perspective of individuals with chronic illness and found supportive evidence for cybercoping—digitalized collective effectuating of individualized health problems. This perspective is significant in that these issues have not been looked at from the *public's* (problem-solving groups of individuals with chronic diseases) perspective in previous health communication or public relations scholarship. Through this study, we found that the more one develops online relational ties, the more enhanced the relational experience and this in turn improves both psychological and physical coping outcomes.

Overall, this study demonstrates that social interactions in digitalized social settings can improve patients' health coping outcomes. Specifically, the results suggest that as chronic illness patients make efforts in forming and maintaining relational ties with others, they could build stronger social and emotional resources, thus allowing them to do collective effectuating—collective problem-solving efforts within social connectedness with fellow problem solvers. Collective effectuating then helps patients enhance affective competence and manageability of their chronic illness.

The findings from the current study support the presence and importance of networked sociability among individuals with chronic health problems and its underlying mechanism. In addition, the findings indicate that simple increase of Internet use among chronic health problem solvers would not yield better cybercoping. Instead, those individuals with chronic health problems will experience better coping outcomes if they make efforts to form and maintain better online relational density—increase ties and interactions—and further if this enhanced relational density increases relational experience within their interactions with their online personal networks.

In the current study, the importance of psychological coping as well as physical coping has been emphasized in the context of chronic diseases, especially because chronic diseases require a higher degree of long-term self-management, compared to other types of diseases. Interactive online platforms facilitated by communication technologies, that is blogging, and their effects as examined in our study have allowed patients to express their views and exchange information more easily than in pre-digitalization eras.

We believe our study findings shed light on the potential benefits of this technology within the health-care community as well as critically assess the boundary conditions for positive effects of digitalization and improving health. Many believe and assert that communication technologies and use of the Internet help decrease health disparities as they increase easy access to various health information. However, this study suggests that positive cybercoping occurs not because of a simple increase of accessibility or use of Internet as surfing tool but because of continual efforts to cultivate and maintain relationships with fellow health problem solvers. Thus, digital communication technologies (e.g., mobile, smart phone) and applications (e.g., social apps for collective efforts) may create and facilitate online communication environments where patients can not only search information, but also form and nurture personal relationships through their communicative behaviors.

Health care providers or policy makers might consider the findings from this study and foster these patients' communicative actions and relationship building efforts through information communication technologies. It is important to note that although people with chronic diseases are often considered to be individuals fighting an individual battle, this study showed support for the benefits of group problem solving and collective effectuating of these supposed individual problems. The communicative actions of information forwarding, information sharing and information seeking as outlined in CAPS (Kim, Grunig,

& Ni, 2010) are of particular relevance to becoming an engaged member of these digital publics and online communities. In addition, the ability to create and maintain these networks and publics through digital technology may signal an important shift from thinking about chronic health as an individual issue to thinking about it as a public organized around and attempting to deal with a common problem. This could in turn prompt new and different collaborations between the fields of public relations and health communication that vastly differ from their current health campaign-based relationship.

ENDNOTE

1   Structural equation modeling (SEM) approach is methodologically superior to conducting a series of separate regression analyses because it can identify and withhold spurious relationships among key variables (Bollen, 1989). In our tests, the SEM test integrates all variables with a new mediator and controlling variables, which enables to assess better conceptual influences among variables.

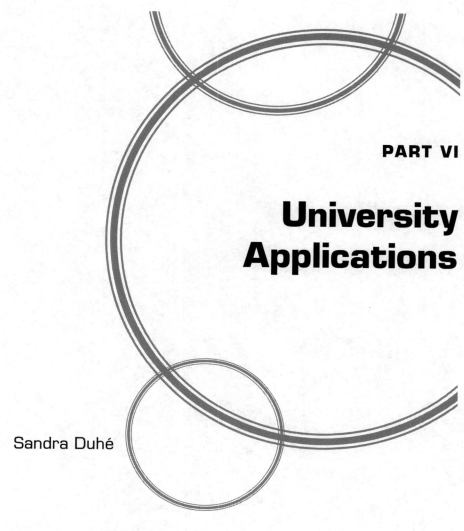

PART VI

University Applications

Sandra Duhé

Overview

Social media use is increasingly becoming standard fare in organizational communication strategies, and universities are no exception. This section examines social media on campus from two different perspectives: regulation and athletics.

Daradirek "Gee" Ekachai and David Brinker provide findings from their content analysis of social media policies enacted by top universities across the US. In the wording of these policies, they found administrators encouraged faculty and staff to use social media platforms and equally acknowledged the potential risks involved. A less equitable balance was struck, however, between conveyance of legal and ethical guidelines, and between the bounds of empowerment and accountability. Gee and David pose relevant questions about who should be involved in the creation of social media policies.

Chang Wan Woo and Wonjun Chung offer insight into sports public relations, an area of practice that is perennially popular with students. Here, they discuss how the relationship-building value of social media was perceived by three interrelated publics in NCAA athletics: athletic staff members, student-athletes, and students. Each group surveyed had an overall positive view of social media, but Chang and Wonjun used relationship cultivation strategies to delve deeper into how those perceptions differed depending on the parties involved in the relationship.

University Social Media Policies

A Content Analysis Study

Daradirek Ekachai
David L. Brinker, Jr.

This chapter investigates the content of internal social media policies published online by U.S. universities. The chapter is composed of three parts. It first examines the academic and professional literature about social media policy. The next section presents a content analysis that yields three categories of dual considerations embedded in social media policies: opportunity and risk, legal and ethical, and empowerment and accountability. Finally, the discussion section examines these categories by critically evaluating the apparent values reflected by the policy language. In particular, the discussion seeks to raise questions about how policies present the opportunities and risks associated with social media use. The ethical and legal discussion revolves around the concerns reflected in the policies, which seem to favor a brand-oriented reputation management approach, to the exclusion of ethical issues related to guidance for faculty and students' conduct online. Finally, the chapter examines how internal stakeholders (faculty and staff) are both empowered and held accountable as public communicators on behalf of the university.

Introduction

A television staff member in Utah, confusing the station's Twitter account for a personal account, lost a job over a tweet that said, "I'm downtown eating. Surrounded by Mormons and repressed sexual energy" (Bergman, 2010, para. 1). Virgin Atlantic Airways fired 13 flight attendants for Facebook messages criticizing its flight safety standards and using inappropriate language to describe passengers (Conway, 2008). Since the advent of social media, there have been instances of job terminations resulting from inappropriate or harmful status updates, images, videos, or comments that could harm the organization's reputation and business interests.

As social media have revolutionized information sharing and changed the way we communicate with each other, organizations have begun to embrace social media to connect, engage, and build relationships with both internal and external audiences. They are leveraging digital and social media to promote brands, increase sales, build reputation, and encourage employees' engagement and productivity. But with this open, free flow of information sharing comes legitimate legal, ethical and business concerns for employers, who want to ensure that their employees use social media legally, ethically, and productively.

Social media policy

A global survey of communication managers in 2007 found the majority of managers believed social media improve employee engagement and internal collaboration (Hathi, 2007). Despite the benefits and opportunities that social media present, management still holds reservations about their possible risks and negative consequences. One concern is that employees' online participation via blogs or other social networks may pose a risk to a company's reputation. Understandably, managers realize that failure to manage social media properly could result in managerial, ethical, and legal liabilities.

Concerns about workplace time and resource management, security, compliance, and reputation management have led some organizations to develop a social media policy to guide, define, and restrict employees' use of social networks. A social media policy serves a dual function of providing internal users with guidelines and boundaries and, simultaneously, empowering them with open access and recommendations or best practices to use social media intelligently and effectively. In light of the need to balance empowerment and accountability, the pressing challenge for organizations is to create a social media policy that respects the rights of employees while protecting their brands online.

University social media policy

Universities, like other business enterprises, employ social media platforms to communicate with a diversity of stakeholders: employees, faculty, current students, parents, alumni, and prospective students, to name a few. Adoption of social media by both faculty and staff continues to grow. A 2010 study, conducted by Babson Survey Research Group, found 59% of faculty respondents had more than one social network account, and nearly one-third used social media to communicate with students and peers ("Sociable Professors," 2010).

Online marketing firm Fathom SEO published a series of white papers examining social media best practices and found common social media standards among colleges and universities (Brady, 2011). Such practices included emphasizing personal relationship building and tailoring content to specific audiences. Brady suggested that a significant challenge facing current university social media practice is "humanizing" the organization to create messages "that connect with an audience on a personal level" (p. 4).

Lavrusik (2009) wrote that university-specific applications of social media included sharing and promoting faculty and student work, emergency notification, and publicizing the expertise of faculty. These specific benefits lent increased utility to social media use in the university context.

Higher education institutions operate under unique ethical and legal boundaries. Such issues as academic freedom, student privacy, FERPA, and HIPAA rules are uniquely important subjects. Universities have an interest in encouraging and maintaining an intellectual community by endorsing openness, freedom of expression, and transparency in their faculty and staff's social media participation. Yet, it is increasingly important for them to establish a social media policy or guidelines that will equip internal publics to represent the institution responsibly.

Addressing the unique legal challenges associated with social media in the educational context, Prairie, Garfield, and Herbst (2010) suggested the increasing importance of formal usage policies at higher education institutions:

> Because technology generally, and social media in particular, evolves at a rapid pace, it can be difficult for large institutions or entire school districts to remain current in terms of policies and monitoring. Like many aspects of the law, it is critically important not to underestimate the challenges that may present without a cohesive strategy that is updated on an annual basis. … [Social media] is also ripe for abuse without the existence of coherent policies. (p. 372)

Moreover, the organizational structures and cultural models of academic institutions may have an impact on the level of openness or locus of control suggested in the social media policy.

This chapter investigates university social media policies because of rapid growth in collegiate social media use. It analyzes the content of published social media policies from 30 U.S. universities to examine how these policies address employees' opportunity, risk, empowerment, and accountability. Ethical and legal areas such as student privacy, confidentiality, and FERPA will also be explored.

Literature review

As a response to concern over "cyberloafing," or lost workplace productivity, abuse of company resources, and potential legal liability, policies governing employees' Internet use have become fairly commonplace and received some attention from scholars (Henle, Kohut, & Booth, 2009; Young, 2010a). Although social media use at work is now routine, managing their use in the workplace has also proved challenging. The need to study policies governing their use is increasingly pressing. A 2011 study of international social media use conducted by employment service firm Randstad found one-half of surveyed employees maintained a social media account, and one-third received some form of guideline or policy on social media professionalism (Randstad Holding, 2011). In the Americas, a Manpower (2010b) study found the United States is the least likely to provide a social media policy, with 24% of employers reporting a formal policy. Perhaps the slow adoption of social media policies is related to the high rate of employers banning social media use entirely. Devaux (2009) reported 54% of surveyed chief information officers said their organizations banned social media, and 19% limited their use to "business purposes only" (p. 20).

Yet, unlike Web browsing behavior addressed by general Internet usage policies, social media are a potentially invaluable communication tool, though their use is associated with both significant advantages and liabilities. Competing literature portraying social media as a possible workplace risk and as a marketing opportunity provided context for this study.

Competing views of social media as risk and opportunity

Social media accounts often blur the line between personal and professional communication (Brus, 2011), and they enable employees to access mass audiences with ease. Rapid transmission and dissemination of content leaves businesses vulnerable to employees acting as informal but potentially influential representatives of the institution. Concerns over reputation management have been noted and studied by workforce professionals (Devaux, 2009; Manpower, 2010a). Some authors have approached social media policymaking as a solution to these problems, and as a means of addressing legal risk (Lyncheski, 2010; Maryott, 2010).

However, the extent to which a social media policy can restrict employee speech to protect employers' interests is unclear. For example, a complaint before the National Labor Relations Board involving a Connecticut paramedic, who was fired over a post on Facebook criticizing her supervisor, was resolved by a settlement in which "the company agreed to revise its overly-broad rules" ("Settlement Reached," 2011, para. 4). Although nothing prevents inclusion of a restrictive clause within a policy, this case suggested that it could be contested.

Opposite these concerns is the recognition that employees' online communications are useful to businesses seeking to engage audiences in a direct and organic relationship (Foley, 2010). The ability for employees to speak directly to mass audiences presents a powerful opportunity for more credible dialogue with audiences than traditional institutional communication outlets, albeit at the cost of the organization's central message control.

Further, social media are proving a valuable platform for strategic marketing (Campbell Laidler, 2010). From a public relations perspective, they are particularly well suited to Grunig's (2001) two-way symmetrical model. Grunig (2009) argued that digital media, properly used, can facilitate this relationship and dialogue-based paradigm: "If the social media are used to their full potential, I believe they will inexorably make public relations practice more global, strategic, two-way and interactive, symmetrical or dialogical, and socially responsible" (p. 1).

This opportunity-risk balance manifests in the language social media policy authors use to frame social media guidelines. Boudreaux (2009) analyzed 46 social media policies from various types of organizations and found that 37% of them portrayed social media as a positive opportunity for their employees and organization, while 15% focused on risk. Employers may find the opportunities enabled by social media becoming increasingly vital and expected, as the cost of entirely blocking social media use increases (Reeves, 2011).

Empowerment and organizations' social media cultural models

Scholars have found links between workplace Internet use practices, internal policy, and organizational culture (Czerniewicz & Brown, 2009; Young, 2010a). The social media-related cultural models examined in this study were proposed by social media thinkers from the Altimeter Group (Li, 2010; Owyang, 2010a). The models classify organizational culture by the arrangement of authority involved in message creation and distribution.

For the purpose of this study, the models were organized into four categories: Organic/Holistic, Coordinated, Centralized, and Multiple Hub-and-Spoke. The organic/holistic culture assumes that any employees in the organization can participate in social media space without formal approval from management. The centralized system uses a single unit, generally the marketing or corporate communication department, as the sole authority to produce, distribute, and control all social media efforts. The coordinated system allows various internal units to produce and distribute social media content but enforces standardization of content and style through training, education, and support. Lastly, the multiple hub-and-spoke model allocates authority to divisions or departments performing autonomously under a common brand.

Through social media policy analysis, this study attempted to examine to what extent universities empower faculty and staff in their social media activities.

Research questions

Petroff (2010) identified some of the most common key messages found in university social media policies: Authenticity and transparency, protecting confidential information, respecting copyrights, developing a social media strategy, respecting your audience, and obeying terms of service on specific platforms.

The present study focused on discovering common values or concerns expressed in university social media policies, and the extent to which the organization's cultural models were reflected in the content of those policies.

Specifically, it examined:

1. The common areas or topics addressed in social media policies.
2. The extent to which the policies addressed the employees' opportunity and risk associated with social media.
3. The extent to which the policies addressed legal and ethical concerns.
4. The extent to which the policies addressed the employees' empowerment and accountability.

Method

The 30 university social media policies examined in this study were selected from a ranking of the top 50 U.S. colleges published by the *U.S. News and World Report* in 2010. Of those, 18 published social media guidelines online. The other 12 policies were obtained via keywords in an Internet search, composing a convenience sample. The unit of analysis is the whole social media policy document. The length of the documents ranged from 320 to 14,276 words, with an average of 1,928 words.

Coding system

Using an inductive data reduction approach, both authors read all 30 policies to familiarize themselves with the content. The authors then organized the data into themes and categories. The categories were exhaustive and mutually exclusive. The authors reviewed and revised the coding categories until mutual agreement was reached. For the purpose of assessing the inter-coder reliability, each author coded one-third of the policies, selected at random. The average percentage agreement across 30 variables was 87%.

Coding themes and categories

Three dual themes emerged from 30 social media policies—Opportunity and Risk, Legal and Ethical Considerations, and Empowerment and Accountability. Opportunity indicators included categories on strategic usage guidelines, tactical usage guidelines, and specific platform guidelines. Risk indicators included categories on image or reputation concerns, acceptable content, loss of productivity, and caution about the public nature of social media.

The second dual theme that emerged from the content of university social media policies was legal and ethical considerations. The legal theme included language that alerted employees not to violate laws on copyrights, proprietary information, confidentiality, Family Educational Right and Privacy Act (FERPA), Health Information Portability and Protection Act (HIPAA), and libel and defamation.

Ethical considerations included required disclosure of university affiliation and roles, and observation of privacy standards. Ethical content guidelines also addressed respect, transparency, honesty, and accuracy.

The third theme involved employees' empowerment and accountability. Empowerment can be observed through the characterization of an organization's cultural model, reflected in its social media policy. Presumably, employees are empowered to the extent that policies authorize them to produce and distribute social media content. Indicators included whether employees were seen as agents or representatives of the university, and whether departments' social media accounts had to be formally approved by a central university authority. Accountability was indicated when policies addressed disciplinary measures for the violation of social media guidelines. It was also implied when universities disclaimed responsibility for online posts or placed liability on the individual user, and when the university retained a right to remove online content.

Findings and analysis

The present analysis was primarily concerned with the value dimensions apparent in social media policies. Descriptive differences in content seemed to fall within three dual social media usage concerns: proactively highlighting opportunities and preemptively mitigating risks, the interests represented in legal and ethical considerations, and the degree to which a policy was interested in empowering or imposing accountability standards on users. These categories elaborated studies of internal policy comparing "proactive" and "reactive" approaches to limiting undesirable Internet use (Young, 2010b). Policies concerned with the opportunities of social media, addressing ethical issues, and creating a culture of empowerment can be characterized as proactive, while risk management, legal concerns, and disciplinary considerations seemed generally reactive.

Opportunity indicators

To encourage social media use, many internal social media policies, guidelines, and handbooks included guidance for use of social platforms. Our research showed both tactical (76.7%) and strategic-level (63.3%) guidelines were commonly provided to give instruction (see Table 1). Strategic uses of social media generally guided users in managerial decisions such as selection of audiences and articulation of problems to be solved with a social media tool.

For example, Dartmouth University dedicated the majority of its policy to this sort of strategic support. The policy offered a six-step strategic plan, encouraging users to "Define Purpose and Goals" (para. 3), "Choose a Tool" (para. 4), "Consider Existing Options" (para. 5), "Commit Resources" (para. 6), "Assign Responsibilities" (para. 7), and "Track Usage" (para. 8). The university framed social media as a communication tool for internal users to employ proactively by recommending that employees set communication goals and measurement standards, and maintain ongoing conversation. By providing a framework for creating a goal-driven social media presence, the university suggested to users that online outreach is encouraged and supported.

Tactical-level guidelines offered general execution tips, such as recommending a timetable for updating content. Although 85.7% of policies analyzed in this study referred to specific social media platforms (e.g. Twitter, Facebook), only 35.7% provided guidance specifically for those outlets—the remainder not distinguishing between platforms. The University of Massachusetts at Boston, for example, suggested messages should be responsive, stating "you should do your best to reply in a timely and appropriate fashion" (para. 4).

Table 1. Opportunity Indicators (N=30)

INDICATORS	N	%
Tactical Usage Guidelines	23	76.7
Strategic Usage Guidelines	19	63.3
Platform-Specific Tactical Guidelines	11	36.7

Risk indicators

Language expressing concern over potential damage to the institution's public reputation was very common (90%) among the examined policies (see Table 2). This reflects, perhaps, the long-standing public relations notion of "uncontrolled" messages. Most (86.7%) of the policies defined unacceptable types of speech, such as obscenity or confrontational language. Social media policies also made frequent (80%) references to the idea that Internet communications are virtually indelible, public, and highly searchable. Tufts University, for example, cautioned users about the potential risks involved with using social media. The policy stated:

> There's no such thing as a truly "private" social media site: search engines can turn up posts years after the publication date, comments can be forwarded or copied and archival systems save information even if you delete a post.... Once you publish something through social media, you lose a degree of control of your message. Be certain before you post something that you are prepared to share it with a potential audience of millions. (para. 8)

Concerns regarding loss of productivity were included in 17 out of 30 policies (56.7%), expressing the need for responsible use of technology/computer resources, time management, or the use of resources for work purposes only. Many policies included similar language addressing respect for university time and property. University of Illinois specifically suggested users "Set time limits as part of your daily routine. Think about what tasks you're not completing while using social media to help ensure your use of time is wise" (para. 4).

Table 2. Risk Indicators (N=30)

INDICATORS	N	%
Image/Reputation Concerns	27	90.0
Defines Acceptable Content	26	86.7
Public Nature of Social Media	24	80.0
Loss of Productivity	17	56.7

Legal indicators

Though social media generally facilitate the rapid exchange of information, the potential legal hazards of online communication are quite apparent. The legal concerns were generally addressed by existing University policies, and social media specific guidelines tended to refer back to these familiar standards.

A majority of the policies examined in this study addressed copyright (76.7%), protection of proprietary information (70%), and confidentiality (60%). One area that did not receive much attention was defamation (46.7%) (see Table 3).

The University of Notre Dame offered a concise summary of legal boundaries associated with social media communication:

> Notre Dame will not tolerate content that infringes on proprietary information, or that is pornographic, libelous, defamatory, harassing, or inhospitable to a reasonable work environment. The Office of Public Affairs will not pre-screen content, but it shall have the right to remove any content that is considered to violate content policies. (para. 17)

Table 3. Legal Indicators (N=30)

INDICATORS	N	%
Copyright	23	76.7
Proprietary Information	21	70.0
Confidentiality	18	60.0
FERPA	16	53.3
Defamation	14	46.7
HIPAA	10	33.3

Ethical indicators

A majority of the policies analyzed addressed ethical considerations in one way or another. Eighty percent of the policies addressed the importance of preserving privacy online, and the importance of disclosing an affiliation with the university. Other ethical concerns involved guidelines for online writing: to respect others (66.7%), accuracy (50%), transparency (46.7%), and honesty (33.3%) (see Table 4).

The social media guide for the University of California at Fresno, recommended honesty, disclosure, and transparency as ethical considerations:

> Your honesty, or dishonesty, will be quickly noticed in the social media environment. If you are blogging about your work at Fresno State, use your real name, identify that you work for Fresno State, and be clear about your role. If you have a vested interest in something you are discussing you should be the first to point it out. (para. 8)

Table 4. Ethical Indicators (N=30)

INDICATORS	N	%
Required Disclosure of Affiliation	24	80.0
Privacy	24	80.0
Respect	20	66.7
Transparency	14	46.7
Accuracy	15	50.0
Honesty	10	33.3

Empowerment indicators

The data analysis revealed that 86.7% of university social media policies empowered internal social media users to be agents or representatives of the university. Further, the majority (53.3%) of social media policies analyzed in this study seemed to reflect the multiple hub-and-spoke model, where the universities allow departments, offices, and student groups to perform and manage their own social media activities as long as they have an approval from immediate supervisors. One-third of the policies followed the coordinated cultural model, where a cross-functional team keeps various units' social media activities organized. Only 13.3% of policies had an open, organic/holistic system that assumes every employee acts as a representative and no formal approval is required. Interestingly, none of the universities followed the centralized model by designating one single department to manage all social media efforts (see Table 5).

University of Michigan's policy is hub-and-spoke oriented because the authority for social media activity is located within sub-units: "Any messages that might act as the 'voice' or position of the university or a school/college/unit must be approved by the university or the director of the school/college/unit or their delegate" (p. 5, para. 2).

By contrast, Albany Law School's policy is an example of a policy coded as a coordinated model:

> Notify the law school: Departments or law school units that have a social media page or would like to start one should contact the Office of Communications and Marketing at Albany Law School to ensure all institutional social media sites coordinate with other law school sites and their content. All institutional pages must have a full-time appointed employee who is identified as being responsible for content. Ideally, this should be the head of the department. (p. 3, para. 3)

This policy suggested that, while the university coordinated the social media activities of its various departments, the responsibility for content and management resided within each autonomous functional unit.

Table 5. Empowerment Indicators (N=30)

INDICATORS	N	%
Users as Agents	26	86.7
Formal Approval	25	83.3
Hub-and-Spoke	16	53.3
Coordinated	10	33.0
Organic/Holistic	4	13.3

Accountability indicators

The findings revealed that only 23.3% of social media policies addressed disciplinary actions against employees who violated social media guidelines. Along the same line, only one-third (33.3%) of the policies stated that the university would remove social media content deemed inappropriate. However, indirect accountability was common, with 63.3% of policies suggesting that users were personally liable, or that the university disclaimed responsibility, for content (see Table 6).

The University of Kentucky's accountability-oriented policy dedicated a full section to "Sanctions" for violations including revocation of rights and privileges, fines, corrective sanctions including termination, and "any combination of disciplinary action, or civil or criminal liability" (p. 5, para. 5).

Table 6. Accountability Indicators (N=30)

INDICATORS	N	%
User Liable for Content	19	63.3
Remove Content	10	33.3
Disciplinary	7	23.3

Conclusion

Opportunity and risk

Social media policies often approach social media use, appropriately enough, as a tool for leveraging stakeholder relationships and accomplishing communication objectives. Such policies are therefore designed to share best practices with internal users to convey opportunities social media represent. Yet, enthusiasm for encouraging social media use is tempered by the very real risks of using social media as a pseudo-official channel. Concerns over loss of control over message and institutional image, as well as legal considerations, can manifest in restrictive policy language. Goldstein (2011) referred to these alternative policy attitudes as the "Please Don't Damage My Brand Approach" and the "Social Can Help Drive Business Approach" (paras. 4, 5). One theme identified in this study was the dual attempt to use social media policies to leverage opportunities while mitigating risks.

Legal and ethical considerations

Organizations offer legal guidelines to protect both the institutions and employees from the explicit threat of legal actions resulting from violations of particular laws (e.g., copyright, defamation).

In this study, concern for the ethical behavior of internal social media users seemed to relate primarily to the idea that individuals represent the organization and can potentially inadvertently affect stakeholder opinion. Unlike legal standards, which are generally well formed and clearly actionable, ethical standards such as "honesty" and "respectfulness" are more akin to reputation issues. These concepts are therefore framed by policymakers as opportunities for more effective communication or as standards of professionalism, rather than mandatory behavioral codes.

This ethical approach seems insufficient in addressing the unique nature of universities. For example, student privacy is a central concern at academic institutions. Identity and attribution are also central to the academic ethos and culture (see, for example, Hyland, 1999). Although social media policies seemed attuned to the unique legal considerations of the university (e.g., FERPA), they seemed to lack a corresponding sensitivity to the ethical standards that internal stakeholders, particularly faculty and students, need for defining certain behavioral boundaries. This imbalance may be a result of the origins of social media policies, often crafted, managed, and enforced by marketing departments. Faculty needs to have an active voice not only in the use of social media, but also in crafting the policy governing it.

A purpose of this book is to identify how new media challenge traditional public relations practices. The social media shift in defining who can act as institutional communicator carries with it significant implications. One of these issues is the need for expanding the understanding of who is involved in the creation of policy. Universities might take under consideration the pedagogical and academic ethical issues that need to be addressed by an academic body. Whereas brand management seemed to be the focus of social media policies, these concerns should not exclude the specific ethical concerns of the university context.

Empowerment and accountability considerations

This study's findings that showed high empowerment indicators but low percentages on accountability seem to indicate that universities are more likely to focus on empowering employees rather than holding them accountable.

For social media success in any organization, striking the right balance of empowerment and accountability is critical, but not easy. If the university only focuses on empowerment, it may no longer control coherent messages it intends to convey to its publics. But if the institution aggressively puts control and accountability measures in place, empowerment is diminished. A good internal social media policy clearly defines what employees can and cannot do online, while protecting the institution by setting boundaries and accountability measures. A good policy also empowers employees, enabling them to use social media tools responsibly and with confidence. Empowering policies can enhance employees' authority and voice, engender loyalty, and may increase work productivity. However, designing social media policy without addressing accountability measures may not deter misconduct online. This study found that most social media policies that embraced coordinated and multiple hub-and-spoke models seemed to empower internal audiences with almost instant authority in social media efforts. Although that level of empowerment is refreshing and welcome, some attention to employees' responsibility and accountability is recommended to protect the university's brand and reputation.

New media have opened the doors to a broader understanding of how news is gathered and distributed, including conversation about the viability of "citizen journalism." Just as the technology has, in a sense, empowered the general public to act as journalists, so it seems to enable an organization's internal publics to act as de-facto public relations representatives. This is a concern from a public relations perspective. This content analysis found that 87% of policies implied employees were agents of the university. The authors suggest this assumption needs more reflection. Is it reasonable to expect that untrained employees will complement the efforts of university public relations professionals, when "citizen journalism" is still not regarded as equivalent to professional journalism? These questions of credibility, public relations objectives, and control of messages are considerations underlying the decision to empower employees as public communicators.

Limitations and suggestions for future studies

Caution is due in interpreting the results of this study because of its sampling technique. The 30 policies analyzed in this study were exclusively published online. It is quite likely that publically available internal policies differ from those that are unpublished. The authors recognize that any relatively small convenience sample is not necessarily representative. Additionally, it should be noted that the language in published social media policies is redundant among policies. For example, the language in the West Virginia University and University of Massachusetts, Boston policies was very similar. Thus, the present study analyzed the language that has circulated, rather than independent attempts to create social media policy. However, this study lays groundwork for an emerging literature that investigates internal social media use.

Although the present study is limited by the infancy of social media policies, future studies may take this chapter as a point of departure for differential analysis of emerging trends in internal policy and social media use. For example, the government, not-for-profit, and corporate contexts may yield categorically different policy themes. The origins of these differences are worth investigating. This study suggests that the creation of social media policy within university marketing departments shapes their content. The typical creator of policies may vary by organization type, resulting in varying emphases within the policy.

Future research should continue examining social media use in the university context. Questions about social media use in the classroom are increasingly salient. How do professors create relationships online with students? How do universities presently handle ethical concerns arising from social media interactions? To what extent is social media communication recognized as a form of speech under the protection of academic freedom? Addressing these questions will become increasingly important as public relations practitioners in academic institutions consider turning over control of messages to internal stakeholders who act as de facto brand representatives.

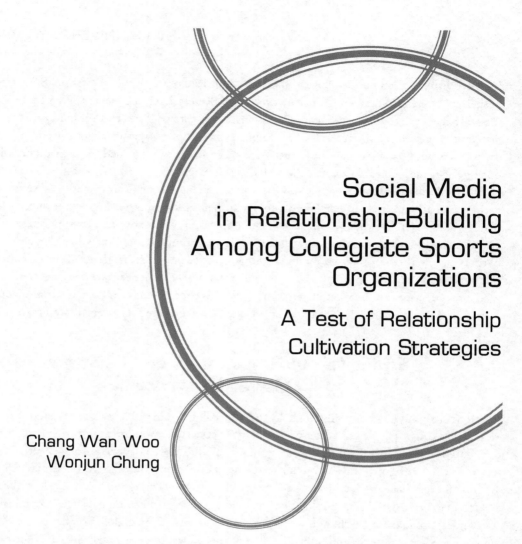

Social Media in Relationship-Building Among Collegiate Sports Organizations

A Test of Relationship Cultivation Strategies

Chang Wan Woo
Wonjun Chung

Today, sports organizations and players use various social media platforms to update their fans and promote parasocial relationships between players and fans. The purpose of this study was to further investigate how various publics around athletic departments value the use of social media as a relationship development tool. A total of 474 respondents (23 staff, 111 student-athletes, and 340 students of three National Collegiate Athletes Association universities) participated in this study. All three publics were favorable to the use of social media by athletic departments and student-athletes as an effective means of relationship management. Particularly, staffs were the publics who regarded the use of social media as a relationship management tactic the most highly among the three groups tested.

Marvin Austin and Greg Little of the University of North Carolina (UNC) were kicked off of the University's football team because of improper contact with an agent ("Marvin Austin Breaks Silence," 2010). The investigation was initiated because their tweets about a party hosted by the agent were seen by the National Collegiate Athletes Association (NCAA). The bad news was that UNC had to give up their best players after the investigation. The good news was that UNC and NCAA were able to recognize one possible illegal action by student-athletes and to investigate the incident by themselves.

Some schools ban players from using social media accounts. Randy Shannon, the former head coach of the University of Miami football team, ordered his players to shut down their Twitter accounts because tweeting is a potential distraction to players' performance ("Miami Hurricanes Players Ordered to Shut

Down Twitter Accounts," 2010). Rick Stansbury, the head coach of the Mississippi State University men's basketball team, also banned the use of Twitter by players after one player used it to criticize his coaching staff and exchange negative comments with a fan (Stevens, 2011). Problems with social media are not limited to college athletes, but occur in professional sport organizations as well. For example, Chad Ochocinco, who is a wide receiver for the Cincinnati Bengals, was fined by his team because he tweeted during the preseason game against the Philadelphia Eagles in 2010 (Star, 2010).

Considering how many people now use social media platforms, it might be wise for sports organizations and athletes to use social media to reach their fans and other important publics (e.g., sponsors, players, staff, etc.). We then wonder if collegiate sports organizations and their publics (student-athletes and students) consider social media to be an effective communication tool when they use it. The purpose of this chapter, therefore, is to offer an examination of social media in terms of the definition of effective public relations and to investigate whether the publics of collegiate sports organizations consider social media an effective means of public relations. We will then discuss the role of social media in collegiate sports organizations based on a survey study of publics of three NCAA affiliated institutions.

Effective public relations and relationship cultivation strategies

Heath (2001) said good organizations do and say well. He said public relations could assist an organization by making it good, which is a prerequisite to effective communication with its constituencies. He also noted that public relations, like any other discipline, does not seek absolute truth, but helps people discover the truth and draw a conclusion using rhetoric or storytelling. This role of public relations within an organization—facilitating conversation between the organization and its publics—is crucial. Pearson (1989) insisted that public relations practitioners need to manage the communication system and set the standard by exchanging dialogue with publics. This dialogic exchange would let organizations share views on issues and intermingle opposing views. In public relations, dialogic exchange would mean, "communicating about issues with publics" (Kent & Taylor, 2002, p. 22). Kent and Taylor said organizations' approach to dialogue requires their commitment to and acceptance of interpersonal relationships among publics.

Grunig, Grunig, and Dozier (2002) tried to find public relations effectiveness within organizations. They said the value of public relations does not come in a short period. It is a long-term process, so softer indicators, such as relationship building, best measure the value of public relations. Hon and Grunig (1999) developed those soft indicators to measure the value of public relations: (a) control mutuality, (b) trust, (c) commitment, (d) satisfaction, (e) exchange relationships, and (f) communal relationships. Hon and Grunig said that organizations often fall short of their goals because they do not build good relationships with their constituencies and that the value of public relations could be measured by the quality of relationships the public relations department helps the organization to establish. They also applied an interpersonal relations study (Stafford & Canary, 1991) to suggest six relationship cultivation strategies: (a) access, (b) positivity, (c) openness, (d) sharing of tasks, (e) networking, and (f) assurance.

Access is defined as "a strategy that a party (either a public or an organization) uses to reach the other party and express or share opinions and thoughts" (Ki & Hon, 2009, p. 6). *Positivity* is an enjoyable relationship. Publics obtain benefits from relationships with organizations, which in turn make relationships more enjoyable. *Openness* indicates sharing of thoughts. To build a satisfying relationship, organizations and publics need to share their thoughts and concerns openly and honestly. *Sharing of tasks* shows how to solve problems related to both organizations and their publics and how to eventually develop mutually beneficial relationships between the two parties. *Networking* means sharing relationships. Organizations can build networks "with the same groups that their publics do, such as environmentalists, unions, or community groups" (Ki & Hon, 2009, p. 9). Lastly, *assurance* reflects the notion of commitment. Organizations need to assure that they will make an effort to handle what their publics care about in order to maintain relationships with their publics.

Relationship cultivation strategies and the use of social media at sports organizations

The relationship cultivation strategies can be applied to the use of social media within sport organizational settings. Social media provide new ways of communicating and relationship building (*networking*). Sports organizations, players, and fans, of course, enjoy this new phenomenon. In this context, *access* can be defined as a strategy that sports organizations or players use to reach their fans and share opinions and thoughts with them. Currently, almost all of the major North American sports leagues such as the NFL, Major League Baseball (MLB), National Basketball Association (NBA), and National Hockey League (NHL) and their players now have stand-alone websites, blogs, Facebook pages, and/or Twitter accounts (Lewis & Kitchin, 2010). For example, by May 1, 2011, Chad Ochocinco had about 2 million followers on his Twitter account (http://twitter.com/ochocinco) and had more than 1.6 million Facebook fans (http://www.facebook.com/OchoCinco).

As an example of *positivity*, Woo, An, and Cho (2008) found that MLB team message boards provided space for fans to enjoy conversation. Ochocinco often shared pictures from the practice field for fun. And, his social media sites contained conversations between him and his fans along with his daily thoughts, which is called *openness*. An example of *openness* and *sharing of tasks* is how former Boston Red Sox pitcher Curt Schilling used his blog when he was attacked by the media about allegedly faking bloody socks during the 2004 American League Championship Series (ALCS) against the New York Yankees. Schilling posted "Ignorance has its privileges" and tried to defend himself to fans instead of talking through traditional media. Sanderson (2010) looked at Tiger Woods' official Facebook page after his marital infidelity crisis and found that fans were expressing anxiety and worries. However, they also defended Tiger Woods' infidelity as a human failing. Sanderson's findings were in contrast to press reports, which usually "reinforced problematic stereotypes and fostered perceptions that athletes' extramarital behavior is newsworthy" (p. 448). Sanderson concluded that social media sites are good public relations tools for athletes.

In terms of *assurance*, sports organizations use social media to respond to fans as well. Pegoraro (2010) followed top-tweeting athletes' Twitter accounts for one week and found that in most sports, more than 17% of tweets were responding to fans. More than 50% of NFL and NBA tweets were fan responses. Furthermore, most athletes frequently wrote about their personal lives, including pop culture. NFL Commissioner Roger Goodell challenged his staff to find new ways of building long-term relationships with fans using various communication channels. As a result, the NFL's vice president of corporate communication began tweeting about behind-the-scene issues (Fisher, 2009).

Collegiate athletic departments also utilize social media. Steinbach (2009) examined the social media revolution in the Big Ten conference and South Eastern Conference (SEC), interviewing several experts who helped those NCAA conferences launch their social media sites. The experts indicated that college sports fans are very passionate and tribal, and they share their passion on social media sites. The SEC was harshly criticized when it banned fans from using social media in its stadiums. However, they quickly compromised to allow fans live blogging, tweeting, and video recording using mobile phones. It brought lots of fans to their social media sites, and the SEC has been successful using social media. Steinbach (2010) said that athletic departments view social media as a form of word of mouth and free advertising. He used the example of the University of Utah football team, which had 500 tickets to sell. The assistant director sent press releases to local newspapers and television stations and posted a notice on the team's Facebook page. The 500 tickets were sold out by the time the press release reached the traditional media outlets. Steinbach also examined the University of North Dakota (UND) athletic department's use of Facebook, Twitter, and YouTube to promote the men's and women's basketball teams. The teams played at the skating rink UND's high profile hockey team used as their home stadium. As such, the basketball teams posted pictures and videos of players riding a bus to the famed arena and bouncing the balls on the ice. This played a big role in selling their tickets.

Although social media can have positive outcomes for sports organizations, there are reasons for universities to be concerned about social media. Some schools have banned student-athletes' use of social media. Similarly, parents worry about potential controversies due to the use of social media (Dinich, 2009). Dinich interviewed directors and coaches of athletic departments of several universities including Penn State University, University of Virginia, and the University of New Mexico. He found many staffs tried to monitor student-athletes' social media sites, implement social media guidelines for their student-athletes, and educate them to be careful about their contents.

Student-athletes serve multiple roles in universities, as their name indicates. They are students who are not only getting an education from their universities, but also representing their universities as athletes. They often spend more than 40 hours a week in practice and still need to manage their classes (Comeaux, 2010). Comeaux said, "Unlike other students, Division I college student-athletes are burdened with many sport demands and expectations that pose grave challenges to their learning and personal development" (p. 391). Student-athletes use social media like their non-athlete classmates. However, their use of social media is often monitored by their coaches (Dinich, 2009; Ruppenthal, 2010). Dinich's interview with former Florida State University safety, Myron Rolle, indicated some characteristics of social media use by student-athletes. Rolle had 4,971 Facebook friends, and he clearly understood the risks of social media. He explained that he was very cautious about his content during his social media activity. Rolle said that when he posted his picture on Facebook, "I'd make sure I'd look at it first and say, 'OK, is this picture OK? Would my mother appreciate this picture?'" (Dinich, 2009, para. 5).

Clapton and Finch (2010) investigated sports team identification of college students at NCAA Bowl Championship Series (BCS) schools. They found students who had higher sports team identification experienced a greater sense of social capital at their universities. That is, the stronger the student fan, the more they trusted the university and felt a greater sense of belonging to the school. They wanted to be Facebook friends with student-athletes, like Myron Rolle, and wanted to be one of the then 99,000 followers of Skylar Diggins (@SkyDigg4), a star basketball player at Notre Dame University. And, like fans of Tiger Woods' official Facebook page, this public would defend student-athletes on their social media sites in crisis situations.

Do social media really help sports organizations to communicate with their publics? Lewis and Kitchin (2010) thought that there is still a long way to go. Favoriso (2008) argued that most sports managers still treat the Internet as a tool to generate short-term revenue and serve as a repository of information. He was concerned about sports managers' mindset of short-term gain rather than long-term relationship building. Taylor and Kent (2010) expressed worries about addressing the effectiveness of social media without specific evidence. We also wondered if collegiate sports organizations could use social media to build and cultivate relationships with their publics.

There are some indirect suggestions in the literature about the contributions of social media to relationship cultivation (see Ki & Hon, 2009). As discussed, sports organizations use Twitter to respond to their fans (access and assurance). Tiger Woods let fans talk about his crisis situation on his official Facebook page (openness and sharing of tasks). Sports fans have spaces such as message boards and Twitter to network with other fans (networking) and share their passions (positivity). However, these concepts of relationship cultivation have not been fully examined within the context of sports public relations. Therefore, we posit our research questions here:

RQ1: To what extent do publics (staff, student-athletes, and students) of collegiate sports organizations consider social media a tool for relationship cultivation?

RQ2: Are there any differences among perceptions of publics of collegiate sports organizations about the use of social media?

Methods

Questionnaire development

The questionnaire for this study was based on Ki and Hon's (2009) work. Their questions were modified to apply to the relationship between sports organizations and their publics. Our questionnaire also reflected all six indicators of relationship cultivation: access, positivity, openness, sharing of tasks, networking, and assurance.

We created different questionnaires for different publics. That is, we designed questions for the staff-player relationship, staff-fan relationship, and player-fan relationship. We also had questions about student-athletes' social media use that measured relationship strategies such as positivity and openness between players (2 questions), players and fans, and staff and players. After making adjustments to reduce the total number of questions, we had 24 questions for athletic department's social media use and 25 questions for each public's (students and athletes) social media use. All questions were measured by a 7-point Likert scale (1=strongly disagree to 7=strongly agree). A March 2011 pretest with students at a large public university in the South produced acceptable Cronbach alpha values higher than 0.70 (with the exception of openness of staff-fan relationship at 0.62)(Baxter & Babbie, 2004),[1] and students had no difficulties completing the survey.

Survey procedure

In April 2011, we collected data from athletic departments, student-athletes, and students at two large southern universities (NCAA Division I) and one Midwest university (NCAA Division III). Students received extra credit for completing the survey.

Sample

A total of 504 participants completed the survey, resulting in 474 surveys being suitable for analysis. Of the 474 participants, 4.9% (23) were staff, 23.4% (111) were student-athletes, and 71.7% (340) were students. There was a diversity of gender, age, and race among respondents in each of the publics surveyed.

Results and discussion

What do staffs think about social media?

All three universities sampled in this study had a Facebook, Twitter, and YouTube account. As Table 1 shows, athletic staffs generally favored the use of social media for relationship cultivation. In particular, staff members thought their athletic department's use of social media would help staff and players to network ($M = 5.8$, $SD = 1.2$) and enhance positivity in the relationships between staff and fans ($M = 5.61$, $SD = 1.3$). Staff members did not strongly agree with the openness of student-athlete's social media use for either staff-player relationships ($M = 4.78$, $SD = 1.8$) or player-fan relationships ($M = 4.57$, $SD = 1.8$). This result is understandable, considering the recent Twitter ban for Mississippi State University men's basketball team (Stevens, 2011) and the University of Miami's football team ("Miami Hurricanes Players Ordered to Shut Down Twitter Accounts," 2010).

Overall, however, staff members agreed that social media use by both their athletic department and student-athletes would help build relationships with their publics, as each relationship indicator scored

Table 1. Report of Means for Staffs' Perception of Use of Social Media ($N = 23$)

ATHLETIC DEPARTMENT'S USE OF SOCIAL MEDIA

	STAFF-PLAYER		STAFF-FAN		D.F.	T
	M	S.D.	M	S.D.		
Access	5.13	1.81	5.00	1.24	22	0.39
Positivity	5.33	1.63	5.61	1.31	22	-0.83
Openness	5.11	1.81	5.39	1.35	22	-0.80
Sharing of Tasks	5.30	1.52	5.43	1.19	22	-0.47
Networking	5.80	1.19	5.46	1.33	22	1.94
Assurance	5.33	1.69	5.37	1.32	22	-0.14
Total (Relationship Cultivation)	5.33	1.47	5.38	1.14	22	-0.17

STUDENT-ATHLETE'S USE OF SOCIAL MEDIA

	STAFF-PLAYER (OR PLAYER-PLAYER)		PLAYER-FAN		D.F.	T
	M	S. D.	M	S. D.		
Access	5.21	1.57	5.13	1.48	22	0.43
Positivity	5.26	1.50	4.91	1.79	22	1.22
Openness	4.78	1.84	4.57	1.84	22	1.39
Sharing of Tasks	5.10	1.74	5.20	1.78	22	-0.38
Networking	5.41	1.60	4.98	1.68	22	1.75
Assurance	5.50	1.47	5.13	1.50	22	1.69
Total (Relationship Cultivation)	5.21	1.51	4.99	1.55	22	1.32

Note. 1 = Strongly disagree, 7 = Strongly Agree

higher than neutral. Paired sample t-tests did not find any statistically significant differences between athletic staff perceptions about the use of social media for staff-player, staff-fan, or player-fan relationships.

What do student-athletes think about social media?

As indicated by the total relationship cultivation scores shown in Table 2, student-athletes thought their athletic department's use of social media would benefit staff-player relationships ($M = 5.13$, $SD = 1.2$) slightly more than staff-fan relationships ($M = 5.01$, $SD = 1.1$), and a paired sample t-test revealed that this difference was statistically significant (t (110) = 2.6, $p < 0.05$). Specifically, student-athletes thought their athletic department's use of social media would enhance accessibility between staff and players ($M = 5.1$, $SD = 1.3$) more so than between staff and fans ($M = 4.95$, $SD = 1.3$). As a result, this difference was also statistically significant (t (110) = 2.3, $p < 0.05$). In addition, student-athletes more strongly agreed that their athletic department's use of social media would help in providing assurance between staff and players ($M = 5.23$ $SD = 1.3$) more than between staff and fans ($M = 5.03$, $SD = 1.3$). This difference was also statistically significant (t (110) = 2.4, $p < 0.05$).

Table 2. Report of Means for Student-Athletes' Perception of Use of Social Media ($N = 111$)

ATHLETIC DEPARTMENT'S USE OF SOCIAL MEDIA

	STAFF-PLAYER		STAFF-FAN		D.F.	T
	M	S. D.	M	S. D.		
Access	5.10	1.28	4.95	1.28	110	2.26*
Positivity	5.30	1.23	5.23	1.18	110	1.18
Openness	5.11	1.25	5.08	1.24	110	0.44
Sharing of Tasks	4.99	1.28	4.83	1.25	110	2.03*
Networking	5.06	1.35	4.99	1.26	110	1.15
Assurance	5.23	1.29	5.03	1.26	110	2.40*
Total (Relationship Cultivation)	5.13	1.17	5.01	1.13	110	2.56*

STUDENT-ATHLETE'S USE OF SOCIAL MEDIA

	STAFF-PLAYER (OR PLAYER-PLAYER)		PLAYER-FAN		D.F.	T
	M	S. D.	M	S. D.		
Access	4.96	1.24	5.05	1.17	110	-1.57
Positivity	5.26	1.22	5.22	1.24	110	0.68
Openness	4.92	1.43	4.96	1.23	110	-0.47
Sharing of Tasks	4.99	1.19	5.14	1.26	110	-2.44*
Networking	5.25	1.19	5.13	1.22	110	1.65
Assurance	5.14	1.18	5.22	1.20	110	-1.03
Total (Relationship Cultivation)	5.09	1.11	5.12	1.11	110	-1.03

Note. 1 = Strongly disagree, 7 = Strongly Agree
* p < 0.05

For student-athlete's social media use, student-athletes generally thought social media would help build relationships with both staff (M = 5.09, SD = 1.1) and fans (M = 5.12, SD = 1.1). The only significant difference found was in sharing of tasks. Student-athletes expressed that sharing of tasks between players and fans (M = 5.14, SD = 1.3) would be more effective for relationship cultivation than the sharing of tasks between staff and players (M = 4.99, SD = 1.2) (t (110) = −2.4, p < 0.05).

Overall, student-athletes did not regard social media for relationship building as highly as staff members did, though they still supported their use. Student-athletes generally favored social media in terms of relationship building with their publics. Especially, student-athletes indicated that athletic departments' use of social media would improve accessibility and assurances in relationships between athletic departments and student-athletes.

What do students think about social media?

Total scores for relationship cultivation in Table 3 show that students thought athletic departments' social media use would benefit their relationship with fans (M = 4.72, SD = 1.3) more so than with players (M = 4.4, SD = 1.3), and this result was statistically significant for each of the relationship cultivation

Table 3. Report of Means for Students' Perception of Use of Social Media (*N* = 340)

ATHLETIC DEPARTMENT'S USE OF SOCIAL MEDIA

	STAFF-PLAYER		STAFF-FAN		D.F.	t
	M	S. D.	M	S. D.		
Access	4.46	1.50	4.79	1.48	339	−5.45*
Positivity	4.46	1.46	4.89	1.43	339	−8.30*
Openness	4.45	1.49	4.87	1.46	339	−7.08*
Sharing of Tasks	4.24	1.44	4.50	1.44	339	−5.71*
Networking	4.46	1.42	4.67	1.42	339	−5.65*
Assurance	4.35	1.45	4.63	1.47	339	−5.03*
Total (Relationship Cultivation)	4.40	1.31	4.72	1.31	339	−8.89*

STUDENT-ATHLETE'S USE OF SOCIAL MEDIA

	STAFF-PLAYER (OR PLAYER-PLAYER)		PLAYER-FAN		D.F.	t
	M	S. D.	M	S. D.		
Access	4.53	1.53	4.91	1.49	339	−5.96*
Positivity	4.79	1.50	4.98	1.48	339	−4.09*
Openness	4.24	1.54	4.74	1.46	339	−7.47*
Sharing of Tasks	4.25	1.43	4.65	1.49	339	−6.88*
Networking	4.74	1.48	4.91	1.50	339	−3.75*
Assurance	4.48	1.43	5.03	1.46	339	−9.23*
Total (Relationship Cultivation)	4.50	1.33	4.87	1.36	339	−9.44

Note. 1 = Strongly disagree, 7 = Strongly Agree
* p < 0.01

strategies (t (339) = −8.9, p < 0.01). The same held true for student-athletes' use of social media, in that students perceived social media having a significantly higher benefit for cultivating relationships between players and fans (M = 4.87, SD = 1.4) than between staff and players (M = 4.5, SD = 1.3) (t (339) = −9.4, p < 0.01). The assurance indicator showed the biggest difference and favored player-fan relationships. Students perceived that when student-athletes use social media, it would influence the level of assurance between player and fans (M = 5.03, SD = 1.5) more so than between staffs and players (M = 4.48, SD = 1.4) (t (339) = −9.2, p < 0.01).

Overall, students had relatively lower expectations of athletic departments and student-athletes building relationships through social media use compared to the other two publics. However, with no indicator scoring less than 4, it can be implied that students have a positive view of social media's ability to cultivate relationships among fans, athletic departments, and players.

Table 4. Summary of Findings

PUBLIC	PERCEPTIONS ABOUT THE EFFECTIVENESS OF SOCIAL MEDIA USE
Staffs	Staffs expect the most effectiveness from social media use in terms of relationship building among three publics. (4.6 < M < 5.9; See Table 1). Staffs think athletic departments' use of social media would best contribute to networking between staff members and players (M = 5.9). Staffs are less favorable toward openness of student-athletes' use of social media, especially between players and fans (M = 4.6).
Student-Athletes	Student-athletes generally expect social media use by athletic departments and players to help build effective relationships among publics (5.0 < M < 5.3; See Table 2). Student-athletes expect athletic departments' social media use to benefit relationships between staffs and players more than between staffs and fans (In every indicator, especially access and assurance). Student-athletes think players' social media use would help fans and players share tasks (M = 5.1).
Students	Students expect social media use by athletic departments and players to help them build relationships with publics, but not as highly as staffs and student-athletes do (4.2 < M < 5.0; See Table 3). Students generally expect social media use would benefit athletic departments' and student-athletes' relationships with external publics more than with internal publics (in every indicator).

Conclusion

All three publics—athletic staffs, student-athletes, and students—were favorable toward the use of social media for relationship cultivation. However, they each regarded the use of social media differently, which answers the second research question, "Are there any differences among perceptions of publics of collegiate sports organizations about the use of social media?" As we discussed and Table 4 shows, staffs expected the most from social media use, except for the openness indicator. Student-athletes also considered social media use positively. In particular, they expected their athletic department's use of social media to benefit the relationship between the organization and themselves. Students saw social media as a communication tool because students agreed that athletic department and student-athlete social media use would help relationships with fans more than with themselves. However, students were the least favorable about the social media's relationship building potential between athletic departments and student-athletes. This makes sense because these particular students, who may be fans, are not athletes and thus not personally involved in the interaction between athletic staffs and players. Therefore, their understanding of that relationship, and how to improve it, is naturally limited in scope.

As Taylor and Kent (2010) argued, the effectiveness of social media is not yet clear. Social media pioneers (e.g., Shankman, 2010; Solis, 2011) have said social media can help organizations with engagement. Solis further claimed that social media brought the "public" back into public relations because social media allow organizations to communicate *with* rather than *to* publics. The dialogic loop suggested by Kent and Taylor (1998) is now more open and immediate, and dialogic communication created by social media would set the standard for public relations as Pearson (1989) suggested. This dialogic, or two-way, communication also contributes to public relations excellence (Dozier, Grunig, & Grunig, 1995; Grunig, 2001; Grunig & Huang, 2000).

This study has some limitations that should direct future studies. First of all, we had a small number of staff members compared to other samples. In addition, in-depth interviews with staff members would make findings more informative. For example, their relatively low level of agreement with student-athlete's social media use in regard to openness could be explained more precisely with in-depth interviews. Furthermore, we did not recruit participants from larger schools. Those colleges belonging to NCAA BCS schools get more media attention and have more revenue for collegiate sports. Lastly, combining these results with data from professional sports organizations would increase applications of the study and serve as an interesting point of comparison.

This study showed that the relationship cultivation measurement developed by Hon and Grunig (1999) and Ki and Hon (2009) still remains valid and is reliable when adapted for sports settings instead of perceptions about organizations' general communication strategies. Even though our study had some limitations, we believe it provides a starting point for utilizing relationship cultivation measures to investigate effectiveness of various strategies and tactics, including social media, used by a variety of organizations.

ENDNOTE

1 Analysis of subsequent survey results revealed Cronbach alpha values between 0.72 and 0.87 for all items tested, thus indicating a higher level of internal reliability in the actual survey compared to the pretest.

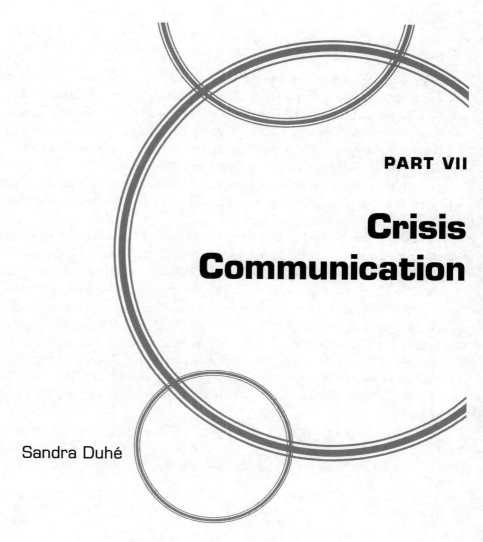

Sandra Duhé

PART VII

Crisis Communication

Overview

Crisis preparedness/response is an interdisciplinary field that requires the expertise of a variety of professionals to oversee the public health/safety, logistical, legal, and resource concerns involved. A crisis response strategy may be optimally engineered, but if affected publics perceive related communications as insufficient or, worse, unreliable, the response is considered a failure. Social media platforms satisfy the desire for immediate information in a crisis but also make the response process more visible. This section addresses how an evolving media landscape affects, challenges, and supports crisis communication.

Brooke Fisher Liu, Yan Jin, Lucinda Austin, and Melissa Janoske introduce the innovative Social-Mediated Crisis Communication Model (an evolution of the Blog-Mediated Crisis Communication Model) that can be used to build relationships with publics before, during, and after crises. Their model helps crisis managers understand how social media publics transmit information and how organizations can respond to that information. They provide a hypothetical case to illustrate their points.

Timothy Coombs introduces and defines the concept of the "paracrisis." Not to be confused with a crisis, a paracrisis still poses challenges to organizations because of its highly visible portrayal on the social Web. In a straightforward, informative style, Tim provides direction on how organizations can respond to a paracrisis, if they choose to do so. He provides several examples of how paracrises have been handled online.

Beyond their educational missions, colleges and universities have legal and ethical obligations to share crisis information with their students. Joe Downing and Mark Casteel provide an overview of new communication technologies that are proving to be more effective than text messages. Joe and Mark emphasize the importance of public and private organizations working together to keep students safe.

Barbara Gainey explains that public agencies are often the first available source of information about a crisis. Barbara highlights how public agencies use Facebook, Twitter, YouTube, and other outlets to share information about preparation, sheltering, and related crisis messages. She also provides references to impressive sites that are designed to guide agencies in the use of social media but are useful for public relations practitioners in any sector.

Ronald Lee Carr, Cornelius Pratt, and Irene Carolina Herrera provide a compelling account (experienced firsthand by two of the authors) of how social media served as a cultural-shifting force in the aftermath of Japan's triple disaster in March 2011. A standout of the chapter is their description of how one mayor used YouTube to reach his constituents in a style that was anything but subtle and indirect.

The last two chapters focus on BP's response to the April 2010 Deepwater Horizon rig explosion, each focusing on a different online platform. Marcia DiStaso, Marcus Messner, and Don Stacks analyzed crisis-related Wikipedia content for more than a year following the explosion. Throughout that time period, Wikipedia remained a top-ranking information source in search engines. Much of the content was negative, though well referenced. Marcia, Marcus, and Don build on their Wikipedia reputation management work in the first edition to once again remind practitioners of the importance of monitoring Wikipedia.

Jinbong Choi studied how BP used Twitter in the 40 days following the explosion. He identified frames and keywords used during that time span and found the company's primary focus was on information dissemination rather than placing blame or defending its actions.

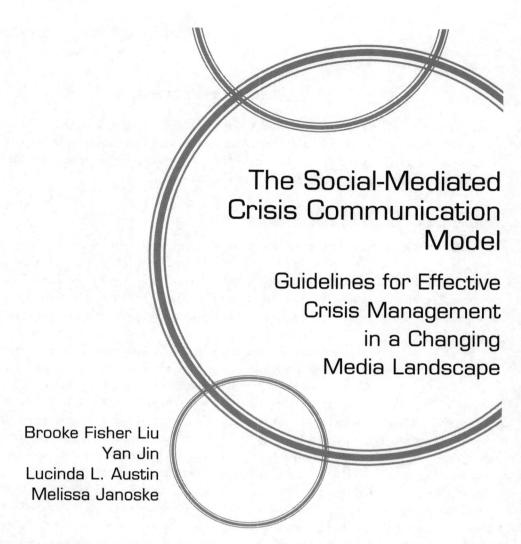

The Social-Mediated Crisis Communication Model

Guidelines for Effective Crisis Management in a Changing Media Landscape

Brooke Fisher Liu
Yan Jin
Lucinda L. Austin
Melissa Janoske

Although public relations professionals recognize the importance of understanding how to effectively engage their publics via new media, crises create unique challenges for relationship building, especially when reputations may be on the line. To address these challenges, the Social-Mediated Crisis Communication (SMCC) model provides guidelines for effectively integrating new media into crisis management. This chapter summarizes how crisis managers can use the SMCC model to build relationships before, during, and after crises. Based on synthesis of 62 interviews with crisis communicators and young adults and experiments with 338 young adults, the chapter's findings provide valuable insights for the changing media landscape.

In 2010, the Gulf oil disaster was the most searched term on Yahoo ("Oil Spill," 2010). In 2011, Google and Twitter partnered to provide Egyptians with a speech-to-tweet service, enabling Egyptians to overthrow President Mubarak through dissent diffused online through networked power (Beckett, 2011; Melanson, 2011). These, and countless other examples, indicate new media must be integrated into crisis planning, response, and recovery. Yet, organizations continue to struggle with how to make the business case for effectively engaging with their publics via new media (Hathi, 2009). To address this struggle, Jin and Liu (2010) proposed the first model to predict how organizations can effectively communicate with key publics via new media, traditional media, and offline word of mouth communication, which we introduce in this chapter.

Literature review

In this section we review research on publics' new media consumption in a crisis context and then turn to research on organizations' new media use in a crisis context. The findings indicate the need for the Social-Mediated Crisis Communication model (SMCC), which synthesizes current best practices and thereby helps crisis managers navigate new media opportunities.

Publics, crises, and new media

Forty-one percent of Americans receive most of their news via the Internet, and 65% of people aged 18–29 cite the Internet as their main source of information (Pew, 2011). For immediate and interactive feedback, users more frequently turn to social networking sites such as Facebook and Twitter than more traditional forms of media such as newspapers or television (Pew, 2010a). During crises, publics turn to social media for a wide variety of information and support, including emotional support (Macias, Hilyard, & Freimuth, 2009; Stephens & Malone, 2009).

Social media play a role in both the construction and deconstruction of crises, as the choice of channel can strongly impact publics' perceptions of organizational reputation in certain circumstances. For example, one study found that when publics receive information from an organization in crisis via Twitter they were less likely to attribute crisis responsibility to the organization (Schultz, Utz, & Göritz, 2011). Certain forms of social media, such as Twitter, Facebook, and individual blogs, are most frequently used for information sharing, especially in the first 12 hours of a crisis, when the most citizen-generated content appears online (Baron & Philbin, 2009; Heverin & Zach, 2010; Wigley & Fontenot, 2010).

Social media can also skirt the schedule of more traditional media and allow publics to access information where and when they see fit (Procopio & Procopio, 2007; Purcell, 2011) as well as mobilize based on that information (Murdock, 2010). Publics with high involvement in crises, however, are more likely to actively engage with media channels such as newspapers and magazines (Avery, 2010). Publics' involvement in crises is also affected by electronic word of mouth (eWOM) communication, which can both amplify and increase organizations' crisis messages (Kozinets, de Valck, Wojnicki, & Wilner, 2010). Publics tend to adapt their desire to spread eWOM through their judgment of the appropriateness or value of information to potential recipients (Sohn, 2009). The Internet moves in both directions, as it can extend and ease information retrieval (decentralizing and accelerating the spread of communication) while increasing the risk of information overflow, which can lead to rumors and increased bias (Bucher, 2002).

Publics who are dissatisfied with an organization may use social media to spread negative word of mouth (nWOM) communication on and offline (Bailey, 2004; Richins, 1983). This can take the form of vindicating worth, gaining revenge, or getting others to act against the organization (Ward & Ostrom, 2006). Employers should also exercise caution with employees who blog, as they may express or expand upon nWOM (Kaplan & Haenlein, 2010). Transparency here, as with all communication, is essential (Burns, 2008) because blogs tend to be more subjective than other forms of online media, especially online newspapers, while publishing more frequently during a crisis (Liu, 2010).

Publics who interact with organizations and others via social media are also more likely to engage in other behaviors and engagement, both on and offline (Dutta-Bergman, 2006). In sum, social media can facilitate information sharing, opinion sharing, and emotional expression (Macias, Hilyard, & Freimuth, 2009) so that more traditional public relations actions (e.g., providing information and bolstering organizational reputation) can now be shared by publics during a crisis (Smith, 2010b).

Organizations, crises, and new media

Eighty-three percent of public relations professionals concur that blogs and social media have altered the way organizations communicate, and 81% believe social media have enhanced public relations (Wright & Hinson, 2010a). Use of social media can empower public relations practitioners, but some are slower to adopt technologically complicated forms of social media, which often are used to communicate to smaller audience segments than more traditional channels of mass communication (Diga & Kelleher, 2009; Eyrich, Padman, & Sweetser, 2008). Further, even when organizations utilize social media channels, traditional media tactics are still seen as a necessity for building organization-public relationship strength (e.g., Pfeiffer & Zinnbauer, 2010; Taylor & Perry, 2005).

Clear communication with publics during crises is a top priority, and handling both primary and secondary[1] stakeholder interactions is a crucial step in surviving complicated and decisive moments (Coombs, 2012). Social media can spread information rapidly through a variety of channels to reach diverse publics (Coombs, 2008). Further, social media encourage participation from diverse publics, including those who are not directly affected by the crisis but have some sort of relationship with the organization, which can help facilitate information flow during crises (Booz Allen Hamilton, 2009). Publics are more likely to read and share information where they already have established relationships (Van Hoosear, 2011). The goal is collaborative or people-powered social media, the continued maintenance of which can be a burden on practitioners. To help reduce this burden, organizations can aggregate their social media usage to save practitioners' time and energy, while maintaining posting and monitoring schedules (Swallow, 2010).

When building online relationships before crises occur, research indicates blogs use more dialogic communication than traditional websites (Seltzer & Mitrook, 2007), because of the conversational human voice that stakeholders attribute to blogs (Kelleher, 2009; Kent & Taylor, 1998). Organizations that utilize dialogic voice are then able to better understand user preferences and expectations (McAllister-Spooner, 2009). Openness to this sort of dialogic communication increases stakeholder engagement, which can lead to an increased positive response to a crisis (Yang, Kang, & Johnson, 2010). Reading an organizational blog can help decrease both public belief that a company is in crisis and the perceived magnitude of the crisis (Sweetser & Metzgar, 2007). Managing blog-mediated crises requires an organizational understanding of engagement with both influential and non-influential bloggers, their followers, and non-blog followers (Jin & Liu, 2010).

Although dialogic voice is a powerful tool, messages transmitted via social media should not wildly differ from messages submitted through more traditional media. Consistency in messaging supports the organization's larger communication plan and decreases the potential for confusion among publics (Booz Allen Hamilton, 2009). This message consistency can also lead publics to continue using social media as important information sources after crises. For example, publics who receive crisis information from Twitter are more likely to continue to follow that organization on Twitter after the crisis (Hughes & Palen, 2009). We now summarize the research on the SMCC model, which explicates how crisis managers can better understand and effectively use technology to engage publics, disseminate crisis information, and successfully emerge from crises.

Introducing the Social-Mediated Crisis Communication Model

The literature review confirms that organizations no longer have a choice about whether to integrate social media into crisis management. When organizations face high-stakes threats, it is especially important for crisis managers to provide evidence-based guidelines to effectively meet publics' expectations for information via a complex network of social media, traditional media, and offline word of mouth channels. However, scarce theories are available to provide the foundation and framework for theory-driven and evidence-based crisis communication practice in this complex media landscape.

To address this need to provide evidence-based guidelines to make the business case for integrating social media into crisis management, researchers developed the SMCC model as a framework for crisis and issues management in the changing media landscape (Jin & Liu, 2010; Liu, Jin, Briones, & Kuch, in press—see Figure 1). Originally called the blog-mediated crisis communication model (BMCC), the model later was expanded and renamed to include all social media: the social-mediated crisis communication model (SMCC) (Liu et al., in press). This model helps managers decide if and when to respond to influential social media through monitoring issues and crises online.

The model is divided into two parts that predict how publics transmit information and how organizations can respond to this information. The first part predicts how three key publics transmit information: (1) influential social media creators, who post crisis information online; (2) social media followers, who consume social media creators' crisis information, sharing this information online and offline; and (3) social media inactives, who receive crisis information via traditional media and word of mouth communication with other social media inactives, social media followers, and/or social media creators.

Two arrow types depict how information is distributed by social media directly (solid arrows) and indirectly (dotted arrows). Finally, the model includes five factors that affect how organizations engage with the three key publics:

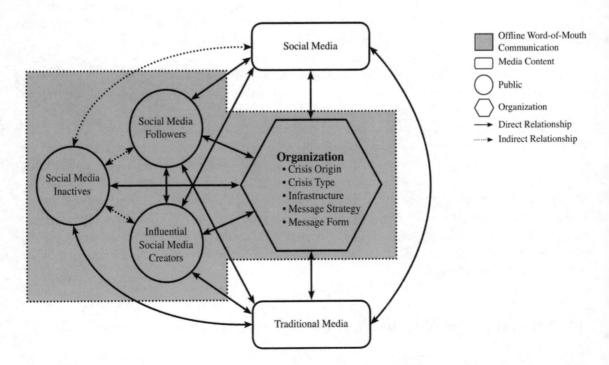

Figure 1. Social-Mediated Crisis Communication Model (Liu, Jin, Briones, & Kuch, in press)

1. Crisis origin: whether the crisis sparked from an internal organizational issue such as mismanaging volunteers or from an issue external to the organization such as a severe weather incident.
2. Crisis type: whether victim, accident, or intentional, which impacts how organizations should adapt their crisis response strategies (Coombs, 2012).
3. Organizational infrastructure (centralized vs. localized): depending on the crisis situation and the social media context, some crises might be best handled according to a central message designed to be communicated by all local affiliates, branches, or chapters as well as organizational headquarters.
4. Crisis message strategy: how the crisis message is communicated (see below and Table 1).
5. Crisis message form: whether the crisis message is distributed via social media, traditional media, and/or word of mouth communication.

The second part of the model combines Situational Crisis Communication Theory (SCCT) (Coombs, 2012) and applied research to provide suggestions for how a given organization should proactively manage social-mediated crises (see Table 1). As the model posits, once these influential social media and issues are identified, crisis managers can apply different response options depending upon the crisis origin and type. Grounded in rumor psychology theory and suggested rumor-quelling strategies (DiFonzo, 2008), Jin and Liu (2010) adapted the three-stage rumor transmission process as the social-mediated rumor

Table 1. Social-mediated crisis response strategies (adapted from Coombs, 2012; Jin & Liu, 2010)

RESPONSE OPTIONS	STRATEGY	RUMOR PHASES			CRISIS RECOVERY
		GENERATION	BELIEF	TRANSMISSION	
Base	Instructing information	X			
	Adapting information: corrective action	X			
	Adapting information: emotion	X			
Deny	Attack the accuser		X		
	Denial		X		
	Scapegoat		X		
	Ignore		X		
Diminish	Excuse		X		
	Justification		X		
	Separation		X		
Rebuild	Compensation				X
	Apology				X
	Transcendence				X
Reinforce	Bolstering		X		
	Ingratiation		X		
	Victimage		X		
	Endorsement		X		
Punish*	Legal action			X	

*Not a recommended blog-response strategy

cycle that communicators should manage at each phase: (1) rumor generation, (2) rumor belief, and (3) rumor transmission.

The SMCC model incorporates the dominant crisis communication management theory, SCCT, into its recommended social-mediated crisis response strategies shown in Table 1. SCCT first advises that organizations prioritize protecting their publics from harm through two strategies: providing instructing and adapting information (Coombs, 2012). Instructing information informs publics about how they can protect themselves from physical threats while adjusting information helps publics cope with any psychological threats (Coombs, 2012). After providing necessary instructing and adjusting information, organizations select from four response options—deny, diminish, rebuild, and reinforce—with various strategies within each option (for a full description of these response options see Coombs, 2012 or Jin & Liu, 2010).

To determine how to monitor issues online, Jin and Liu (2010) proposed an influence matrix for evaluating blogs' influence for issues and crises, which also applies to other social media types (see Table 2). The matrix defines influential social media as ones that organizations monitor as part of their regular issues management process or new ones that emerge during a specific issue or crisis. Organizations cannot respond to all issues and must prioritize issues through environmental scanning, monitoring, forecasting, and assessment (Hallahan, 2001a).

Table 2. Matrix for evaluating social media influence (adapted from Jin & Liu, 2010)

MEASUREMENT TOOLS		SOCIAL MEDIA CHARACTERISTICS					
		CREDIBILITY	CONSISTENCY W/ KEY PUBLICS' ATTITUDES	FREQUENCY OF UPDATES	CONSISTENCY W/ ORGANIZATION'S CRISIS RESPONSE	DIALOGICAL SELF-PORTRAYAL	INTERACTIVITY
Outputs	Number of posts about crisis by valence		X	X	X		
	Number of comments about crisis by valence		X	X	X		X
	Number of unique visitors	X					
	Number of RSS subscribers	X					
	Number of crisis-related links to and from other sites	X					X
Outtakes	Search engine blog rank	X					
	Third-party endorsements	X					
	Business/media affiliation of the social media content creator	X					
Outcomes	Key publics' awareness of social media channel	X				X	X
	Key publics' post-crisis trust of social media channel	X					

The matrix is organized by three measurement types: outputs, outtakes, and outcomes (shown on the left in Table 2), which can be measured to highlight the strength of varying social media characteristics (the columns to the right). Outputs measure how many people are paying attention to social media, outtakes measure social capital and social networking, and outcomes measure how social media affect people's behaviors and relationships. Depending on which characteristics are most important to organizations, such as frequency of updates for timely information or interactivity with publics, organizations can identify social media that are most important to monitor and pitch for coverage.

Given the large number of SCCT response strategies (shown in Table 1), more research is needed on the way in which the crisis message is conveyed affects publics' responses. Recent research on the SMCC model indicates how crisis information form (traditional media, new media, or word of mouth communication) and source (organization or third-party) affect publics' response to organizational crisis strategies, which we now discuss.

Summary of research on the Social-Mediated Crisis Communication Model

To date, the SMCC model has been tested through interviews with American Red Cross communicators, which informed recommendations for crisis monitoring and responding. Additional testing through interviews and experiments with college students informed how crisis information is transmitted and shared. These interviews and experiments explored how factors such as crisis origin and crisis information form and source affect information seeking, crisis communicative tendencies, acceptance of crisis response strategies, and emotional responses to crises.

Interviews with American Red Cross communicators

The SMCC model, originally titled the BMCC model, was renamed after interviews with 40 American Red Cross communicators highlighted the increasing influence and importance of a variety of social media platforms in crisis communication (Liu et al., in press). This research suggested crises are often ignited online through multiple types of social media as well as through offline word of mouth communication. Findings from this study suggested the importance of increasing focus on social media, such as Facebook and Twitter, compared to traditional blogs, which Red Cross communicators did not view as useful for crisis communication.

Exploring publics' crisis communication behavior

Following the Red Cross interviews, additional research was conducted to provide insight into publics' motivations for media use in crises and their communicative tendencies. A series of qualitative, in-depth interviews was conducted with college students at a large East Coast University to explore motivations for seeking crisis information, information seeking behaviors, and communication about crises via varying types of media (Liu, Jin, & Austin, in press). Prior to the interviews, prospective participants completed an online questionnaire, and the researchers invited 22 students to participate in the interviews based on their media consumption habits.

Interview findings revealed that participants used social media during crises for insider information seeking and checking in with family/friends, and they used traditional media for educational purposes.

Convenience, involvement, and personal recommendations encouraged social and traditional media use, and information overload discouraged use of both. Humor and attitudes about the purpose of social media also discouraged use of social media as a method of information seeking, while credibility encouraged traditional media use (Liu, Jin, & Austin, in press). For example, participants who believed that social media were typically a lighter, more humorous form of communication, designed to build relationships and keep in touch, were less likely to use social media to seek information in times of crises.

In regard to communication behaviors and information sharing, interviews revealed that humor value and having/seeking insider information motivated participants to communicate online about crises as predicted by the SMCC model (Liu, Jin, & Austin, in press). For example, individuals who believed information was humorous were more likely to share information via social media, and individuals were eager to share insider information with others via social media as well. Other factors affecting media type usage were social norms and privacy concerns. Participants indicated that social norms surrounding crisis situations often dictated their communication behaviors; for example, if most of the individual's friends were using a particular type of media to communicate, the participant was more likely to use that form of media as well. Privacy concerns affected communication in that some individuals were hesitant to use social media to communicate about crises out of concern that their information could be seen by others.

Overall, interview participants' preferred methods for communicating about crises were in-person communication, followed by text messaging and Facebook. Interview participants frequently mentioned discussing crises with friends and family, indicating that these groups were influential for the participants.

Effects of crisis information form and source

Based on these interviews with college students and the prior Red Cross interviews, researchers designed an experiment for 162 college students at a large East Coast university to test the effects of essential components of the SMCC model (Liu, Jin, & Austin, in press). This study tested the effects of crisis information form (traditional media, social media, or offline word of mouth) and source (third party or organization) on publics' acceptance of crisis response strategies and publics' crisis emotions.

Using findings from the prior interviews with college students, six crisis situations were selected which interviewees suggested were the most relevant and important for college students at the particular university (Jin, Liu, & Austin, 2011). For crisis information form, results from this study revealed that publics were more likely to accept defensive, supportive, and evasive crisis responses via traditional media than via social media or word of mouth. For information source, publics were more likely to accept an organization's defensive and evasive crisis responses when these came directly from the organization experiencing the crisis. However, publics were most likely to accept supportive responses when these came from a third party.

Interaction effects were also relevant, as publics were most likely to accept accommodative crisis responses when these came from the organization experiencing the crisis via word of mouth, while they were least likely to accept these responses when delivered by the organization via traditional media. This finding indicates that if an organization in crisis decided to use accommodative responses (e.g., taking full responsibility, asking forgiveness, offering compensation to the affected), it is important to ensure these responses are conveyed via word of mouth instead of solely relying on traditional media channels. This may sound counterintuitive to public relations practitioners who often recommend traditional media for reaching out to a broad audience. However, it is important for practitioners to keep in mind that in crisis situations, publics are eager to learn more from the communication channels they have more direct access to and interaction with. Our results suggested an accommodative gesture from the organization tends to be more positively received when communicated via word of mouth. The significant interaction effects of crisis information form and source indicated that the source of the crisis response moderated publics' acceptance of crisis messages distributed via traditional media, social media, and word of mouth.

The role of crisis origin in the effects of crisis information form and source

Expanding upon this research, researchers conducted additional analyses to explore the effects of crisis origin in combination with form and source through a 3 (crisis information form) x 2 (crisis information source) x 2 (crisis origin) mixed-design experiment with 338 college students at a large East Coast university (Jin, Liu, & Austin, 2011). This study sought to explore publics' acceptance of crisis response strategies depending on the crisis origin (whether the crisis sparked from an internal organizational issue or from an issue external to the organization), crisis information source (the organization or a third-party), and the crisis information form (social media, traditional media, or word of mouth communication).

Internal crisis origin, which participants perceived the organization as highly responsible for, led to stronger crisis emotions and higher acceptance of more accommodative organizational crisis responses. External crisis origin, which participants perceived the organization had low responsibility for, increased likeliness to accept an organization's defensive responses. Interaction effects revealed that publics were more likely to accept evasive crisis responses when the crisis origin was external and information came from the organization. In addition, if the crisis information came from a third party through social media, publics' attribution-dependent emotions such anger, contempt and disgust were intensified or aggravated when the crisis origin was internal. Findings indicate the key role of crisis origin in affecting publics' preferred information form and source, which impacts how publics anticipate an organization should respond and what crisis emotions they are likely to feel when exposed to crisis information.

Overall findings from studies testing the SMCC model show the importance of considering crisis origin, information form, and information source, as well as crisis type and message strategy. These findings also clearly indicate the importance of strategically matching crisis information form and source when organizations respond to crises, which currently are not considered in dominant crisis theories such as SCCT.

Conclusion

Imagine you are the chief public relations officer of a state university, charged with responding to the following crisis. After a major athletic victory at your university, more than 2,500 students have gathered downtown to celebrate. As students rock parked cars, take down street signs, and set fires, the University calls in riot police to intervene. Shortly before midnight, town police spray water and use tear gas to calm the crowd, eventually arresting 31 students. The chaos is captured on thousands of cell phones, with students immediately uploading videos and pictures of the riot to Facebook and YouTube, including blood on the sidewalk and police seemingly beating students. The following morning, local news stations pick up the videos and pictures, including students' allegations of police brutality, which run on the two local broadcast stations and the front page of the town's newspaper.

Confronted with a crisis situation like this, the SMCC model would instruct you to first identify the key publics in the crisis. To identify these key publics, your public relations team should employ the measurement tools highlighted in Table 2. For example, frequent posters of YouTube videos of the riots would be influential social media creators that would need to be carefully monitored. Also, online content endorsed by journalists should also be carefully monitored as this content was produced by influential social media creators. Further, these journalists should be monitored as key social media followers. Finally, social media inactives such as significant donors should be considered for outreach to make sure the riots have not damaged the University's relationship with these donors.

Next, the SMCC model advises examining how the crisis origin, crisis type, and your organization's communications infrastructure affect the selection of the most desired message strategy and message form. Given that the crisis emerged from within a single organization and is considered a victim crisis because the University is not responsible for staring the riot, the University should start with the base response options featured in Table 1. Further, given that the accusations of police brutality could increase perception of the university's crisis responsibility, the public relations team also should consider using some of the rebuild options if the accusations prove to be accurate. The most desired crisis message form should resonate with the key publics in this crisis, which include the influential social media creators, social media followers, and social media inactives.

SMCC model research has found that the base response options group into supportive strategies, and publics are most likely to accept supportive strategies when the crisis information comes from third parties such as journalists via traditional media than from organizations in crisis (Jin, Liu, & Austin, 2011; Liu, Austin, & Jin, 2011). For accommodative strategies, however, publics are most likely to accept these strategies when the crisis information is conveyed from the organization in crisis and least likely to accept these responses when delivered by the organization via traditional media (Jin, Liu, & Austin, 2011). Therefore, in handling the outlined crisis, the public relations team should consider working with partners to communicate supportive response strategies indirectly to key publics, but accommodative strategies are more effective when communicated directly to publics. It is also important for the public relations team to keep in mind that if the crisis is perceived as originating internally, publics will develop stronger emotions (Jin, Liu, & Austin, 2011).

Finally, continuing to use the measurement tools highlighted in Table 2, the public relations team would be able to determine when the crisis has ended because of decreasing online and offline coverage of the crisis. At this point, the team would evaluate how well the crisis was managed, using lessons learned to improve the management of future crises.

While the popular media tools of the day will change in the future, the need for crisis managers to effectively engage with key publics via diverse channels will not. Crisis managers will continue to work on achieving a balance between engaging publics and measuring what impact such engagement has on their organization-public relationships. The SMCC model can help strike this balance, especially as additional research is conducted to test, refine, and expand the model.

ENDNOTE

1   Primary stakeholders include those whose actions can directly harm or benefit the organization such as customers and employees. Secondary stakeholders include those who can affect or be affected by the organization such as media and competitors.

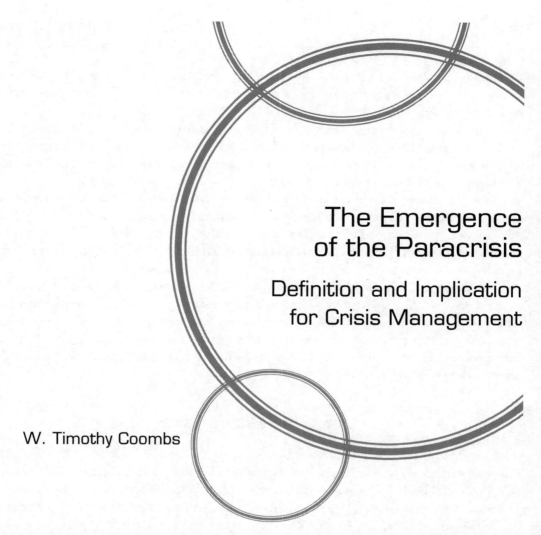

The Emergence of the Paracrisis

Definition and Implication for Crisis Management

W. Timothy Coombs

Social media are making crisis prevention efforts more visible to people outside of the organization facing a potential crisis. These visible crisis threats have been called paracrises. People watch to see if and how management responds to the crisis threats. How people evaluate the organization's paracrisis response can impact their perceptions of and relationship with the organization. This chapter examines the challenges presented by paracrises, the various communication response options for handling them, and the role of social media in the process. The key yield from the chapter is a greater understanding of paracrises and advice on how to construct effective crisis communication strategies for addressing paracrises.

Crisis communication is a rapidly growing discipline with strong ties to public relations. The focal point of the crisis communication research has been the crisis response, what organizations say and do after a crisis hits. This focus is natural given that what stakeholders usually see is the crisis response. However, the crisis response is just the tip of the crisis communication iceberg. Crisis communication is involved throughout the crisis management process, not just during the crisis response (Coombs, 2010).

The Internet has begun to reveal more of the crisis communication iceberg to stakeholders. Of particular note is how crisis prevention is becoming more visible due, for the most part, to social media. Moore and Seymour (2005) observed: "This process [crisis management] is being changed by the increasing ability of technology to communicate, act and influence perceptions during a crisis, as well as the more recognized potential to cause a crisis" (p. 30). Their point is that the Internet has increased an organization's vulnerability to crises. Moreover, the Internet is now scanned closely for the emergence of potential crises. No organization would consider its crisis scanning efforts complete if it was missing an Internet component.

The focus of this chapter is the way the Internet, featuring social media, has altered the nature of crisis prevention. Scanning locates potential crises. Once located, managers assess the crisis threat and what actions to take to reduce the threat (Coombs, 2012). Prior to social media, crisis prevention action was largely a private matter. Managers located the threat, and the evaluations were completed outside of public view. Rarely did stakeholders even know there was a threat or how the organization chose to respond to the threat. Through social media, crisis threats are becoming public spectacles. Stakeholders who are aware of the crisis threat can judge how the organization chose to respond. Just as in the crisis response, what the organization says and does in relation to the crisis threat can affect stakeholder perceptions of, interactions with, and attitudes toward the organization.

The term paracrisis is used to encapsulate the public management of crisis threats. The first section of the chapter explains the conceptualization of paracrisis including why a new term is even necessary. The second section examines the communicative response strategies most commonly used to address a paracrisis and the potential outcomes for each response. The third section reviews the lessons crisis managers can learn from paracrises. The chapter ends with a general discussion of the paracrisis' role in crisis management and crisis communication.

Conceptualizing paracrisis

The term "paracrisis" was introduced at the 2010 European Public Relations Education and Research Association (Euprera) convention in Jyväskylä, Finland. That was followed by its inclusion in the third edition of *Ongoing Crisis Communication* (Coombs, 2012). With the term in its infancy, it is important to explain why paracrisis was developed and what it can contribute to public relations. The conceptualization of the paracrisis defines the term, links it to social media and specific types of crises, and denotes its connection to reputation.

Defining paracrisis

A paracrisis is like a crisis but is really a "public" warning sign that a crisis is emerging. The term "para" can mean "like," hence the term paracrisis. Like a crisis, paracrises require strategic action designed to lessen the potential harm from the situation (Coombs, 2012). There is confusion between a crisis and a paracrisis. People discussing the event will often call a paracrisis a crisis when it is not. The organization has not defined the situation as a crisis and summoned its crisis management team to address the situation. Stakeholders are not facing imminent peril from the event. In most cases, the risk is to the organization's reputation and does not involve public safety. However, the event has the potential to evolve into an actual crisis. It is imprecise to call such a situation a crisis. The term paracrisis helps to eliminate some of that confusion.

Another option would be to simply call it a warning sign. However, the paracrisis is referring to a very specific class of warning signs. A warning sign, what Fink (1986) termed a prodrome, indicates that a crisis could be emerging. Part of crisis prevention involves the identification and evaluation of warning signs along with remediation efforts when necessary. Remediation, also known as mitigation, involves efforts to reduce the likelihood of a warning sign becoming an actual crisis. What makes paracrises a special class of warning signs is their public visibility.

Paracrises, social media, and crisis types

Historically, crisis prevention efforts have transpired behind the scenes. Crisis managers worked to prevent crises with little recognition or visibility for their efforts. It is not that crisis managers where purposefully hiding their actions. Instead, people were simply unaware of what actions were being taken if they were not directly involved in the effort to reduce the crisis threat. Crisis managers were unsung heroes in their efforts to prevent crises. Social media have altered the private nature of some crisis prevention efforts. Social media is a broad category of user-created content on the Internet covering blogs, vlogs, microblogs, content sharing sites, and social networking (Safko & Brake, 2009). The "brand names" of social media include Twitter, Flickr, YouTube, and Facebook. Crisis warning signs frequently emerge in social media. Examples include a YouTube video by Domino's employees for a potential product harm crisis, a tweet from a Ketchum employee that angered a major client by insulting the client's hometown, and the hijacking of Nestlé's Facebook site to pressure the company into responsible sourcing of palm oil.

Organizations can face a wide array of crises including acts of God, terrorism, product tampering, product harm, industrial accidents, and management misconduct. Fink (1986) argued that every crisis will produce warning signs. The difficulty for crisis managers is to find the warning signs. The more visible public warning signs tend to be associated with rumor and challenge crises.

Rumor crises are when false information about an organization is circulated among stakeholders. Challenge crises are when stakeholders claim the organization is acting immorally, unethically, or irresponsibly and should change those behaviors (Coombs, 2007; Lerbinger, 1997). Prior to the Internet, rumors and challenges were more private. Rumors spread from person-to-person through conversations and letters. Challenges were directed toward the organization through direct contact or letters.

Email and websites began the transition of making the early signs of rumors and challenges more public. People outside of the rumor or challenge conversation could now watch the crisis develop. Someone did not have to send you a message about a rumor. You could "see" the rumor online. An early Internet rumor was that Febreeze kills pets. Given one major use of the product is to address pet odors, the rumor was serious. Emails and discussion group postings helped to make the rumor visible to stakeholders. Similarly, the challenge was no longer just between the challenger and the organization. In 2010, stakeholders visiting Facebook and YouTube could learn about Greenpeace's charge that Nestlé was irresponsibly sourcing palm oil for its chocolate. Other stakeholders could witness the challenge unfold online. The social media dramatically accelerated the public accessibility and viewing of rumors and challenges. The variety of communication channels, ease of posting messages, and potential to reach a wide range of stakeholders made social media the perfect "medium" for cultivating rumors and challenges.

Rumors always had a somewhat public element as people needed to pass the inaccurate information on to others. Organizations were concerned about rumors because it was inaccurate information that harms the organization in some way. A common recommendation for rumors is that an organization should confront them directly and publicly if the rumors are damaging (Day, 2003; DiFonzo & Bordia, 2000). The Internet and social media have raised the stakes by making it easier to spread rumors to large numbers of people. Furthermore, people can easily find rumors as they search for other information online. There are websites devoted to debunking rumors, and Starbucks once had a section of its website devoted to rumors. New technology has made rumors easier to breed but has not altered how to address rumors.

The increasingly public nature of challenges has altered the way crisis managers address challenges. Prior to websites and especially social media, challenges were predominantly private matters. The interactions were between the organization and the challengers. Occasionally, challengers might be able to attract news media coverage and convert the challenge from a private to a public matter. But public challenges were more an exception than the rule because the news media have limited interest in challenges (Ryan, 1991).

Social media have reinvented the challenge. By posting challenge information for public viewing, the private becomes public. The challengers no longer need the news media for visibility and can bypass their gatekeepers. Through social media, challengers are their own media outlets. Figure 1 visually presents the shift. The darker circle on the left indicates the initial private nature of challenges while the lighter circle on the right indicates the more public nature of today's challenges. It is the public nature of the challenges that creates new demands. Now crisis managers are handling challenges in public view. When their actions are more visible, there is greater scrutiny and an accompanied pressure to be effective. Other stakeholders will be evaluating and often commenting on the crisis prevention efforts in the social media. Missteps are amplified and can intensify the reputational threat posed by a challenge. Moreover, there is a danger of *not* responding to a challenge, a legitimate response option, because people will complain about the lack of attention. The entire challenge process transpires in the social media, altering the prevention dynamics to a degree and what constitutes an effective response.

Challenges are effective when they punish an organization is some way. The challenge messages create negative reputational messages. I resist using the phrase "negative publicity" because social media do not fit the traditional definition of publicity, hence the phrase "negative reputational messages." Management wants to remove the potential reputational threats, so they take actions to address the challenge if the challenge is deemed a threat.

Past Challenges   Present Challenges

Figure 1. Social Media and Visibility of Challenges

Strategic responses and reputations

When faced with a challenge, managers develop a strategic response. The basic strategy is to remove the threat. The strategic response to challenges involves the need to: (1) assess the situation, (2) select a response, and (3) execute the response. The assessment has not changed. Managers need to determine if making the requested changes is consistent with the organization's strategy and is cost effective. Organizations should not change behaviors if it is counter to their strategy and long-range plans. Cost effective means there can be some benefits to making the changes. For instance, when McDonald's changed its packaging in the 1980s due to activist concerns, the chain became an environmental leader in the fast food industry. To this day, McDonald's draws reputational benefits from that change. Or the change simply silences critics. The organization does not derive reputational benefits but does eliminate negative publicity that can be eroding its reputation. What has changed about assessment is the public scrutiny. When YUM! Brands agreed to directly pay the Coalition of Immokalee Workers (CIW)[1] a penny per pound of tomatoes, the company ended the negative attention the CIW was drawing toward the company.

The CIW challenge to YUM! Brands demonstrates how seemingly powerless and marginalized stakeholders can create an effective challenge—one that changes organizational behaviors. When the challenge began, the CIW was not considered a powerful stakeholder by YUM! Brands. For the first three years of the challenge, the CIW was largely ignored by YUM! Brands. The CIW used communication skills to enhance its power and potential impact of its threat. The CIW sent speakers around the country to college campuses and worked with campus organizations to publicize those events. A website (http://ciw-online.org) was created to help raise awareness of the CIW along with a DVD about the cause. These public relations efforts enhanced the power of the CIW by making their challenge a greater reputational threat to YUM! Brands (Coombs & Holladay, 2010a).

Clearly, reputation is a critical element of challenge crises. The corporate reputation is recognized as a valuable, intangible asset by managers. Favorable reputations have been linked to recruiting, investing, and sales (Davies, Chun, da Silva, & Roper 2003; Fombrun & van Riel, 2004; Money & Hillenbrand, 2006). Reputation can be defined as how positively or negatively stakeholders perceive an organization (Argenti & Druckenmiller, 2004; Dowling, 1994). Reputations are evaluative and created through direct and indirect contact with an organization (Brown & Roed, 2001). Indirect contact can include news stories, websites, and social media content (Deephouse, 2000). Stakeholders frequently rely upon only small pieces of salient information when making reputation judgments (Carroll & McCombs, 2003). Organizations devote considerable time and effort to cultivating and maintaining favorable reputations.

Crises are recognized as threats to reputations (Barton, 2001). A cursory review of the crisis communication literature reveals a strong focus on protecting the organization's reputation during a crisis (e.g., Benoit, 1995; Coombs, 2007). When a crisis hits, an organization's reputation suffers some damage (Coombs & Holladay, 2006). Managers take action on challenges when they are perceived as a reputational threat. Negative comments appearing in the news media and/or social media can erode an organization's reputation. One motivation to act on a challenge is to remove the corrosive reputation element —the challenge. At times, managers alter organizational behaviors simply to remove the reputational threat so long as the change has a lower perceived cost than the reputational threat. This is similar to when organizations settle a lawsuit because of the expense and stipulate they did nothing wrong. The settlement is a pragmatic move designed to reduce costs, not an admission of guilt.

Typically, crises threaten to disrupt the organizational routine and damage reputations (Coombs, 2007). Reputational threats alone are unlikely to disrupt organizational operations. Once more, it is more accurate to not call reputational threats crises. The term paracrisis more accurately captures the reputational threats posed by challenges. Precision is important when applying the term crisis. Paracrisis denotes a less important situation but a situation that still requires a carefully crafted strategic response in order to ensure that a crisis does not occur.

Communication response strategies for paracrises

Just as in crisis communication, those managing paracrises can choose from a range of potential communication response strategies. The selection of the paracrisis response strategy is driven by the strategic evaluation of the situation discussed earlier. Managers have three basic response options when handling a paracrisis: reform, refute, and refuse. Each response option is designed to help reach specific outcomes for the paracrisis' resolution.

Reform means that the organization accepts the challenger's position that the organization is acting inappropriately and changes the undesirable behavior. The reform is accompanied by an account—why the organization made the change. The account may or may not acknowledge the challenge. Common reform accounts include good business sense, accommodate stakeholder demands, and apologies for not making the change sooner. The good business sense account explains that the change will help improve the organization's competitive position in the marketplace. The accommodate stakeholder account suggests that stakeholder expectations have changed and the organization wishes to respect those changes. The messages are unlikely to specifically state changing expectations but by talking about the changes the organization is at least tacitly acknowledging changes in stakeholder expectations. An apology recognizes the challenges, the validity of the challenge, and expresses remorse for not acting sooner on the matter. A reform account could utilize one, two, or all three accounts. The desired outcome is to end the challenge and perhaps eventually to strengthen the organization's reputation.

In 2008, Johnson & Johnson (through its McNeil Consumer Healthcare unit) launched an edgy ad for their Motrin brand pain reliever. The center of the campaign was an Internet ad that talked about how using baby carriers can create back pain for young mothers. However, the message noted how carrying a baby was like a badge of honor and people should understand if a young mother looks tired and is cranky. Many young mothers were offended by the ad and used social media, especially Twitter, to complain about the advertisement. The Twitter storm began on a Friday and by Monday the advertisement was removed with the following reform message:

> With regard to the recent Motrin advertisement, we have heard you.
>
> On behalf of McNeil Consumer Healthcare and all of us who work on the Motrin Brand, please accept our sincere apology.
>
> We have heard your complaints about the ad that was featured on our website. We are parents ourselves and take feedback from moms very seriously.
>
> We are in the process of removing this ad from all media. It will, unfortunately, take a bit of time to remove it from our magazine advertising, as it is on newsstands and in distribution.
>
> Thank you for your feedback. Its [sic] very important to us.
>
> Sincerely,
> Kathy Widmer
> Vice President of Marketing
> McNeil Consumer Healthcare (Gates, 2008, paras. 8–12)

In addition, Johnson & Johnson sent this reform statement to bloggers who had posted about the offensive advertisement:

> I am the Vice President of Marketing for McNeil Consumer Healthcare. I have responsibility for the Motrin Brand, and am responding to concerns about recent advertising on our website. I am, myself, a mom of 3 daughters.
>
> We certainly did not mean to offend moms through our advertising. Instead, we had intended to demonstrate genuine sympathy and appreciation for all that parents do for their babies. We believe deeply that moms know best and we sincerely apologize for disappointing you. Please know that we take your feedback seriously and will take swift action with regard to this ad. We are in process of removing it from our website. It will take longer, unfortunately, for it to be removed from magazine print as it is currently on newstands and in distribution. (Gates, 2008, paras. 2–3)

Johnson & Johnson was recognizing the challenge, the validity of the stakeholder challenger, and apologizing for the action.

The Nestlé response to Greenpeace charges of irresponsible palm oil sourcing included significant changes to its sourcing and a renewed focus on the environment. As indicated in excerpts from their change statement, Nestlé was recognizing the importance of the issues but not the Greenpeace challenge itself:

> Nestlé views destruction of tropical rainforests and peatlands as one of the most serious environmental issues facing us today.... Together with TFT, Nestlé has established Responsible Sourcing Guidelines and has committed to ensuring that its products do not have a deforestation footprint. This is the first time any company has made such a commitment.... Action plans are being developed with each supplier to ensure traceability and sustainability. This includes technical assistance to ensure full legal compliance of all plantation activities and to identify and protect forests of High Conservation Value, peatlands and high carbon value forests. ("Update," 2010, paras. 1, 3, and 4)

The changes by Nestlé granted legitimacy to the challenge even though the Greenpeace challenge is not mentioned directly. Hence, the Nestlé response qualifies as acknowledging stakeholder demands. Greenpeace recognized its demands were being met by ending its campaign against Nestlé following this statement of changes.

Refute means that the organization maintains its actions are appropriate and does not change. The refute strategy makes the organization's rationale for its decision public and open to scrutiny. The desired outcome is to blunt the reputational threat by revealing that it really is not a threat. The argument is that the organization is right and should not change its behaviors—there is nothing wrong. Managers need to be willing to explain and defend their assessments of the challenge. This is a departure from private challenges. Within the social media, many stakeholders will be viewing and evaluating the refutation statements. If stakeholders dislike the rationale for the refutation, the crisis threat could be re-energized and again escalate toward becoming a crisis.

"Skins" is a television show about teenagers that includes teen issues of drug use and sex. The show was first aired in the United Kingdom. In 2011, MTV began to air a US version of the show. Many conservative groups were offended by the show. The Parents Television Council (PTC) began a campaign against the show claiming it was child pornography and urged companies to drop their advertising by placing the show on their "do not buy" list. The PTC even speculated as to whether or not "Skins" was the most dangerous show ever. A number of companies did drop their advertising from "Skins" including General Motors, H&R Block, Wrigley, and Taco Bell. Taco Bell stated the show "was not a fit for our brand" (Stelter, 2011, para. 2). Wrigley said the company did not want "to endorse content that could offend our consumers" (Chen, 2011, para. 7). Both companies utilized a reform with good business sense.

MTV, on the other hand, choose to refute the PTC. Here is the refutation from MTV:

> Skins is a show that addresses real-world issues confronting teens in a frank way. We review all of our shows and work with all of our producers on an ongoing basis to ensure our shows comply with laws and community standards. We are confident that the episodes of Skins will not only comply with all applicable legal requirements, but also with our responsibilities to our viewers. We also have taken numerous steps to alert viewers to the strong subject matter so that they can choose for themselves whether it is appropriate. (Nededog, 2011, para. 9)

MTV was refuting that the show was child pornography by noting it complied with all laws and community standards. Moreover, MTV maintained it was responsible to viewers by warning them of potentially offensive content.

Refuse means that the organization simply ignores the challenge and offers no formal response. The desired outcome is that other stakeholders will ignore or dismiss the challenge. The idea is that any response will draw unnecessary attention to the challenge. Without attention, it has no power and will eventually disappear. When Green America, in early 2011, called on Hershey to change its practices of buying cocoa from suppliers known to use child and slave labor, the challenge went unnoticed by Hershey.

Hershey did not respond on its website, did not release statements, and was not active in addressing the challenge in the social media. One belief is that by not repeating the challenge, you reduce the attention given to the challenge—a response keeps the challenge alive. This strategy assumes the challenger will not keep promoting the challenge. Refuse strategy has risks of its own. Other stakeholders may question the wisdom of not responding to the challenge. Moreover, the challenge can continue to circulate and spread to more stakeholders if the challengers keep pushing the message. As with refute, the crisis threat could become re-energized and thereby create a need to re-evaluate the organization's response strategy. Both refute and refuse response strategies are subject to increased scrutiny due to the social media's ability to publicize the challenge. The "public" scrutiny alters the dynamic of the refute and refuse responses from when they were just "private" responses.

Lessons learned from paracrises

To highlight key lessons about paracrises, it is instructive to consider an extended case study, Nestlé and Greenpeace. In March 2010, Greepeace launched a campaign requesting that Nestlé change its palm oil sourcing practices. The concern was that Nestlé did little to ensure that its palm oil did not come from sources linked to the destruction of the rainforests and prime orangutan habitat. The campaign was called "Ask Nestlé to give rainforests a break." The campaign began on March 16, 2010, with a report by Greenpeace detailing irresponsible sourcing by Nestlé. Immediately Nestlé ended its ties with Sinar Mas Group, the most irresponsible of palm oil producers. On March 17, 2010, Greenpeace posted a video to YouTube that was a parody of a Nestlé Kit Kat commercial entitled "Have a break?" The video showed a man eating a Kit Kat at work but the candy bar was orangutan digits that dripped blood as he ate them. The scene shifted to orangutans in trees as the forest around them was being cut down.

Nestlé tried to have the video removed from YouTube. The effort to remove the video was termed censorship and the story began circulating in the social media. Nothing draws comments from social media writers like efforts to sensor social media. As you might guess, interest in the video skyrocketed. The setting then moved to Facebook on March 19, when Greenpeace began posting to the official Nestlé Facebook page. The posts included co-opting the Kit Kat logo to say Killer instead of Kit Kat. Nestlé stated: "We welcome all comments, but please don't post using an altered version of our logo as your profile pic. And please read our statement to answer many questions." People on Facebook did not appreciate the post. Here is one comment, not from Greenpeace, "You NEED to change the tone in your Facebook response. You're committing social media suicide." The response from Nestlé was: "Oh please…it's like we're censoring everything to allow only positive comments" (Leonard, 2010, para. 15). The deleting of posts with inappropriate logos sparked another round of censorship discussions in the social media. Informal tracking showed social media coverage of Nestlé was increasing and becoming more negative (Leonard, 2010).

After a few more "nasty" comments, Nestlé posted: "This (deleting logos) was one in a series of mistakes for which I would like to apologise. And for being rude. We've stopped deleting posts, and I have stopped being rude" (Leonard, 2010, para. 19). For two days, Greenpeace and its supporters controlled the content of the Nestlé Facebook page as Nestlé became a bystander at its own social media site. Two months after the campaign began, Greenpeace declared victory. In May 2010, Nestlé announced it would take action to locate and remove all irresponsible palm oil suppliers from its supply chain. Moreover, Nestlé was entering into a partnership with The Forest Trust to ensure that all of its products were free from any links to deforestation ("Sweet," 2010; "Update," 2010).

The Greenpeace-Nestlé case highlights four important, interrelated lessons about social media and paracrises:

1. Have a sense of humor—do not overreact
2. Know the etiquette
3. Accept lack of control
4. Be where the action is

Challengers will use humor to provoke their targets. Managers need to have a sense of humor and to accept that challengers will create parodies and have a right to do so. Overreacting (not having a sense of humor) will create the impression of censorship and of a larger organization trying to bully stakeholders.

Nestlé showed a lack of social media etiquette, especially in its Facebook postings. If you are going to utilize a social media, understand its rules of conduct. Organizations can response to content but must do so within the boundaries of the social media. Etiquette is related to lack of control. Social media is user-created content. Organizations cannot hope to and should not attempt to control it. Attempts at control will fail and once more bring charges of censorship. In turn, censorship charges will ensure additional negative discussions of an organization in the social media. Finally, be where the action is means that organizations must be part of the social media they are monitoring. Ideally, a response to a paracrisis should include the social media in which it originated. Clearly managers should post the response in a number of channels, including where it all began. If the organization has not been involved in the particular social media prior to the paracrisis, stakeholders are unlikely to notice or to appreciate the message. Still, many social media experts argue better late than never. However, do not expect much return on a social media response to paracrises if the response is your first foray into a specific social media channel (Coombs, 2012).

Conclusion

Social media are moving some crisis threats from the private to the public realm. Nowhere is the trend more evident than in challenges. Social media allow other stakeholders to see a challenge (the charge that an organization is acting irresponsibly) and the organization's response to that challenge. Particularly for challenges, the management of crisis threats is more of a public matter. This chapter has outlined the role social media have played in creating paracrises. Paracrises are crisis threats that look similar to crises because the threat is public, threatens the organization's reputation, and requires a strategic response to prevent further damage and escalation into a crisis.

The public nature of their management separates paracrises from other crisis threats. Paracrises require special attention because the responses—efforts to manage them—are subject to close public scrutiny. Efforts to manage paracrises are evaluated critically, and mistakes can intensify the damage done by the paracrisis and result is a true crisis. The chapter has focused on paracrises related to challenge crises. Three potential paracrises responses are outlined, along with when conditions favor the use of each strategy and the benefits and risk of using each strategy. In the end, I hope the reader has an appreciation for paracrises and the way that crisis communication can be used affectively to address these threats.

Crisis prevention is closely related to relationship management. Addressing crisis threats, such as paracrises, try to prevent an escalation of relationship damage between the organization and the stakeholders challenging the organization. The social media give voice to stakeholder concerns. Managers need to hear that voice and decide how best to respond to what is being said.

Paracrises are part of the evolution of crisis communication being facilitated by social media. Basic crisis communication strategy still matters, but the social media expand crisis communication by creating additional channels and new pressures for crisis managers. Social media usage will only increase over time. It is safe to say that the social media are a social force that cannot and should not be ignored. It behooves us to understand how social media are shaping crisis communication if we hope to improve its practice.

ENDNOTE

1 The CIW is a community-based organization that represents about 4,000 mostly low-wage workers in Florida. The members of the organization are mostly immigrants that are Latino, Mayan Indian, and Haitian. Their primary work is picking tomatoes and citrus. The CIW is not a powerful union but rather a represents a group of workers that are frequently exploited and marginalized by large industries ("About CIW," n.d.).

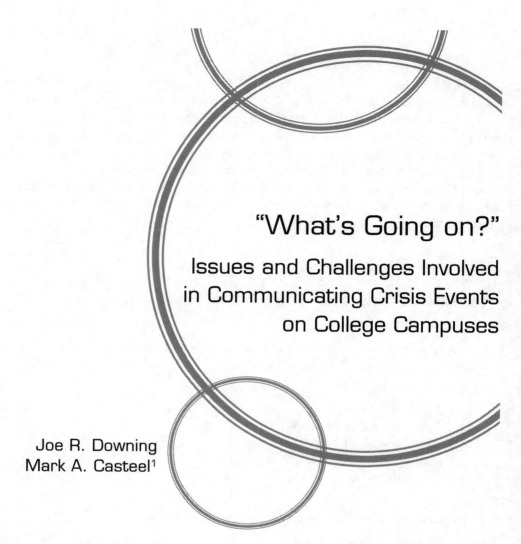

"What's Going on?"

Issues and Challenges Involved in Communicating Crisis Events on College Campuses

Joe R. Downing
Mark A. Casteel[1]

When a crisis occurs, crisis and emergency communicators from both the public and private sectors face a daunting task. Not only must they get the message out quickly, but the message must also be accurate and provide specific information about a plan of action. Communicating crisis information quickly is especially important for college officials, who have both legal and ethical responsibilities to their multiple constituents. In this chapter, we focus on the technological advances that have recently created new opportunities for college officials to communicate with their stakeholders. Although traditional text messaging (SMS) has been championed as a viable alerting tool, it suffers from several technological limitations that severely limit its effectiveness. We discuss the limitations of traditional text messaging plus other commonly used warning media (e.g., sirens, email, MMS) and then summarize the newly emerging Cellular Broadcast Service (CBS) capability that will allow brief, authenticated alerts to be sent to cell phones that have the necessary technology. There are two significant advantages to distributing warnings via CBS technology: (a) the warnings can be targeted to individuals within range of a specific cell tower; and (b) the messages do not negatively affect voice and SMS network traffic. We also discuss the federal government's implementation plans for this new warning technology. We conclude the chapter by examining future trends, placing much emphasis on the need to more fully incorporate social networking into the crisis management field.

On April 27, 2011, an EF-4 tornado with winds up to 190 mph hit Tuscaloosa, Alabama, home of the University of Alabama. The tornado caused massive damage to the Tuscaloosa area, with over 65 fatalities and 1,000 injuries ("Historic Tornado Outbreak," n.d.). Although the tornado missed the University's campus, five students sadly lost their lives. The devastation to the community was so severe the University cancelled final exams and postponed spring commencement until August (Reeves, 2011).

This profoundly sad event highlights the pressing need facing any government, business, or school official to get information about a crisis out in a timely manner. Other recent high profile crises like the Virginia Tech shootings in April 2007, illustrates the urgent need for emergency officials to notify their constituents and the public of the need for immediate action. Notably, however, when Virginia Tech officials used their email warning system known as *VT Alerts* to warn of gunfire (later determined to be from a nail gun) that occurred on November 13, 2008, only a portion of the messages were actually received (Carter, 2008b).

Essential Elements of Effective Warnings

For warnings, there is a sizable body of social science research on what constitutes effective warning messages. These key elements are listed below.

The message should be –

- Clear, with no jargon;
- Specific in its instructions;
- Accurate and truthful;
- Certain, using authoritative language; and
- Consistent, also explaining any changes from past messages.

The message should contain the following information:

- What exact action should be taken;
- When to take action, including when action should start and when it should be completed;
- Where the hazard is taking place, in order to define clearly to whom the message is directed;
- Why protective action is necessary including consequences if action is not taken; and
- Who is sending the message.

The message should be confirmed as follows:

- Through frequent repetition, and
- Through issuance over multiple communication paths.

Source: Drawn from the work of Dennis Mileti and colleagues provided to committee members as background information prior to the workshop.

Figure 1. Summary of social science research on effective warning messages. From "Public Response to Alerts and Warnings on Mobile Devices: Summary of a Workshop on Current Knowledge and Research Gaps," 2011, p. 10. Reprinted with permission from the National Academies Press, Copyright 2011, National Academy of Sciences.

Given the emphasis on quick message distribution, all communicators should know the essential components of an effective crisis response message. Figure 1 includes a summary of relevant social science research on this topic. Mastrodicasa (2008) argued that university and college crisis communicators, in particular, are placed in a unique position with regard to the dissemination of crisis and warning messages. The speed with which they disseminate information following a crisis will be the primary means of evaluating the efficacy of their response (Mastrodicasa, 2008). Not only do they have to consider multiple stakeholders (e.g., students, employees, parents, board members), but they have both ethical and legal responsibilities. For instance, in July 2010, the Jeanne Clery Disclosure of Campus Security Policy and Campus Crime Statistics Act (1990), which is commonly referred to as the *Clery Act*, was amended to include, among other requirements, that college administrators must test their campus's emergency response procedures yearly (34 C.F.R. § 668.46.g.6).

In today's age of social networking and cell phones, many colleges and universities still rely on cold war era warning sirens. Radcliffe Institute, for instance, upgraded their siren warning capabilities as a direct result of the Virginia Tech shootings (Jonsson, 2007). Sirens are also popular in the Midwest and South for tornado warnings (League et al., 2010). Sirens, however, can only sound an alarm. Given this narrow application, universities are now installing or upgrading public address systems on their campuses, which have the ability to actually broadcast specific messages as well as warning tones ("Cisco Helps," 2008). Some of these systems can even be heard over one mile away (Lenckus, 2007). Strobe lights and digital signage are also commonly used technologies.

Today, universities and colleges must avail themselves of the best technologies available to communicate crisis information. Until recently, this meant using Reverse 911 systems (Sanders, 2008) or distributing campus wide emails. However, Reverse 911 systems often are tied to a landline phone number (Schmitz, 2011). Cell phones can be called, but their numbers must first be manually entered into a database. Surprisingly, most cell users choose not to sign up for emergency notification messages ("Residents Shun Cell Phone," 2011). Given the decreased number of landline phones—especially among college students—and the tendency to keep one's cell phone number private, Reverse 911 systems are of limited utility. Plus, landline phone networks can get bogged down during a crisis. Alternatively, email consumes much less bandwidth than a voice call: A five minute voice call consumes as much bandwidth as around 4,000 emails of similar length (Noam, 2001). However, like Reverse 911 systems, email is not readily accessible to students on the move (Rae, 2011). Further, email is not a popular media choice for students (Richtel, 2010). Sending stakeholders short message service (SMS) messages (i.e., text messages) to their cell phones is popular, but, as we will discuss at length in this chapter, is fraught with problems.

College officials use wireless pager devices as another stand-alone alerting tool. For instance, IntelliGuard Systems markets its RAVENAlert™ wireless pager device to colleges. Colleges, in turn, distribute the device to faculty, staff, and students ("RAVENAlert," n.d.). The device, which is essentially a key fob, lights up, emits a loud alert and vibrates when campus police send stakeholders a message. Also, the device has limited messaging capacity—much like wireless pagers used in hospitals and restaurants.

As we will argue, there are limitations to each of these single alerting tools. For instance, in the case of the pager, students must carry yet another device. As such, the current trend on college campuses is to use a multipronged approach to alerting—an approach that relies on a combination of different alerting tools. A technology that has become widely available and heavily used by many institutions of higher education since the Virginia Tech shootings is the *mass notification service* (MNS).

Mass Notification Service (MNS)

An MNS is a commercial alerting service that provides broad distribution of alert and warning messages across a variety of platforms (e.g., cell phone SMS, email, voicemail, wireless pagers), depending upon the needs of the organization. With regard to cell phones, MNSs are popular given the high usage of cell phones among college-aged students. In December 2010, it was estimated that over 302 million Americans owned a cell phone, approximately 94% of the population ("Background on CTIA," n.d.). For college students, the estimate was an astonishing 99.8% as of June 2010 (Ransford, 2010). Given this high rate of cell phone ownership, it only makes sense that universities use this readily available resource.

Numerous technological and logistical problems exist, however, in implementing an MNS system, especially for universities where the subscriber rate can number in the tens if not hundreds of thousands. Probably the largest challenge is building and maintaining a valid distribution list (Latimer, 2008). Some institutions utilize an opt-in procedure whereby potential users must sign up their cell phone number. Other institutions, such as Penn State, require subscribers to provide a verification code after initially signing up (a double opt-in procedure). Of course, some subscribers forget or neglect to respond with the appropriate validation code, effectively reducing the number of individuals who should (and want to) receive the warning message (Latimer, 2008). A more effective procedure is opt-out, where the institution automatically loads all subscriber information into the database after which individuals may choose to remove themselves from the subscription. When Notre Dame switched from a double opt-in procedure to opt-out, their student participation rate rose from 62% to 99.8% (Latimer, 2008). Once the distribution list is built, an additional challenge is maintaining the database so that all stakeholders' contact information is valid (Downing, 2011).

MNSs have been shown to be effective in a number of high profile tests. Virginia Tech, for instance, tested their system on October 8, 2008. Approximately 36,000 email messages were delivered in less than 8 minutes, and approximately 24,000 voicemail and SMS messages were delivered in slightly over 19 minutes. Follow-up surveys revealed that 98% of the campus community was aware of at least one of the test messages (Roberts, 2008). Penn State also tested its MNS on February 23, 2011, and successfully delivered SMS messages to 88,771 subscribers from 23 campuses in just over 18 minutes (Mountz, 2011).

Although the use of an MNS might be seen as a panacea for disseminating crisis messages, some recent and high profile failures of MNSs have led some to question its effectiveness (Downing, 2011). One important consideration is that not all subscribers receive the test warning messages. For instance, in a test of the University of Florida's system, about 3% of over 41,000 messages did not get delivered, for undetermined reasons (Sanders, 2008). Some reasons that were suggested for the 3% failure rate were full inboxes, users not subscribed to an SMS plan, and an outdated database. Traynor (2008) also noted that cellular networks were never designed to withstand the avalanche of SMS calls generated during a crisis. Given these limitations, the success rates of the recent tests of MNSs are therefore somewhat encouraging.

Technological limitations in delivering SMS alert messages

In addition to some of the practical and implementation problems with MNSs discussed, other technological limitations inherent in SMS also present challenges. First, SMS is limited to 93–160 characters of plain text per cell phone screen (depending on carrier and cell phone). Although sending multiple screens of information is possible, some older cell phones do not support the "concatenation" technology responsible for reassembling the message in the correct order. This increases the likelihood that end users will receive a message out of order, or only part of a message, as occurred during the Malibu wildfires of

October 2007 (Latimer, 2008). The challenge, then, is for communication researchers to figure out how to design an information-rich disaster/crisis message using limited characters.

Second, wireless networks do not deliver SMS messages in real time, increasing the chances that messages could build up in a user's message queue, especially if one's cell phone is turned off (Traynor, 2010). Of course, this also increases the likelihood the messages will be delivered out of order. Although Meng, Zerfos, Samanta, Wong, and Lu (2007) found the vast majority (91%) of SMS messages were delivered within 5 minutes, 5.1% of the messages were never delivered.

Third, sending large numbers of SMS messages simultaneously creates a tremendous burden on the wireless network. During the September 11 attack, for instance, call volume on the Verizon network increased up to 100% while it increased 1,000% on the Cingular network (Traynor, 2010). Likewise, during the London bombings on July 7, 2005, text message use increased 20% while voice calls increased 67% ("Mobile Networks," 2005).

Fourth, SMS servers are susceptible to security breaches. Since no protocol exists to authenticate the sender of a message, cases of SMS spoofing have been reported (Lo, Bishop, & Eloff, 2008). A related issue is the message sender's name often is not included in the SMS, leading individuals to question the credibility of the message and potentially weakening the impact of a warning (Perry & Lindell, 1991).

Clearly, relying solely on SMS to deliver crisis warning messages in a timely and efficient manner is not warranted given the limitations and challenges just mentioned. This conclusion should not be surprising, however, given the original intent of SMS to function as a means of communication from one individual to another individual (Latimer, 2008).

One recent technological advance that has been heralded by some (Palen & Liu, 2007) as an improvement over SMS for crisis communication is the use of SMS with multimedia, known as multimedia messaging service (MMS). Simply put, an MMS message, unlike an SMS message, can include embedded graphics and video, and therefore provide for richer message content. As should be readily apparent, however, MMS messages actually create more problems than they solve. The primary problem is the bandwidth required to send graphics and video. During a crisis event, the mass distribution of MMS messages would simply exacerbate the network load issue already caused by SMS messages.

Moving beyond the point-to-point SMS network

Integrated Public Alert and Warning System (IPAWS)

Since 1994, the United States government has relied on the Emergency Alert Service (EAS) to communicate time sensitive messages to the public. The EAS uses one-way analog communication channels like NOAA weather radio and television and radio broadcasts to deliver these emergency messages.

In 2006, Congress passed the Warning, Alert, and Response Network (WARN) Act that tasked the Department of Homeland Security (DHS) to upgrade the existing EAS to meet the communication needs of a mobile society. For instance, one limit of the EAS is that officials cannot "target" their message to a specific location (e.g., to a particular county) (Traynor, 2008). Since 2006, the DHS—specifically FEMA—has overseen the creation of its new Integrated Public Alert and Warning System (IPAWS). IPAWS will allow federal, state, tribal, and local government officials to deliver three types of emergency alerts to the public:

- Presidential Alerts
- Imminent Threats (e.g., tornado warnings)
- Child Abduction Emergency (AMBER) Alerts. ("Commercial Mobile," n.d.)

CAP 1.2/IPAWS Profile XML Message

```
<alert>
   <identifier>MSU-e46aeg08-459c-889c-c596-p14e857dc524</identifier>
   <sender>BaylorCountyEOC</sender>
   <sent>2010-06-12T09:23:00-05:00</sent>
   <status>Actual</status>
   <msgType>Alert</MsgType>
   <source>DemoOnly _ WatchOfficer@BaylorCounty.tx.us</source>
   <scope>Public</scope>
   <code>IPAWSv1.0</code>
   <info>
      <language>en-US</language>
      <category>Safety</category>
      <category>CBRINE</category>
      <category>Other</category>
      <event>Hazardous Materials Warning</event>
      <responseType>Evacuate</responseType>
      <urgency>Immediate</urgency>
      <severity>Severe</severity>
      <certainty>Likely</certainty>
      <eventCode>
```

Based on the message above, one can see that

- The *Watch Officer* from *Baylor County's* Emergency Operations Center sent the *Alert* message to the *Public* on June 12, 2010, at 9:23 a.m. CDT
- The message, which is in English, concerns *Safety*: The county has issued a *Hazardous Materials Warning* message and they want the *Public* to *Evacuate Immediately*
- The CAP message uses IPAWS version 1.0 code.

Figure 2. CAP message (fictitious) that uses XML formatting. Message adapted from "EAS, IPAWS evolve to meet government mandates," by R. Brown, 2010, Broadcast Engineering.

Broadcast/cable providers in the United States were required to purchase the equipment necessary to receive IPAWS-enabled messages by September 10, 2011 (Timm, 2011). If not, they faced fines levied by the FCC. The FCC also requires that IPAWS messages be delivered to the public in 10 minutes or less (Traynor, 2008).

Commercial Mobile Alert System (CMAS)

In 2008, the Federal Communications Commission (FCC), in collaboration with other government agencies like DHS/FEMA, Department of Justice, National Weather Service, and other government and industry groups, began developing the Commercial Mobile Alert System (CMAS). CMAS standards allow officials to send authenticated text-based emergency messages to wireless providers. Wireless providers, in turn, will deliver CMAS-type messages to their subscribers' cell phones. The federal government is touting IPAWS in general—and CMAS in particular—as its "Next-Generation EAS."

Common Alerting Protocol (CAP)

CMAS uses a standardized message template called the Common Alerting Protocol (CAP). CAP is an open-standard, Extensible Markup Language (XML) based system designed to provide a simplified warning process (Gerber & Ferree, 2011). This allows different alerting systems to exchange data with each other. Within this defined XML structure, officials fill in the contents of different sections with XML "tags" in the message template (see Figure 2 for an example). Thus, officials can personalize a warning message around the type of message, the time the message should be sent (and when it should expire), and the specific geographical location where the message should be sent (Gerber & Ferree, 2011). CAP also authenticates the sender of the text message, which addresses the security issues inherent with sending SMS messages through the existing point-to-point wireless network. Figure 3 illustrates what, from the mobile users' standpoint, a CAP message looks like on a cell phone screen.

CAP relies on a new network standard called *short message service-cellular broadcasting* service (CBS) ("National Research Council," 2011) to deliver CMAS messages. The FCC has asked individual wireless providers in the United States to participate voluntarily as Commercial Mobile Service providers. To date, most major wireless providers have agreed to participate. Commercial Service providers must have the capacity to deliver CMAS messages to their mobile subscribers by April 7, 2012 ("FCC's Public Safety," 2009).

To receive CMAS messages, mobile users must have a CMAS-capable handset that can read a CAP message. While only select high-end smartphones (like the latest iPhone) currently can receive a CAP message, wireless providers have agreed to make future consumer handsets CAP-enabled (Wyatt, 2011).

Cellular Broadcast Service (CBS)

To deliver CMAS messages, Commercial Service providers must use CBS technology. CBS, which is already in widespread use across Europe (Klein, n.d.), allows wireless carriers to send one-way CAP messages to every phone that is within range of an individual cell site ("National Research Council," 2011). Officials' use of CMAS messages, created with CAP standards and delivered through CBS technology, will address two limitations of sending emergency SMS text alerts. First, CBS messages use a separate data channel than the existing point-to-point network to communicate messages. As such, CBS technology is not affected by network congestion. Second, CBS technology allows emergency officials to target a message to a specific cell tower ("National Research Council," 2011), such that only individuals within range of the tower (and therefore the ones for whom the warning is intended) will receive the message.

Unlike a traditional SMS message where mobile users have to open the message, a CMAS message pops up on a mobile user's screen. Further, the CMAS message has a distinct audio tone and vibration cadence ("Personal Localized," 2011). Although all mobile users will receive Presidential Alerts, individuals can opt out of receiving Imminent Threat and AMBER Alert-type messages. Wireless customers will not be charged to receive CMAS messages.

Making CMAS less technical: Rebranding CMAS into the Personal Localized Alerting Network [PLAN]

On May 10, 2011, the FCC announced that FEMA, in partnership with the four major U.S. wireless providers (AT&T, Sprint, T-Mobile, and Verizon) would roll out CMAS messages in New York City and Washington, D.C. by the end of 2011 (Wyatt, 2011). Interestingly, the government also appears to have begun its public awareness campaign by rebranding CMAS. That is, instead of referring to the Next-Generation EAS by the relatively meaningless acronym CMAS, the government now refers to CMAS as the *Personal Localized Alerting Network*, or PLAN. ("Personal Localized," 2011). During the May 10, 2011, press conference, New York City Mayor Michael Bloomberg remarked:

> Every weekday our population of 8.4 million residents swells to more than 12 million as commuters and tourists come to town. Under the PLAN program, we'll be able to broadcast (emergency) messages to any of them who are within our target area. ("New Emergency System," 2011, para. 4)

FEMA and its partners said that the full deployment of PLAN would take place by Congress's April 2012 deadline.

Challenges for PLAN

Government officials face three significant socio-technical issues as they roll out the PLAN initiative in the United States. First, the current version of PLAN only supports English text messaging alerts.[2] Second, CMAS standards, based on the recommendation of the Commercial Mobile Service Alert Advisory Committee, by design do not allow embedded hyperlinks to be included in the message ("National Research Council," 2011). The Commercial Mobile Service Alert Advisory Committee made this recommendation for fear that once individuals receive the CAP message, individuals will browse

Figure 3. AMBER Alert message (fictitious) delivered to the Waco, TX, area. Message adapted from "New Opportunities for Public Alerting," by M. Gerber and J. Ferree, 2011. Cell phone template retrieved July 3, 2011, via Yahoo! Developer Network (http://developer.yahoo.com/ypatterns/about/stencils/). Creative Common Attributions (http://creativecommons.org/licenses/by/2.5/).

the Web from their smartphone to get more detailed information about the crisis. This, in turn, might overwhelm wireless providers' 3G and 4G networks. Third, the FCC only requires wireless providers to send text-based alerts that have no more than 90 characters ("Commercial Mobile Alert System," n.d.). Thus, a significant problem with PLAN is how to include the relevant message elements (see Figure 1) in 90 characters or less[3] without a hyperlink to richer communication channels.

Also, government officials should work closely with colleges to define what communication roles each organization will play during a crisis. The college should clearly state in its Crisis Communication Plan who should communicate directly with local emergency management officials in the event of an Imminent Threat. Indeed, in the pre-crisis communication phase, the college must work with local officials to establish what type of campus crisis constitutes an Imminent Threat.

What lies ahead for crisis communication on college campuses?

Leveraging cellular handset features

Perhaps the most popular innovation in handsets is that GPS capabilities are integrated into the phone. Indeed, for several years U.S. regulators have required handset manufacturers to include location-tracking technology in their phone so that during an emergency public safety officials can track where people are (Sharma & Vascellaro, 2008). With GPS, campus officials can now send targeted messages about school closures and during a school lock down (Craig, 2007). In the future, campus stakeholders would receive location-specific messages/maps, including evacuation routes, sent directly to their phone.

One revolutionary advance that has recently become popular is the use of a cell phone's camera to provide overlays of information in real time. This technology is known as *augmented reality* (AR). One application of AR that is relevant to emergency/crisis communicators is the use of real-time weather information (e.g., weather, radar, and current warnings) superimposed on the phone's camera image. Applications currently exist for both the Android (Wise, 2011) and iPhone operating systems (Trance, 2011). Although still in their infancy, AR applications have the ability to provide real time information in threatening situations.

Delivering these maps and creating Augmented Reality applications will likely require the increased bandwidth speed promised by wireless providers' 4G network. Further, and equally important, is that an organization leverages the power of social networks to help distribute its crisis messages.

Decentralized crisis response structures on college campuses

Large, complex organizations like the federal/state/local government and colleges have traditionally used a "command and control" model to centralize coordination during a crisis ("What Technology," 2007, para. 7). Indeed, this chapter largely has focused on organizations' use of one-way communication channels (like PLAN) during a crisis. Colleges, like the government, frequently rely on one-way communication channels between campus safety officials and their primary stakeholders (e.g., faculty, staff, students, parents, local emergency management officials).

An exciting new area that combines the scholarly study of crisis situations with the practical consideration of who actually uses and reacts to crisis warnings is an area known as *crisis informatics*. Crisis informatics adopts a more sociological perspective, and in the words of Palen, Vieweg, Liu, and Hughes (2009) "views emergency response as an expanded social system where information is disseminated within and between official and public channels and entities" (p. 469).

Within the crisis informatics model, emergency response officials increasingly leverage peer-to-peer, ad-hoc networks (also called *smart swarms* and *word-of-mouth meshes*) and use social network sites like Facebook and Twitter to help distribute crisis messages ("What Technology," 2007). College campuses are especially conducive to social networks since, according to a 2010 Pew Internet & American Life Project study, 18% of college students use Twitter. This is a significantly higher percentage of Twitter users than other segments of the general population (Young, 2010).

Moreover, innovations in CMAS-enabled cellular handset technologies, coupled with faster wireless networks (e.g., 4G networks) will enable two-way communication between campus officials and their stakeholders. For instance, the University of Maryland (UMD) announced in early 2011 that it has begun to test its V911 app. V911 allows individuals to interact with campus officials during an emergency. When the app is launched, the user is immediately voice-connected to a dispatcher in the university's department of public safety (Hickey, 2011). While the app is being developed for Android, the campus plans to expand to the iPhone and BlackBerry platform, too.

Mobile apps like V911 also highlight the central tension that exists for many Americans to be safe yet still maintain their privacy. According to a recent Nielsen report, over 50% of respondents reported that they had privacy concerns with location-based services (Bilton, 2011). Currently, PLAN messages are somewhat broad in that government officials can only target their message to specific cell towers and not to individual phones. However, in the future it could be possible for government officials to send more precisely targeted messages to an individual's cell phone (assuming that individual subscribes to a particular location-based service). The question, however, remains whether individuals will accept this technology given its potential to invade their personal privacy.

Conclusion

The crisis warning responsibilities facing college administrators have changed dramatically since the Virginia Tech shootings. Rather than relying solely on rather antiquated technologies, college officials today utilize a variety of mass notification tools and technologies. They also are beginning to integrate social networks and other interactive tools into their crisis response plan, based largely on the fact that this is the reality of many of today's college students. In the future, college administrators must work closely with government officials to establish the communication roles between the government's PLAN messages and an individual campus's mass notification service. The rapid-paced advances in both technology and online social connectedness that likely will continue almost demand that public and private organizations continue to work collaboratively to secure the public's safety.

ENDNOTES

1 The authors thank Katie Kosalek for her help with this chapter.

2 As of this writing (May 2011), the FCC is currently assessing whether to provide alerts in foreign languages as well as English ("Commercial Mobile Alert System," n.d.).

3 Technically, CMAS-enabled handsets can receive up to 15 screens (90 characters each) of text. The handset then concatenates the CAP message into a single message on the user's screen. CMAS-enable handsets also can determine if a CAP message is new or if that message has already been received by the handset ("SMS Over SS7," 2003).

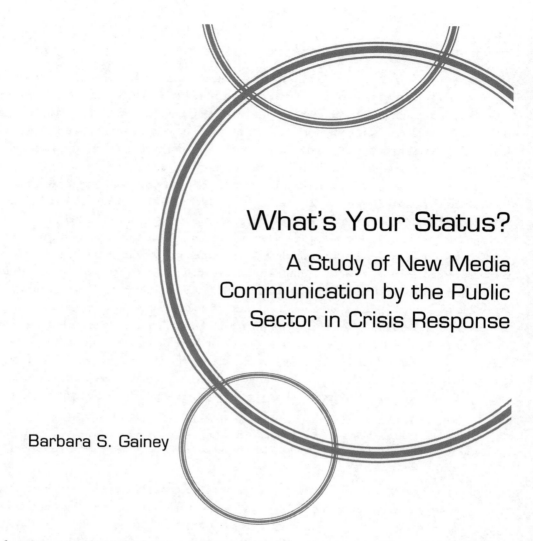

What's Your Status?

A Study of New Media Communication by the Public Sector in Crisis Response

Barbara S. Gainey

The reality of crisis management/crisis communication in the new century is that citizens, the traditional and online media, and the for-profit sector typically turn to the public sector for the initial response in a crisis. For example, public agencies are looked to for guidance in tornadoes, floods, and health threats. Communities turn to schools and civic centers as emergency shelters. Whether responding to acts of terrorism, natural disasters, or health emergencies, public-sector agencies from the Centers for Disease Control to the local health department are front-and-center in initiating initial and often long-term crisis responses. To be most effective, these public agencies are becoming adept at using new media channels to reach stakeholders. These public organizations are, in many ways, setting a new standard for online communication through the use of social media such as Facebook and Myspace, microblogging sites such as Twitter, and visual communication sites such as Flickr and YouTube. This chapter will examine the role of new media in the public sector's crisis communication response efforts.

Crisis response in the new century must take into account that crises of any magnitude often overlap boundaries, extending beyond a single organizational setting or isolated local event, and prompting the immediate attention of the traditional media and the ever-present social media network. We expect the attention of the international community to follow the military killing of a terrorist kingpin or catastrophic natural disasters that result in thousands of lost lives and threats to the safety of a nuclear power plant. Ripples of concern also follow a volcanic eruption in Iceland that affects air traffic in Europe. A deepwater oil rig calamity off the coast of Louisiana affects a corporation based in Great Britain, tourism in states along the Gulf of Mexico, and the broader seafood industry. Likewise, it sparks new discussion of reliance on oil and the safety of deep-water drilling. The H1N1 pandemic spreads worldwide, creating concerns about communal gatherings, crowded classrooms, global travel, and possible work slowdowns due to sick employees. The competitive nature of a 24/7 news cycle means that any event can catapult from local-interest item to mass media or Internet fodder. For organizations, the potential for a crisis to spread to a national or global audience becomes alarmingly real.

Citizens, the media, and the for-profit sector often turn to the public sector for that all-important initial crisis response. Whether confronted with a health crisis, a natural disaster, infrastructure failings, or terrorist threats, federal, state, and local emergency disaster agencies and other governmental units are looked to for guidance, assistance, and shelter. It is essential that public-sector organizations are crisis-ready to provide what is often the initial and, sometimes, the long-term response in crises. That crisis response most often now includes communication through social network channels, in addition to the mix of traditional media outlets. The communication response turnaround has been drastically reduced from a possible 12- hour-to-24-hour window to a more rapid response as soon as information can be confirmed and posted. The demands of a society that is increasingly mobile and electronically connected has pushed the public sector to get up-to-speed quickly with new technologies and social networks. Public agencies, especially at the federal level, have moved from passive websites that exist mainly to provide an online presence, to various stages of interactivity, online transactions, and, perhaps ultimately, to shared decision making (Chun, Shulman, Sandoval, & Hovy, 2010, p. 1). As stated by the Knight Commission (2009):

> The advent of the Internet and the proliferation of mobile media are unleashing a torrent of innovation in the creation and distribution of information....Political leaders and many government agencies are staking out ambitious agendas for openness. The potential for using technology to create a more transparent and connected democracy has never seemed brighter. (*Introduction*, pp. 4–5)

Public-sector organizations by their very nature are often required to have crisis or emergency plans in place. Adding social media and mobile technologies to their response plans have enabled public-sector organizations to step to the front in initial crisis response. This is not to imply that all public agency crisis response bases are covered 100% of the time, but that the viral direction has been set, as have new standards for the practice of public relations and crisis communication.

Crisis response best practices

Regardless of the organization, most crisis response best practices are similar. To position themselves to be able to respond most effectively, public-sector agencies must establish planned, systematic, two-way communication with key constituencies before a crisis event occurs. Like other organizations, public agencies must be able to identify key stakeholder groups and build a clear, consistent, and credible communication relationship, using communication tools to facilitate engagement and interactivity whenever possible. Public organizations are expected to be prepared to communicate with increasingly diverse publics who speak different languages; represent a variety of cultural and economic backgrounds; and have interests and needs that span different generations, races, and lifestyles.

Public organizations are also expected to have some type of warning system in place to alert agencies to potential crises in the months and years ahead. Because of their charge to protect lives and property, public organizations often try to discern threats and hazards that may emerge in the distant future. Such potential crises include natural disasters such as hurricanes, floods, earthquakes or tsunamis, health threats such as viruses and influenza pandemics, financial crises, technology attacks in cyberspace, unrest that transcends political and geographic boundaries, terrorism, infrastructural decline, and industrial and environmental disasters (Barton, 2008; Boin, Hart, Stern, & Sundelius, 2005; "Know your enemy," 2006). According to W. Craig Fugate (as cited in Achenbach, 2011), director of the Federal Emergency Management Agency, "You don't get to pick the next disaster....We plan for the things we know, but we also plan for the things we don't know" (para. 5).

Other best practices include:

- Ensure top management support for crisis management efforts (Dodd, 2006; Pauchant & Mitroff, 1992).
- Maintain a current and comprehensive crisis management plan to guide the organization in a crisis and a trained crisis team to lead crisis response efforts (Lee, Woeste, & Heath, 2007).
- Rely on a mix of traditional (newspapers, broadcast media, town forums) and new (websites, blogs, email, social networking sites) media to communicate with stakeholders before, during, and after a crisis.
- Evaluate the crisis response effort, make appropriate crisis plan and organizational changes based on what has been learned, and plan for future crisis response efforts.

Public engagement and self-efficacy

A key aspect of relying on social media in a crisis is the opportunity to engage one's stakeholders, involving them in two-way communication that is both purposeful and meaningful. Engagement and meaningful conversation demonstrate that the organization is seriously participating in a dialogue, sharing information, listening, and responding to feedback. The organization is willing to consider modifying its behavior based on this dialogue. Stakeholders have an opportunity to influence the organization, bring about change, and even share in the responsibility for making decisions that may affect them.

Engaging the public has been cited by one researcher as one of the critical issues facing public schools of the 21st century (Marx, 2000). Marx defined public engagement as: building public understanding and support, developing a common culture, building a sense of community, creating legitimate partnerships and collaborations, capitalizing on the community as a source of support, developing parent participation, and building a sense of "we" versus "us and them" (p. 88). Former director of communications for the Rockwood School District in Glenco, Missouri, and consultant on public engagement projects with the Annenberg Institute for School Reform at Brown University, J. S. Arnett (1999) said that communication with the community should be viewed as an asset rather than an expense: "Like an old-fashioned barn raising, confidence is an organization's strength, and the ability to withstand the elements is often proportionate to the number of people who participate in its construction" (p. 27). Rich Bagin, executive director of the National School Public Relations Association, indicated engagement involves a deeper level of involvement when he wrote that public engagement participants are "engaged *with* people and issues, not *by* them" (1998, p. 8).

Engagement is important because it builds trust and "habits of cooperation" (Knight Commission, 2009, p. 52) that allow people to work together for a common purpose. According to Longstaff and Yang (2008):

Perhaps the most important and least understood role for policymakers is insuring that emergency communications can be trusted by other emergency responders and by the public. This seems to mean that at a minimum, they provide for ongoing communication that helps to build trust and that they mandate plans for communications in times of [unexpected crises]. (*Introduction*, para. 8)

In addition to engagement, social media are used by public organizations to promote self-efficacy, or the ability of citizens to protect themselves and others in a crisis. Springston and Weaver-Lariscy (2007) noted that during a health crisis, stakeholders are worried about gathering information and reducing uncertainty. To this notion, Johnson-Avery (as cited in Springston & Weaver-Lariscy, 2007) added: "Reputations are going to be based NOT on how well the organization protects itself but instead on how well it protects and informs its publics" (p. 91). Stakeholders want to be empowered to protect themselves and their loved ones and may respond positively to organizational messages that promote self-efficacy (Heath, 2006).

What's your status? Social media case studies

The status of social media usage among, as well as social media resources designed for, public-sector organizations continues to evolve. A U.S. government how-to site (www.howto.gov/social-media) discussed a myriad of social media formats, including blogs, microblogs such as Twitter, wikis, video, podcasts, discussion forums, RSS feeds, and mobile apps.

Another site (http://govsocmed.pbworks.com/Web-2–0-Governance-Policies-and-Best-Practices) detailed social media 2.0 governance and best practices in a wiki format and provided a variety of social media policies in an online database (http://socialmediagovernance.com/policies.php).

FEMA, the Federal Emergency Management Agency, coordinates the federal response to disasters with other federal, state, and local agencies, and uses blogs, Twitter, and YouTube to communicate in crises. In fact, each of FEMA's regional offices has its own Twitter account (Tinker, Dumlao, & McLaughlin, 2009). When Administrator Fugate (2010) launched FEMA's first blog, he called attention to the agency's Facebook page, Twitter page, and new mobile website. He asked readers (whom he referred to as "the rest of our team," para. 4) for ideas, tips, and feedback on the agency's new communication endeavor.

The U.S. Department of Energy website (www.eere.energy.gov/socialmedia/) offered various ways for publics to connect with the agency and share information. At the time of this writing, applications available included RSS feeds, blogs, Facebook, Twitter, YouTube, widgets and gadgets, and social bookmarking. The agency had multiple Facebook pages, including Energy Savers (http://www.facebook.com/energysavers.gov), to discuss ways to conserve energy and save money. The American Red Cross, on its social media page (http://www.redcross.org/connect/), stated it was working hard to help citizens prepare for, prevent, and respond to "life's emergencies" (para. 5) through blogs, a disaster online newsroom, flickr photostream group, Twitter, Facebook pages and groups, YouTube videos, podcasts, and LinkedIn.

Clearly the established and emerging social media channels have tremendous communication potential in an actual crisis. In mid-2011, one of the more sophisticated social media sites among public agencies belonged to the Centers for Disease Control and Prevention, or the CDC (http://www.cdc.gov/socialmedia/). The H1N1 (swine flu) influenza pandemic was the first public health threat communicated, in part, through social media sites to share information and key messages about prevention and control of the disease (Taylor & Stephenson, 2009). During the H1N1 outbreak, the CDC website provided daily updates on the spread of the disease, as well as consumer health information, general public health information and FAQs, and instructions for at-risk groups. Key messages included guidance on proper hand-washing and coughing techniques to avoid the unnecessary spread of germs, among other good health habits.

The CDC's (2010b) impressive and comprehensive social media toolkit explained how the agency worked with the U.S. Department of Health and Human Services beginning in April 2009 to inform, educate, and interact with citizens about the H1N1 and seasonal flu outbreak. Tools provided included widgets that allowed interested readers to display CDC content on their own Web pages, share the CDC's messages, and become health advocates in the process. A total of 11 flu-related widgets were used in the campaign, incorporating messages about flu prevention, seasonal updates, an interactive quiz, and a national flu activity map. The CDC flu widgets were viewed more than 5 million times. Related graphics, messages, and tools were made available in English or Spanish.

The CDC launched a text messaging pilot program to provide current H1N1 alerts. Other related options offered were e-cards, which could be sent to friends and family, Twitter feeds, podcasts, RSS feeds, mobile information, email updates, and online videos. The CDC created a database of subscribers to its "Tip of the Week" messages during the flu campaign, which it was able to use to send power outage safety and evacuation information to those affected by Hurricane Ike on the Texas Gulf coast in September 2009 (Tinker, Dumlao, & McLaughlin, 2009).

Social media have quickly established a reputation as the "go-to" sites for collaboration and information sharing when telephone and email are rendered unreliable by natural disasters. After the January 2010 earthquake in Haiti, there were more than 1,500 Facebook status updates per minute containing the word "Haiti," according to one account (Keen, 2010). The American Red Cross turned to Twitter to get out messages about donating to Haiti relief efforts (Miller, 2010). The Red Cross fan page and a state chapter of LinkedIn communicated crisis information and donation information (Miller, 2010). At the same time, speed does not guarantee accuracy in the chaotic aftermath of a disaster. Social media incorrectly reported that 800 Doctors Without Borders staff were missing in Haiti (800 were working in Haiti) and a CNN Twitter hoax reported that American and JetBlue were flying medical personnel to Haiti for free (Miller, 2010).

Citing statistics from the Pew Research Center's Project for Excellence in Journalism, *PRSA Tactics* contributor Beaubien (2011) highlighted the pervasive use of social media in the aftermath of a crisis:

> Within days of the [2011] Japan earthquake and tsunami, 64 percent of blogs links, 32 percent of Twitter news links and the top-20 YouTube videos carried news and information about the crisis. An estimated 4 billion people worldwide—and 84 percent of Americans—now use mobile phones that can access social-media sites. (para. 3)

In an eight-month Pew Internet and American Life Project study of three communities, Rainie and Purcell (2011) found social media are emerging as "key parts of the civic landscape and mobile connectivity is beginning to affect people's interactions with civic life" (para. 3). Likewise, findings revealed that those who think local government shares information well are also more likely to be satisfied with other parts of civic life.

Conclusion

The sign of the times—or in Facebook-speak, the posts on our walls—leaves no doubt that social media are changing our world. Our personal, work, and community lives are affected by this new interconnectivity that is bridging geographic, economic, and philosophical divides. As organizations try to prevent or survive crises, communication with affected stakeholders must include social media in the communication mix. As organizations try to maintain or rebuild reputations touched by crisis, social media are playing a role in helping organizations reconnect through transparent, honest, and credible communication with stakeholder networks vital to their success. Often at the forefront of crisis communication, public-sector organizations must strive to overcome bureaucratic obstacles, turf issues, and government-speak to provide reliable, effective content. A number of federal, emergency preparedness, health-related, and educational institutions are leading the way. As public agencies, these organizations are sharing their experiences, as well as their expertise, through online resources.

The public sector has often been the proverbial "red-haired stepchild" of crisis literature. Much of the early writings were on the for-profit sector and were published in journals housed in business-college libraries. (Readers will likely be familiar with the gold-standard crisis response of Johnson and Johnson in the Tylenol crisis of 1982 and the case studies of other corporate crises such as Exxon, Pepsi, and Wendy's, just to mention a few.) However, the reality of catastrophic societal crises in the 21st century—natural disasters and health and terrorism threats—is often not confined to one location or one business. Transboundary crises, those that cross political, jurisdictional, or geographic boundaries, illustrate the need to rely on first responders, who in most cases represent the public sector. For now, social media are serving as relatively new tools for crisis communication. Social media serve many functions, such as providing life-saving information and enabling survivors and helpers to assist others (Palen, Vieweg, Liu, & Hughes, 2009).

Applying lessons learned from a crisis is one of the most underdeveloped aspects of crisis management (Boin, Hart, Stern, & Sundelius, 2005). The data gathering and tracking abilities of social media may facilitate a more effective learning process for organizations involved in crisis, which could be a fruitful area for future research. Ultimately, crisis response may move from a traditional centralized response structure to a more "distributed network of vast information sources and skills, including those collective skills and products generated by the public" (Palen et al., 2009, p. 478). The potential for such a change in long-standing, bureaucratic crisis response structures bodes well for the more democratic, community-building function of social media. If crisis communication is to become a more *collective* process (as indicated by recent trends), the quality of stakeholder relationships is all the more important for first responder agencies. In this spirit, Merni Fitzgerald (as cited in American Red Cross, n.d.), director of public affairs in Fairfax County, Virginia, proposed that information distributed through social media networks should be collected and shared among public agencies because disaster response "is a shared responsibility" (p. 18).

To better engage stakeholders, public organizations must continue to experiment with more interactive Web features, make liberal use of video and audio clips to enhance websites, appeal to a more tech-savvy demographic, and make existing sites easier to navigate. In a crisis, access to information should be simplified and not frustrated by sites that require time-consuming searches and multiple clicks to get to crisis-related, and potentially life-saving, information (Gainey, 2010). Crisis response plans that incorporate clear and credible social media messages as part of strategic communication efforts are likely to render benefits for responding agencies and their publics over the short- and long-term. These "innovative social practices and powerful technologies" (Knight Commission, 2009, p. 62) provide new and dynamic ways of reaching diverse audiences. Public, private, and for-profit organizations communicating together with communities and their citizens can be a mighty force when confronting the stresses and challenges of a major crisis.

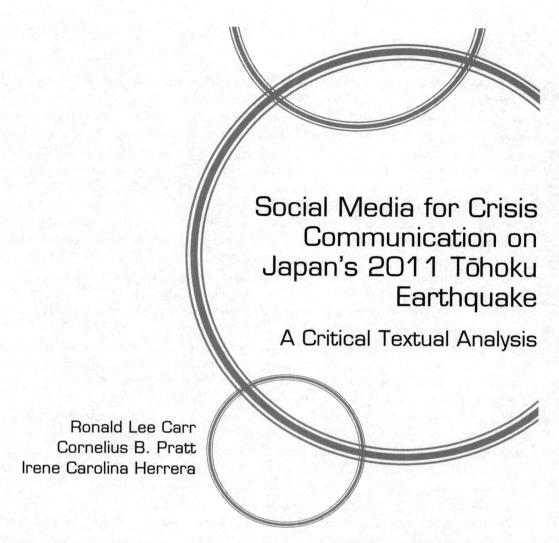

Social Media for Crisis Communication on Japan's 2011 Tōhoku Earthquake

A Critical Textual Analysis

Ronald Lee Carr
Cornelius B. Pratt
Irene Carolina Herrera

Videos (on YouTube), tweets (on Twitter), and posts (on Facebook) are redefining communication and its process. In 2010, Facebook's chief executive officer, Mark Zuckerberg, in alluding to an evolving communication revolution, wrote, "People have really gotten comfortable not only with sharing more information and different kinds, but more openly and with more people." This new communication norm, while antithetical to the Japanese traditional penchant for privacy and for indirect, nuanced communication, has major implications for relationships among the Japanese. For one thing, social media platforms, in the aftermath of Japan's Tōhoku Earthquake, a triple disaster, were used to fulfill several activities: on YouTube, to express a plea for help, to confirm information, and to hurl criticism at constituted authority; and on Twitter, as a medium of public communication, of risk communication, and of persuasion. For another, they are redefining Japanese cultural values and assumptions vis-à-vis communications—those that no longer depend solely on conventional outlets and cultural practices, but are being expanded to incorporate alternative media platforms as channels for the public good as well as for public vehemence and outrage.

Instead of marching in the streets, the Japanese have taken to social media.

—Birmingham, 2011, para. 4

Social networking sites (SNS), including Twitter, Facebook, and YouTube, are redefining the boundaries of electronic communications through a convergence of the Web, traditional communication, and everyday social interaction among humans. Thus, they are also increasingly used by organizations to communicate with and to engage their socially aware publics on a variety of issues (Fieseler, Fleck, & Meckel, 2010; Mishra, 2006); to cultivate relationships by projecting a culture of transparency, by being relevant and valuable to stakeholders, and by interacting with them (Waters, Burnett, Lamm, & Lucas, 2009); to engage and inspire networked people and empower them to participate in social change (Aaker & Smith, 2010; Kanter & Fine, 2010); and to gather e-commerce behavior information for market segmentation, in light of increasing evidence that, even though there are different motivations for using SNS, "users of these platforms are not only receptive to viral modes of ecommerce information but that these people act upon this advice" (Jansen, Sobel, & Cook, 2011, p. 134). Fan pages that connect to a social graph[1] have emerged worldwide. A number of these pages were developed in the aftermath of the 2011 earthquake-tsunami in Japan and were geared toward increasing awareness about radiation levels (and their consequences) and supporting victims in Japan through fundraising. Their use is growing significantly. As Facebook's chief executive officer, Mark Zuckerberg (as cited in Kirkpatrick, 2010) said, "People have really gotten comfortable not only with sharing more information and different kinds, but more openly and with more people" (para. 5).

The purpose of this chapter is threefold. First, it presents some of the major traditional crisis-response strategies that organizations and government agencies use to limit the consequences of crises.

Second, it identifies some of the evolving directions of those strategies, particularly those that focus on the growing use of social media like YouTube and Twitter that have become, as Muralidharan, Rasmussen, Patterson, and Shin (2011) noted, indispensable communication-management tools for organizations in crises. It accomplishes that purpose through a critical textual analysis of videos and tweets of two select social media in the aftermath of Japan's triple disaster of March 11, 2011. In essence, it seeks to answer a two-part question: What symbolic messages, as thoughts and feelings, did social media evoke about Japan's Tōhoku Earthquake in their readers, and to what extent did they indicate the presence of known categories of crisis communication strategies? Our emphasis here on social media contributes to the limited literature on their use to respond to crises (e.g., Coombs, 2008; Liu & Kim, in press; Sweetser & Metzgar, 2007).

Third, it presents the implications of its findings for identifying ways in which the use of social media can indeed be a well-organized vehicle for planned, systematic organizational responses to crises.

Japan's Tōhoku earthquake: A triple disaster

It is 2:46 p.m. Friday, March 11, 2011, in Japan. The ground shakes, tall buildings sway like tall palm trees swept by gusting winds, structures buckle, dust reduces visibility to near zero—all occurring particularly in and around Sendai.[2] Everyone in Tokyo is glued to an iPhone or to some other hand-held device, even as people's collective instincts lead them to a calm dash for the door to seeming safety. No shrieking and panicking. Some simply take cover. In some instances, a subdued cacophony ensues. There are screams, there are yells, there is quick footwork. What is happening is nothing new to the Japanese, whose lifelong experience reminds them of their high-context culture that requires instant stoicism, reasonable calm, and calculated discipline in all stressful situations. Resilience, courage in adversity, and loyalty are the bedrocks of their coping strategy. Most have experienced firsthand a series of temblors. "Is this just another one or *the* big one?" some may ask. This one is palpably different in intensity, they realize within minutes: It is taking much longer than is usual, and its severity is gnawingly troubling.

Where is the epicenter? How bad is it after all? It is widely known that when an earthquake strikes in Japan, almost everyone looks at a hand-held device at about the same time. That ritual is undertaken in search for answers—or guidance—from the newsrooms at Nippon Hōsō Kyōkai, Japan's only public radio and television corporation, whose emergency crisis team brings the public up to date on developments in a fast-developing event.

An occurrence that meets all four standard criteria for a crisis—unexpected, nonroutine, uncertain, threatening (Ulmer, Sellnow, & Seeger, 2011)—is unfolding. This time, however, with inexplicably dire consequences. Japan has been dealt its worst calamity since 1945, when the United States dropped atomic bombs on Hiroshima and Nagasaki, decimating both cities, annihilating humans, desecrating wildlife, and endangering the environment.

Initially, traditional news media were the purveyors of information on the natural disaster. Henceforward, Twitter became a main deliverer of updates: It is a 9.0-magnitude earthquake whose epicenter is off the coast near Sendai, a northeastern industrial city; a tsunami is occurring in its wake; reactors at the Fukushima Dai-ichi Nuclear Power Station are in danger of a meltdown; some coastal villages along a 273-mile (440-kilometer) stretch are being swept to sea; more than 13,000 lives are lost, more than 14,000 missing, and more than 136,000 are displaced; property is damaged and reconstruction is pegged at $300 billion; roads are impassable. There is urgent need to cool the four Fukushima nuclear reactors, to contain the spread of radioactive substances, and to monitor radioactivity levels.

Social media platforms are a tool for responding to a crisis. Japan's then-prime minister, Naoto Kan, and several of his high-profile cabinet members step to a podium in the full glare of the nation, take a full, synchronous bow to apologize to the nation for what some citizens have interpreted as the failure of the government to wrest control of the disaster. Such a public gesture is rooted in a hallmark of transformational leadership that (a) uses emotional activation and appeals to motivate stakeholders to accomplish higher-level needs at above-expected levels (Bass, 1990); and (b) produces a public expectation of a more accommodative stance than transactional leadership and a more accommodative stance when exposed to democratic leadership than to autocratic leadership (Hwang & Cameron, 2008). And it is also indicative of findings that a leader's appropriately expressed emotions in response to a crisis can inspire and motivate followers and can affect the leader's public standing (Madera & Smith, 2009); that an ethic-of-care approach is most effective in addressing stakeholder reactions to unintentional harm (Bauman, 2011); and that a leader's expressing both anger and sadness or sadness alone is perceived as a more effective leader than a leader's expressing anger alone (Madera & Smith, 2009).

Globally, social media wield significant influences on both interpersonal and organizational communications. On the interpersonal front, they are used to create and share information (Zhong, Hardin, & Sun, 2011). Organizationally, they are used by employees to better understand organizational roles, to encourage group cohesiveness and improve work processes, and to develop strong professional and personal ties (Baehr & Alex-Brown, 2010; Baehr & Schaller, 2010). Outside of organizations, employees such as those of British American Tobacco (BAT), a key marketer in China, promote their company on Facebook by posting photographs of BAT events, products, and promotional items and moderating discussion fora (Freeman & Chapman, 2010). Politically, social media are also being used to engage constituents in a dialogue on key political issues; to empower them politically (England, 2011; Grant, Moon, & Grant, 2010; Ifukor, 2010; Keane, 2009; Posetti, 2010); to foment opposition to unpopular government policies through open discussions (Dareini, 2009); and to manage grassroots political campaigns (Levenshus, 2010).

There should be no presumption that traditional media no longer have a crucial role in crises. Far from it. In fact, it must be emphasized that, for the most part, technologically savvy audiences tend to be enamored with social media, whereas older, tradition-bound audiences still find traditional media useful. Such media have advantages over social media. First, they usually are well entrenched in communities, making access to them routine. Second, even the traditional media are not entirely so anymore, as they are co-opting some of the elements of the virtual world in order to reach wider audiences. Nearly all major news outlets are accessible electronically. U.S. news networks (e.g., CBS, ABC, and NBC) update their news clips periodically online, making them as virtual as they can be. Third, they tend to have institutional credibility. *The New York Times*, for example, still has a reputation as a newspaper of record. In the United Kingdom, that attribute is widely associated with BBC News. And, to the degree that professionalism is the hallmark of traditional media, their fare has the cachet that is unusually not the province of social media, which tend to be tied to personalities and to an individual's circumstances.

The importance of this chapter is underscored by the growing communications value of these new media, which enable users to seek, create, and disseminate information, and by the increasing need to make crisis response tools more effective in light of complex socioeconomic and competitive challenges organizations face worldwide. This chapter examines how the use of such burgeoning media can help organizations to better communicate in and about crises. And, the analysis will be presented against the backdrop of the dominant characteristics of Japan's high-context communication culture, whose Confucian roots emphasize harmony and patience—and even self-deprecation.

Crisis communication strategies

Communicating effectively in a crisis requires that immeasurable time be spent thinking proactively and planning meticulously for the unforeseen. Crisis planning, therefore, requires an antithetical response to a mind-set of *ex nihilo nihil fit* (i.e., nothing comes out of nothing). It is important, for example, that crisis communication practitioners understand, predict, or anticipate public opinions as well as estimate an organization's stance on a situation so that they can more effectively counsel management on it, can better manage it, or can avoid further conflicts with specific stakeholders (Hwang & Cameron, 2008).

Strategies for communicating crisis are guided by two primary theories: Benoit's (1995, 1997) theory of image restoration or repair, and Coombs's (2007) situational crisis communication theory. Benoit's (1995, 1997) image-restoration theory assumes that because communication is a goal-oriented activity it is a vehicle for maintaining a favorable image. Even though this theory appears instantly suitable for resolving human-induced crises (e.g., Blaney & Benoit, 2001; Blaney, Benoit, & Brazeal, 2002; Harlow, Brantley, & Harlow, 2011; Kauffman, 2008; King, 2006), it is likewise relevant to addressing those caused by natural disasters (e.g., Benoit & Henson, 2009). Benoit's (1995, 1997) five general categories of self-defense strategies in a crisis are denying charges, evading responsibility, reducing the severity of the offensiveness of a wrongful act, taking corrective actions, and admitting wrongdoing and begging for forgiveness.

Coombs's (2007) situational crisis communication theory (SCCT) categorizes crises into three clusters: victim (low attributions of crisis responsibility); accidental (moderate attributions of crisis responsibility); and preventable (high attributions of crisis responsibility). Each requires specific organizational responses grouped into three postures: deny, which indicates low concern for crisis victims; diminish, which seeks to reduce organization's vulnerability to the crisis; and deal, which indicates high concern for victims and high acceptance of responsibility.

There is also a critical ethical, managerial component of crisis communication strategies—a rational and morally driven response to events and "a sustainable and ethically attractive method for addressing crises and providing ethical outcomes for stakeholders" (Snyder, Hall, Robertson, Jasinski, & Miller,

2006, p. 373). Bauman (2011) identified three general ethical approaches that leaders use in managing a crisis: virtue ethics, ethic of justice, ethic of care. Virtue ethics focuses on "how an agent's character and dispositions guide decision making. A person's character takes precedence over strict ethical codes or merely considering the consequences of one's actions" (Bauman, 2011, p. 283), as utilitarian ethics requires. An ethic of justice "avoids taking the feelings of others into account and impartially considers the rights of all parties involved before making a decision. [It] is characterized by using objective standards, making impartial judgments, and resolving conflicting rights" (Bauman, 2011, p. 284). The decision of the government of Japan to evacuate *all* residents of areas within a 12-mile (20-kilometer) radius of the Fukushima Dai-ichi Nuclear Power Station is a lucid illustration of a leadership grounded in an ethic of justice. Finally, whereas an ethic of care does not consider an individual's feelings, but those of the collectivity, considering the feelings of others is critical for building, maintaining, and strengthening stakeholder relationships and for fulfilling an organization's responsibilities to others.

It is clear that both image restoration and SCCT theories, cast within an ethical framework, were indicated in Japan's crisis communication in the aftermath of the triple disaster. For one thing, the government denied on several occasions that it was not doing enough to put its full weight behind its rescue and recovery efforts. For another, the government apologized publicly for its limitations in bringing the Fukushima Dai-ichi nuclear station under control within a more reasonable time. And, the government showed on several occasions that its ethic of care was strengthening and deepening its relationships with its citizens.

Critical textual analysis

Crisis communication managers strive to accomplish two tasks: to share information with their stakeholders and to persuade them in the process. Social media, as platforms of crisis communication, are used as vehicles in a rhetorical discourse of the issues associated with a crisis. The importance of such platforms is accentuated by the reality that whenever a threat (e.g., that posed by the Fukushima nuclear reactors) cannot be perceived by human senses, dependence upon others for information increases. Therefore, in Japan, public education became even more important in bridging the gap between experts and publics in radiation risk perceptions and ratings (Weisæth, Knudsen, & Tønnessen, 2002).

What themes of information dissemination were indicated in those platforms? And in what form? To provide an answer to that question, this chapter presents a textual analysis of the Tōhoku triple disaster of March 11, 2011, as presented on YouTube, Twitter, and Facebook. In the context of this study, textual analysis enables a researcher to decipher the overarching themes based on the sum of mini-messages on SNS. The analysis is systematic and invokes a broader process of critical reflection on various interpretations of messages and ideological streams embedded in text. The assumption here is that readers, as active consumers of texts, ascribe meanings and specific ideologies to their themes and subthemes.

Textual analysis is an essential component of critical discourse analysis in that it draws upon different, even though particular, discourses, views, and perspectives, presenting opportunities for articulating a structured whole (Fairclough, 2003). This analysis drew upon McKee's (2003) post-structural approach, Hall's (1975) critical analysis of text, and Lester-Roushanzamir and Raman's (1999) three steps that focus on defining text, reading delimited text, and identifying audience interpretation. The critical component of textual analysis empowers researchers to examine how an audience questions assumptions and focuses their discourse on the implications of messages and on the actions they imply. There were several occasions in which rallies and protests were directed against Japan's constituted authority: Tokyo Electric Power Company and the government of Prime Minister Naoto Kan.

Selected texts

We began our analysis by reviewing, over eight weeks (March 11–May 7, 2011), videos (on YouTube) and tweets (on Twitter) on the Tōhoku earthquake. We then identified their relevance to the disaster and viewed/read them closely, specifically for two overarching variables: content (i.e., themes) and form (i.e., use of explicit versus implicit language).

What we learned

Our analysis of government-citizen relationships and of grassroots discourse in the aftermath of the disaster indicates an added role for the government as a boundary-spanning crisis manager: a manager who serves as both an advocate for and a conscience of the community.

From YouTube

At 9 p.m. Thursday, March 24, 2011, Katsunobu Sakurai, the mayor of Minami-soma, in the Fukushima prefecture, which is close to the epicenter of the March 11 earthquake, sat in front of a small digital camera manned by local resident Kenichiro Nakata ("SOS from Mayor," 2011). He spoke to the world through a medium with which the 55-year-old mayor wasn't himself particularly familiar. Shortly thereafter, Sakurai's 11-minute, 13-second message was on YouTube. By May 5, the site had received nearly 450,000 hits ("SOS from Mayor," 2011), and *Time* magazine in 2011 named Sakurai one of the world's 100 most influential people (Beech, 2011). The content and form of the broadcast indicated to Sakurai's most elderly constituents that he had abandoned Japan's traditional politesse, that there was an evolving communications pattern that was uncannily sublime, and also that political rhetoric was no longer confined to traditional news outlets such as print media and television.

The mayor's YouTube discourse had three themes: a plea for help, a confirmation of factual information, and a criticism of authority. After expressing appreciation to all those who had helped in various communities during Japan's national strife, he asked for volunteers to help bring in goods. Most heart-rending was the mayor's plea for petrol, without which citizens could not evacuate or even find their loved ones, some of whom had been swept to sea.

The importance of the mayor's last-ditch effort to help his community cannot be overlooked. Most significant was his criticism of various organizations, including government, private industry, and the domestic media. Mayor Sakurai used a direct, uncharacteristically vehement, no-holds-barred style in criticizing the government and TEPCO, the world's fourth-largest utility company and owner of the nuclear plants. "And with the scarce information we can gather from the government or TEPCO, we are left isolated." Mayor Sakurai then informed his audience that the people of Minami-soma were evacuating "voluntarily" and the "city administration" was offering its support. The mayor dismissed the national government's attempts at managing the disaster and taking over the duties of a local government. The mayor implied that it was the national government's orders to residents that they "stay inside" that ultimately affected their ability to survive. That contrarian stance taken by a local mayor not only challenged the orders of Kan's administration, but also implied that its directives had contributed to the present problem.

On the video, Mayor Sakurai showed a poignant disdain for the Japanese Fourth Estate. On YouTube, he complained that Japanese journalists were simply "gathering information over the phone," rather than going to the stricken coast and seeing firsthand what was happening. The mayor claimed that without "stepping foot into the area" there was no way the media could ever understand developments in his community. In other words, the conventional press has failed to perform its most vital function: to inform its viewers of the facts. Such direct, even incendiary, criticism, made publicly, was antithetical to the Japanese traditional communication style that is consistent with that of high-context cultural values and assumptions.

Mayor Sakurai's 11-minute, 13-second plea to the world established him as a breed apart in the Japanese cultural landscape that reveres using succinct, indirect, implicit communication at all times, even in instances of national strife. But Sakurai shared the facts as he knew them and also challenged authority in his attempt to return the management of his city to its rightful owners: the people of Minami-soma city.

From Twitter

Two days after the earthquake, the Japanese prime minister's office launched a Twitter account (@ KanteiSaigai) to announce live updates of the crisis as events unfolded. Three days later, an English-language version of that account (@JPN_PMO) was launched to provide real-time, 140-character updates to the international community. Within a week, TEPCO also inaugurated an account in Japanese. As of this writing, it had more than 307,000 followers and 73 tweets, but the utility did not have an English-language account.

Government-managed tweets indicated three themes: public information, risk communication, and persuasive communication. A journalistic-style timeliness of the information disseminated publicly was geared toward bringing both domestic and international audiences up to date on internal developments. Initially, the government provided enough information to the public, empowering it to sidestep traditional media outlets. By May 8, 2011, the Japanese-language account had more than 330,000 followers and 888 tweets; the English-language version more than 25,000 followers and 170 tweets.

The Twitter account was used mostly to tweet breaking news and to quote high-level government officials. To reinforce its online presence, the prime minister's office also launched a Facebook account (http://www.facebook.com/Japan.PM), which, as of this writing, had more than 8,500 fans. This initiative illustrates the government's interest in having a ubiquitous presence on the Web. It also demonstrates the intricate interdependence between and convergence of traditional and Web-based social media.

Two weeks after its first tweets, a mock TEPCO CEO account (https://twitter.com/#!/TEPCO_CEO) also hit the Twitter viral stream, posting ironic and satirical texts as a form of protest against and criticism of TEPCO's inability to give concrete, precise, and transparent information on the status of the Fukushima nuclear plant.

The second apparent theme was risk communication. The national government provided information to the public, enabling people to assess their own susceptibility to radioactive materials from the explosion at the Fukushima Dai-ichi nuclear plant, and to the nuclear-engineering community with both an interest and an expertise in reducing such risks.

The third theme was persuasive. The national government exhorted the public to be supportive of the nation's energy-conservation practices, including its acceptance of rolling blackouts. The English-language account had imperative tweets that urged all citizens to conserve electricity and allowed that "it was understandable" for foreign governments to require that their nationals adhere to a much larger evacuation zone of 50 miles (80 kilometers), significantly more than the 12-mile (20-kilometer) zone suggested initially by Japanese officials.

Twitter and Facebook were not only used by the prime minister's office and TEPCO to share first-hand information with citizens, but also served as platforms for grassroots participation in a public discourse. People on the street, volunteer groups, and nongovernmental organizations expressed their thoughts on the situation. Japanese Twitter users indicated discontent with TEPCO's radioactive materials-containment plan, protested and exchanged eyewitness accounts, obtained information on developments, and searched for loved ones.

Conclusion

Ledingham (2001) noted that relationship management emphasizes the importance of identifying and constructing programs that are based on shared interests and measuring specific outcomes. It holds that publics expect mutuality in their relationships with an organization. It is this reciprocity, which occurs within the framework of relationship management, that sustains and nurtures a community relationship, which is related to personal relationships. Ledingham (2001) wrote, "When shared interests are the basis for public relations initiatives grounded in a commitment to mutual benefit, and when those initiatives are designed to accommodate differing interests, then community *can* be the result" (p. 292). Wilson (1994) described it as a shift in management styles from the tradition of power and influence to corporate community-building, cooperative relationships, and relational responsibility. That said, the lessons of this chapter lead to four major implications.

First, social media are redefining communication patterns among Japanese in crisis situations. To the degree that the mayor of Fukushima used YouTube to communicate and maintain relationship with his constituents, most of whom are middle-aged or older, suggests that the traditional communication modes are playing second fiddle to SNS in communities, regardless of the age of the encoder and decoder. The Japanese tradition of interdependence and harmony requires that they communicate in an indirect rhetorical style, which encourages the use of implicit and ambiguous words and the avoidance of leaving an assertive impression (Gudykunst & Ting-Toomey, 1988; Okabe, 1983). The redefinition of such a style points to robust prospects for the increasing use of SNS in crisis situations in a manner that could encourage crisis communication managers to reconsider their professional communications within existing contexts of extant theoretical frameworks: Benoit's (1995, 1997) image-restoration strategies and Coombs's (2007) situational crisis communication theory.

Second, SNS are not only communication tools, but also have a therapeutic value. These platforms reflected the lives, attitudes, and beliefs of the everyday Japanese, who, at once, vented anger, frustration, and grief. Thus, the Japanese penchant for privacy was undermined by the importance of public expression of anger in a crisis of the magnitude experienced on March 11, 2011 (Glionna & Hall, 2011). For the Japanese, and perhaps for other Asian societies, such public expressions hold indicators of relationship building and maintenance in a period of unusual national challenges. A local government's ethic of care, demonstrated in its use of, say, YouTube, strengthens and deepens its relationships with its citizens.

Third, an implication of the findings of this analysis for the nature of Japanese government-citizen relationships is that the dynamic of such relationships is as much a function of traditional communication patterns as it is that of evolving communication technologies grounded in globalization. Even though the government strives to engender a caring, citizens-focused relationship, that relationship will be anchored on the evolving characteristics of the nature and intensity of disasters to which such relationships are subjected, and against which they are evaluated. Those possibilities, therefore, point to the perceived level of credence of a government agency and to the agency's credibility among its publics. Again, the use of an SNS by the mayor of Minami-soma could suggest an emergence of SNS as a medium of citizen dissent, even in societies noted for harmony in all forms of government-public relationships.

Finally, and consistent with the third purpose of this chapter, this analysis points to the relevance of Coombs's (2007) SCCT as a framework for managing communications with constituents directly affected by a natural-technological crisis. Consistent with Coombs's (2007) SCCT framework, the government of Japan fell into the "victim cluster" (p. 137), in which it had very weak attributions of crisis responsibility and was perceived as a victim of the natural disaster. Yet, the government exercised an ethic of care, positioning itself to maintain and strengthen its relationships with constituents. That public position may, however, have been undercut by the yakuza, Japan's "organized crime syndicates" (Kingston, 2011, p. 227), which, even though viewed by the government as "a social evil" (p. 236), has nevertheless responded quickly to new risks and challenges of the 21st century by, for example, being among the first to donate relief supplies to disaster victims in the hardest-hit prefectures of Fukushima, Iwate, and Miyagi (McCurry, 2011), just as it did during the Kobe earthquake of 1995, when it opened the first soup kitchens for displaced survivors (Kingston, 2011). In any event, this analysis supports the premise that a relationship with constituents can be better nurtured within a "victim cluster" that is enabled, as it were, by videos, tweets, and posts on a government's stance and its disaster response.

ENDNOTES

1 A social graph, a network of people with shared interests, seeks to build and maintain loyalty and trust among participants of that network.

2 The first author, a Tokyo resident, recounted his experiences shortly after 2:46 p.m. on March 11: "'This is it,' I thought. Even so, I calmly helped students out of Azabu Hall on our Tokyo campus. I watched the construction site of a near-completed, high-rise building across the street. The building was standing intact and looked secure and I thought for one fleeting moment how solidly engineered Japanese buildings were."

And the third author, also a Tokyo resident, observed: "I was in a supermarket at the time it struck. I quickly realized it was a much stronger earthquake after things began falling off the shelves. A Japanese girl and I ran to an open area clear of any glass windows or tall buildings. Even though it was shocking, it was nothing compared to what I later saw on TV once I got home, as a tsunami swept the whole coast of Tōhoku and aftershocks continued. Fear and panic definitely increased the next day when smoke gushed from nuclear plants in the northeast. For a moment, Japan had collapsed."

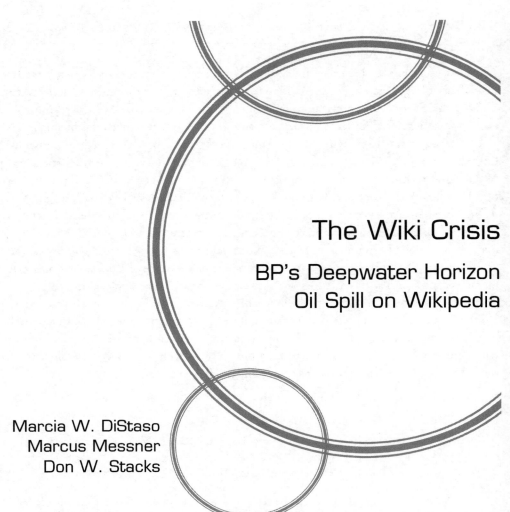

The Wiki Crisis

BP's Deepwater Horizon Oil Spill on Wikipedia

Marcia W. DiStaso
Marcus Messner
Don W. Stacks

In times of crisis, people turn to the Internet and the use of social media increases. In fact, social media coverage is believed to have a higher level of credibility than traditional news media coverage. On April 20, 2010, the explosion on the Deepwater Horizon oil platform in the Gulf of Mexico caused an existential crisis for BP and information about it quickly became available on Wikipedia. By analyzing the Wikipedia content for this crisis over a year from the day of the explosion, this study found that the crisis information at the fingertips of the public was a top loading in search engines, had a higher number of views and more edits than the main BP article, predominately contained negative content, and had high levels of news media references.

Introduction

Long gone are the days when content on the online encyclopedia Wikipedia was regularly criticized for its inaccuracies. Initially, the "wiki wars" and "Wikipedia politics," in which hoaxes were created purposefully to damage reputations, caused headaches for public relations practitioners (Boxer, 2004; McCaffrey, 2006). Today, however, it is the meticulous referencing and high reliability of Wikipedia that constitute the challenge for practitioners. Every scandal and legal issue, no matter how far in the past, is documented and fermented in the articles on major corporations. Changing statements on Wikipedia that are referenced and fact-based has become a difficult and time consuming task (DiStaso & Messner, 2010).

Through its collaborative nature, Wikipedia gives the public an opportunity to shape the knowledge about any important issue of our time, including the image of major corporations like BP. This often leads to the publication of a variety of topics—many of which corporations would rather not see as prominently. Subsequently, the information is widely used by the media and the public, often appearing right next to the company website in online searches (DiStaso, Messner, & Stacks, 2007; Messner & South, 2011). During a crisis like BP's Deepwater Horizon oil spill in the Gulf of Mexico during the spring and summer 2010, the Wikipedia article is updated immediately and continuously. Information is critical in times of crisis, and often the damage from a crisis occurs not because of the incident itself, but rather because of the response to it and the lack of information (Coombs, 2012).

The purpose of this chapter is to incorporate academic research and practical advice to explain why Wikipedia is a social medium that companies must pay attention to during crises. Tips on how to properly handle it are provided. This includes an analysis of Wikipedia content before, during, and after the BP Deepwater Horizon crisis. As the findings show, monitoring Wikipedia should be a part of all social media plans, especially in a crisis. Public relations practitioners cannot afford to forget about Wikipedia and its effect on the public.

Literature review

Social media and the online crisis

Fearn-Banks (2002) identified a crisis as "a major occurrence with a potentially negative outcome affecting the organization, company, or industry, as well as its publics, products, services, or good name" (p. 2). More broadly, it can be looked at as an event that "creates an issue, keeps it alive, or gives it strength" (Heath & Palenchar, 2009, p. 278). The perceptions of stakeholders are what define an event as a crisis, so if stakeholders believe an organization is in crisis, it is (Coombs, 2012). The need for truthful and timely information is magnified in times of crisis. Often a company's reputation is damaged more from the way the company handles a crisis than from incident itself (Galford & Drapeau, 2003).

In times of crisis, people turn to the Internet (Discovery, 2011) and the use of social media increases. Social media are often seen as a way to identify warning signs about a potential crisis, but they can also serve as an effective means for providing updated information about a crisis (Coombs, 2008). At times, social media coverage is believed to have a higher level of credibility than traditional media coverage (Sweetser & Metzgar, 2007) because social media provide current, unfiltered communication (Johnson & Kaye, 2010). Ultimately, the Internet provides the public with a plethora of information that in a crisis is much more difficult or even impossible to get through other means (Bucher, 2002).

For organizations, however, "in a crisis, social media is a game changer" (Wright & Hinson, 2010a, p. 22) because news moves quickly over the Internet, and companies can no longer take time to develop message strategies before acting.

Despite the increased use of social media in crises, existing research is still limited, although growing. The majority of current crisis research focuses on blogs (e.g., Liu, 2010; Oyer, 2010; Scoble & Israel, 2006; Sweetser & Metzgar, 2007) and websites (e.g., Bucher, 2002; Perry, Taylor, & Doerfel, 2003; Taylor & Kent, 2007), but some researchers have explored microblogging (e.g., Schultz, Utz, & Göritz, 2011) and social networking (e.g., Tyma, Sellnow, & Sellnow, 2010). A common conclusion is that individuals use social media to seek immediate and in-depth crisis information.

According to The Holmes Report (as cited in Holmes & Sudhaman, 2011), BP's Deepwater Horizon oil spill in the Gulf of Mexico was the top corporate crisis of 2010. The crisis developed after an April 20 explosion on the oil platform Deepwater Horizon and the oil spill that followed, which was not contained until July 15. The completion of the relief well and thereby the end of the immediate crisis took until September 19. The company faced continuous public criticism for causing the spill but also for its crisis management. The entire Gulf region still struggles today with the economic and environmental damages from the spill (Harlow, Brantley, & Harlow, 2011; Sellnow, 2010).

Wikipedia and corporate reputations

Wikipedia is one of the most popular websites in the world and, thereby, greatly influences Internet users (Alexa.com, 2011). Started in 2001, it today has more than 3.6 million articles in English and 18 million articles across all of its 281 language editions. As of May 2010, 53% of American Internet users used Wikipedia to look for information (Zickuhr & Rainie, 2011). Initial criticisms of hoaxes and inaccuracies caused by the collaborative nature of the online encyclopedia's contribution and editing process have quickly turned into praise for its breadth, timeliness, and reliability (Cohen, 2011; Crovitz, 2009; Giles, 2005; Guentheroth, Schoenert, & Rodtmann, 2007; Leung, 2009). Wikipedia's high usage rates have also moved its articles high on the search results of Web search engines (DiStaso & Messner, 2010; Langlois & Elmer, 2009; Stross, 2009). Messner and South (2011) found that Wikipedia has been framed as credible and accurate in traditional news media coverage.

Subsequently, public relations practitioners have increasingly paid attention to Wikipedia. Wright and Hinson (2010a) stated that companies pay more attention to the impact of social media on their reputations, and Dizikes (2008) observed that practitioners actively engage in editing of Wikipedia entries about their companies. Based on DiStaso and Messner's (2010) findings, companies should pay close attention to their Wikipedia articles. Their longitudinal study on the tonality of entries and framing of *Fortune 500* companies found that negative content and mentions of legal issues and scandals increased over time on Wikipedia. The findings also demonstrated that corporate entries are heavily edited by a great variety of contributors. Dizikes (2008) stated that for companies, "Wikipedia has become something of a battleground for the truth, or, at least, a kind of operating history" (para. 4).

Though studies have analyzed the framing of corporations on Wikipedia, scholars have yet to examine the influence of references on the encyclopedia's content. As McCombs (2005) argued, sources can have strong agenda-setting effects on content. Fernando (2010) stressed the importance of references in the discussions within the Wikipedia community: "An article needs to be well referenced or it will be taken down" (p. 9). According to Fullerton and Ettema (2009), Wikipedia editors regularly engage in discussions over authoritative information and settle disputes through referencing, especially to reports in mainstream media. In major breaking news situations, Wikipedia is often updated quickly with references to media reports from around the world (Melanson, 2010; Siegel, 2007).

Research questions

Based on this discussion, the following research questions were explored:

RQ1: What potential influence do the BP Deepwater Horizon crisis Wikipedia articles have and is it different than that of the main BP article?

RQ2: How is public opinion formed on BP Deepwater Horizon crisis Wikipedia articles and is it different than that formed by the main BP article?

RQ3: How did the amount of BP Deepwater Horizon crisis Wikipedia content change over time?

RQ4: What is the tone of the BP Deepwater Horizon crisis Wikipedia articles and is it different than that of the main BP article and the crisis content on the main article?

RQ5: How many references are used in the BP Deepwater Horizon crisis Wikipedia articles and are they different than those in the main BP article and the crisis content on the main article?

RQ6: What type of references are used in the BP Deepwater Horizon crisis Wikipedia articles and are they different than those in the main BP article and the crisis content on the main article?

RQ7: What is the tone of BP Deepwater Horizon crisis Wikipedia content with BP references?

Method

This study builds on previous research by DiStaso and Messner (2010) who analyzed the Wikipedia entries of the top ten *Fortune 500* companies over time. Although they found that most company Wikipedia articles contained content about corporate legal problems, scandals, and crises, they did not focus on specific crisis content or look beyond the main articles. The goal of this study was to determine how a major crisis is portrayed on Wikipedia over time and across different articles. A content analysis was conducted for the top corporate crisis in 2010: BP's Deepwater Horizon oil spill in the Gulf of Mexico (Holmes & Sudhaman, 2011).

The main Wikipedia article for BP was created on August 27, 2002, and not too long after the explosion on April 20, 2010, two supplemental crisis Wikipedia articles were started: Deepwater Horizon Oil Spill (henceforth called "Oil Spill") and Deepwater Horizon Explosion (henceforth called "Explosion"). All three Wikipedia articles were analyzed sentence-by-sentence.

The analysis was conducted for three separate time points. The main article was analyzed on April 19, 2010 (a day before the explosion), October 20, 2010 (six months into the crisis), and April 20, 2011 (the one-year anniversary of the explosion). The Oil Spill crisis article was analyzed on April 21, 2010 (the day it was created), October 20, 2010 (six months into the crisis), and April 20, 2011 (the one-year anniversary of the explosion). The Explosion crisis article was analyzed June 6, 2010 (the day it was created), October 20, 2010 (six months into the crisis), and April 20, 2011 (the one-year anniversary of the explosion). The coding was completed by two coders, with an intercoder reliability coefficient of .91 using Scott's (1955) *pi*.

Potential influence

Similar to how potential influence was analyzed by DiStaso and Messner (2010), this study identified the search loadings for BP and the Deepwater Horizon oil spill crisis. The most common search engines at the time of this writing are: Google, Yahoo! and Bing (Alexa.com, 2011). Unfortunately, the analysis was not conducted during the crisis, but instead on April 27, 2011—one year and seven days following the BP Deepwater Horizon explosion. A search for BP was conducted along with a search for the crisis using the terms "BP Oil Spill" and "BP Explosion." Page views were also calculated for each of the articles using the Wikipedia article traffic statistics tool.

Formation of public opinion

The people who write and edit Wikipedia articles form public opinion on those topics (DiStaso & Messner, 2010). This is no different in a crisis. Therefore, this study also looked at rigor (total number of edits) and diversity (total number of unique users). This essentially identifies the scrutiny an article receives (the more edits, the more scrutiny) and how many people participated in the formation of public opinion on the topic (an active public).

Tone

As in previous research (e.g., DiStaso, Messner & Stacks, 2007; Kweon, 2000; Michaelson & Griffin, 2005), tone was analyzed by coding each sentence as positive, negative, or neutral. For example, a sentence was coded as positive if it said something like "BP currently invests over $1 billion per year in the development of renewable energy sources, and has committed to spend $8 billion on renewables in the 2005 to 2015 period." An example of a negative sentence is "In 1991 BP was cited as the most polluting company in the US based on EPA toxic release data." An example of a neutral sentence is "BP has operations in over 80 countries, produces around 3.8 million barrels of oil equivalent per day and has 22,400 service stations worldwide."

References

The references were collected from each article, themes were identified, and the data were categorized.

Results

This study analyzed the Wikipedia content and statistics for BP's Deepwater Horizon oil spill in the Gulf of Mexico. This included sentence-by-sentence coding for three different Wikipedia articles at three different times: BP, Deepwater Horizon Oil Spill, and Deepwater Horizon Explosion.

RQ1: What potential influence do the BP Deepwater Horizon crisis Wikipedia articles have and is it different than that of the main BP article?

Even a year after the Deepwater Horizon explosion, the main crisis article was a top loading when a search for the crisis was conducted (see Table 1). The main article for BP loaded after the company website(s) but was still the second listing in Yahoo! and Bing (see Table 1).

Table 1. Search Loadings

	MAIN ARTICLE	CRISIS ARTICLE*
Google	5	1
Yahoo!	2	1
Bing	2	1

* *The search results linked to the Deepwater Horizon Oil Spill article when searching for both the oil spill and the explosion.*

The Oil Spill crisis article was viewed a total of 3,662,815 times from its creation on April 21, 2010, until April 20, 2011 (the one-year anniversary of the explosion). May and June 2010, the months right after the explosion and beginning of the oil spill, were the busiest months for viewing.

The Explosion crisis article was viewed a total of 197,171 times from its creation on June 6, 2010, until April 20, 2011 (the one-year anniversary of the explosion). July and September 2010 were the busiest months for viewing.

The main Wikipedia article for BP was viewed a total of 2,323,199 times from April 19, 2010 (the day before the explosion), until April 20, 2011 (the one-year anniversary of the explosion). May and June 2010 were the busiest months for viewing.

The Oil Spill crisis article was the most viewed out of the three BP articles during the analysis time period, with 1,339,616 more views than the main BP Wikipedia article; the Explosion crisis article was the least viewed out of the three.

RQ2: How is public opinion formed on BP Deepwater Horizon crisis Wikipedia articles and is it different than that formed on the main BP article?

From its creation on April 21, 2010, until April 20, 2011, (the one-year anniversary of the explosion) the Oil Spill crisis article had a total of 5,846 edits (rigor) made by 1,717 editors (diversity). This was an average of 3.4 edits per person.

From its creation on June 6, 2010, until April 20, 2011, (the one-year anniversary of the explosion) the Explosion crisis article had a total of 284 edits (rigor) made by 154 editors (diversity). This was an average of 1.84 edits per person.

From April 19, 2010, (the day before the explosion) until April 20, 2011, (the one-year anniversary of the explosion) the main BP Wikipedia article had a total of 1,619 edits (rigor) made by 776 editors (diversity). This was an average of 2.09 edits per person.

The difference between the Oil Spill crisis article and the main BP article was 4,227 edits and 941 editors.

RQ3: How did the amount of BP Deepwater Horizon crisis Wikipedia content change over time?

Since the main BP article was analyzed the day before the crisis, the first date found no content about the crisis, but 17.8% of the main BP article was about the Deepwater Horizon crisis on the six-month anniversary of the explosion ($n=46$) and 14% on the one-year anniversary ($n=34$).

The Oil Spill crisis article was started on April 21, 2010. At the end of that day, the article had 25 sentences. On the six-month anniversary of the explosion, there were 522 sentences, and 689 at the one-year anniversary.

The Explosion crisis article was split from the Oil Spill article on June 6, 2010. At the end of that day, the article had 83 sentences. On the six-month anniversary of the explosion, there were 104 sentences, and 115 at the one-year anniversary.

RQ4: What is the tone of the BP Deepwater Horizon crisis Wikipedia articles and is it different than that of the main BP article and the crisis content on the main article?

The Oil Spill crisis article and the Explosion crisis article both had high percentages of negative content (see Table 2). Over time, both articles saw an increase in the percentage of positive sentences, but also an increase in the percentage of negative sentences. Although the Oil Spill did start out with 23 neutral sentences and 22 negative sentences, as it grew it rebalanced the tone to contain low percentages of neutral content and high percentages of negative content. By the one-year anniversary, the Oil Spill article was 74.7% negative, but did contain 10.2% positive content. The Explosion article was 68.7% negative and 7.0% positive at the one-year anniversary. The full main BP article contained 33.1% negative content and 15.7% positive, and the crisis content in the main BP article was 55.9% negative and 17.6% positive.

RQ5: How many references are used in the BP Deepwater Horizon crisis Wikipedia articles and are they different than those of the main BP article and the crisis content on the main article?

The Oil Spill crisis article had 16 references (0.64 references per sentence) on the day it was created. On the six-month anniversary of the explosion it had 419 (0.80 references per sentence), and 501 at the one-year anniversary (0.73 references per sentence) (see Table 3).

The Explosion crisis article had 84 references (1.01 references per sentence) on the day it was separated from the Oil Spill crisis article. On the six-month anniversary of the explosion it had 103 (1.01 references per sentence), and 110 at the one-year anniversary (0.96 references per sentence).

The main BP Wikipedia article had 77 references (0.44 references per sentence) on the day before the explosion. On the six-month anniversary of the explosion it had 158 (0.61 references per sentence), and 152 at the one-year anniversary (0.63 references per sentence). This is fewer references on a sentence basis than both of the crisis articles.

The crisis content on the main BP article was much shorter than the crisis articles, but the number of references was very similar to the number of references in the crisis articles. There was no crisis content on the day before the explosion, and at six months there were 40 references in the crisis content (0.89 references per sentence), and 32 references at the one-year anniversary (0.94 references per sentence).

Table 2. Tone

	FULL MAIN			CRISIS IN MAIN			OIL SPILL			EXPLOSION		
	Early	Mid	Last	Early	Mid	Last	Early	Mid	Last	Early	Mid	Last
Positive	12.1% $n=21$	10.9% $n=28$	15.7% $n=38$	0	17.4% $n=8$	17.6% $n=6$	8.2% $n=4$	10.7% $n=56$	10.2% $n=70$	2.4% $n=4$	7.7% $N=8$	7.0% $n=8$
Negative	27.0% $n=47$	33.7% $n=87$	33.1% $n=80$	0	58.7% $n=27$	55.9% $n=19$	44.9% $n=22$	69.5% $n=363$	74.7% $n=515$	66.3% $n=109$	65.4% $N=68$	68.7% $n=79$
Neutral	60.9% $n=106$	55.4% $n=143$	51.2% $n=124$	0	23.9% $n=11$	26.5% $n=9$	46.9% $n=23$	19.7% $n=103$	15.1% $n=104$	31.3% $n=52$	26.9% $N=28$	24.3% $n=28$

RQ6: What type of references are used in the BP Deepwater Horizon crisis Wikipedia articles and are they different than those in the main BP article and the crisis content on the main article?

Both crisis articles predominately contained news media references. The Oil Spill crisis article went from 81.3% at its start to 73.0% at the six-month anniversary and 73.9% at one year. The Explosion crisis article went from 77.4% to 80.6% and then to 79.1% at the one-year anniversary (see Table 3).

Although the highest number of references was found for the news media in the main BP article and the crisis articles, the main article only had 41.6% to 51.3% of news media references whereas the Oil Spill article ranged from 73.0% to 81.3% and the Explosion article from 77.4% to 80.6%. The difference in the main article was made up by higher percentages in BP corporate references.

The types of references in the crisis content on the main page were very similar to the crisis articles, although they were lacking in governmental references and had between 85% and 87.5% news media references.

RQ7: What is the tone of BP Deepwater Horizon crisis Wikipedia content with BP references?

When combining all of the crisis content from the main BP article and the two crisis articles, 59.1% of the references were for neutral sentences ($n=13$), 31.8% of the references were for negative sentences ($n=7$), and 9.1% of the references were for positive sentences ($n=2$).

Table 3. References

	FULL MAIN			CRISIS IN MAIN			OIL SPILL			EXPLOSION		
	Early	Mid	Last	Early	Mid	Last	Early	Mid	Last	Early	Mid	Last
BP	15.6% $n=12$	15.2% $n=24$	13.8% $n=21$	0	5.0% $n=2$	3.1% $n=1$	6.3% $n=1$	2.1% $n=9$	1.8% $n=9$	1.2% $n=1$	1.0% $n=1$	1.8% $n=2$
News Media	41.6% $n=32$	51.3% $n=81$	51.3% $n=78$	0	85.0% $n=34$	87.5% $n=28$	81.3% $n=13$	73.0% $n=306$	73.9% $n=370$	77.4% $n=65$	80.6% $n=83$	79.1% $n=87$
Government	1.3% $n=1$	2.5% $n=4$	2.6% $n=4$	0	0	0	0	8.4% $n=35$	8.6% $n=43$	6.0% $n=5$	6.8% $n=7$	6.4% $n=7$
Activist Org.	11.7% $n=9$	10.8% $n=17$	11.8% $n=18$	0	5.0% $n=2$	6.3% $n=2$	0	9.1% $n=38$	7.8% $n=39$	7.1% $n=6$	4.9% $n=5$	4.5% $n=5$
Scholarly/ Reference	13.0% $n=10$	6.3% $n=10$	7.2% $n=11$	0	0	0	0	3.8% $n=16$	3.4% $n=17$	0	0	0
Other	16.9% $n=13$	13.9% $n=22$	13.2% $n=20$	0	5.0% $n=2$	3.1% $n=1$	12.5% $n=2$	3.6% $n=15$	4.6% $n=23$	8.3% $n=7$	6.8% $n=7$	8.2% $n=9$
Total	77	158	152	0	40	32	16	419	501	84	103	110

Conclusion

This chapter analyzed a major corporate crisis and its repercussions for corporate reputation online. The Wikipedia article on the BP Deepwater Horizon Oil Spill was the top source of information in all three major search engines. The findings show that the crisis article on the spill was viewed substantially more often than the main Wikipedia article on BP, which included a couple of paragraphs on the crisis. The notion that a crisis or scandal can be effectively "outsourced" on Wikipedia into a supplemental crisis article that only few read does not hold up. The research also demonstrates that Wikipedia has moved beyond its initial inaccuracies and reliability issues. The content on BP, and especially the one on the oil spill, underwent a rigorous editing and revision process. Thousands of editors provided input on the issue and documented it with hundreds of references. The increase in the amount of content created over the course of one year is staggering. It seems as if every statement of experts, critics, and the company itself was documented on Wikipedia. The content changed over the course of the year and was adjusted for the latest developments in real time. Almost every sentence on the crisis had a reference, with media reports widely used for the documentation of crisis events, much more so than in the main article on BP.

Although it is not surprising that the content on a crisis of this magnitude is mainly framed negatively, it should catch public relations practitioners' attention that even corporate press releases were used as references for negative content in the crisis articles. It should also be noted that the two crisis articles were the main links from the main BP article to more information on the Deepwater Horizon crisis. Nevertheless, there were additional links in the crisis articles to additional information on topics such as the economic and political consequences of the disaster, the litigation of the oil spill, and the offshore drilling moratorium. This additional content was not included in this analysis. It can be safely assumed that the amount of negative content overall is even larger than documented in this analysis.

The findings of this study are a call to action for the public relations practitioner. They demonstrate that Wikipedia expands the conventional practice of public relations by providing another medium to monitor. Not monitoring Wikipedia regularly is a dangerous neglect. While social media networks like Facebook and Twitter are fast-moving platforms, on which online crises come and go, Wikipedia's collaboratively created content constitutes greatly what is considered "the truth" in today's online environment. It is essential for practitioners to be part of that process before debates over the framing of content are settled by the collective of editors.

Social pressures are driving these new media approaches to public relations, as the reputation of a company is now influenced by the public on Wikipedia. While traditionalists still lament over the rise of Wikipedia, the online encyclopedia attracts great parts of the public on a regular basis. Currently, 53% of American Internet users look for information on Wikipedia (Zickuhr & Rainie, 2011), while 14% of worldwide Internet users access Wikipedia on a daily basis (Alexa.com, 2011). By monitoring Wikipedia as part of their environmental scanning, companies can have a better idea of the information available about them and therefore a basis for engaging with the online public. Although monitoring Wikipedia is a must, active engagement in debates over crisis content is even more effective.

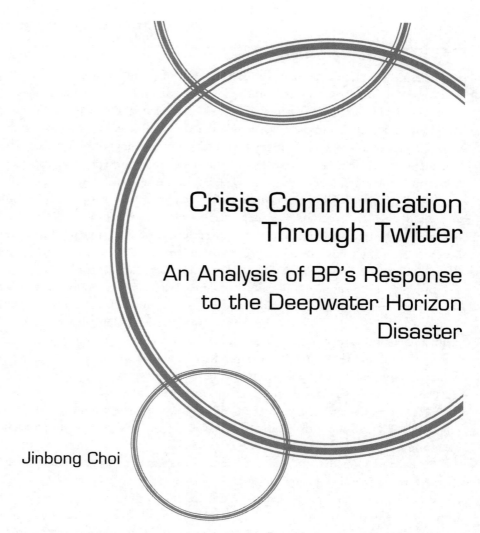

Crisis Communication Through Twitter

An Analysis of BP's Response to the Deepwater Horizon Disaster

Jinbong Choi

This study aimed to analyze how BP used social media (specifically, Twitter) as a crisis communication tool to address an oil spill crisis. Since Twitter's appearance in 2006, public relations practitioners have adopted and used it for their public relations campaigns because of its ability to keep a constant two-way flow of communication with publics. To explore how BP used Twitter in the aftermath of the oil spill, this study identified frames and keywords used in BP's official Twitter account (@BP_America). Findings from an empirical analysis of the tweets BP posted revealed that BP used five frames (information, update, social responsibility, attribution of responsibility, and all that can be done) and 11 keywords (response, update, latest, effort, claims, information, operation, BP CEO, picture/photo, volunteer, and shoreline). The majority of the collected tweets were found to be using the "information" frame and the "response" keyword, reflecting BP's focus on disseminating information about their response efforts while avoiding issues such as "attribution of responsibility."

Introduction

The Deepwater Horizon oil rig, leased by British Petroleum (BP), exploded in the Gulf of Mexico on April 20, 2010, killing 11 rig workers and injuring 17 workers (Canfield, 2011). Adding to the devastation, the well was unable to be capped for nearly four months, allowing for an estimated 4.9 million barrels of oil, methane gas, and pollutants to pour into the Gulf (Canfield, 2011). The Gulf states' $20 billion per year tourist industry was severely affected by the spill and its subsequent news media coverage (Bush, 2010). As of March 2011, an estimated 256,000 individual and business claims have been made against BP (Nelson, 2011). The massive loss of wildlife and wildlife habitats—both on and off land—is devastating. To make matters worse, many Gulf residents rely on wildlife and their habitats to make a living and support their families. Commercial fishermen, among others, were displaced from their jobs, and hotels and tourist attractions were nearly empty in the months following the spill.

BP's brand image has been negatively affected because of damage to the Gulf of Mexico, wildlife and wildlife habitats, and Gulf residents' displacement from their jobs. BP's perceived complacency and lack of urgency to take action to contain the spill triggered negative perceptions about the company. It is now "a company whose name has become synonymous with corporate incompetence, [and] the words BP and public relations no longer fit well together" (Luce, 2010, para. 4).

In an effort to support and gain confidence from the communities affected by the oil spill, BP set up a $20 billion trust to be paid to individuals, businesses, and local government entity claimants ("Compensating," n.d.). To make the claims process easier, BP set up a toll-free number and established claim and community outreach centers. These resources were intended to make the claims process run smoothly and allow those affected to have a first-hand look at what BP was doing to help Gulf coast residents. BP additionally set up a $100 million trust to assist rig workers displaced from their jobs during the U.S. government's moratorium on deep water drilling.

BP's efforts to contain and cease the oil spilling from the well included planning and execution of relief wells and working with experts from various sectors to develop multiple strategies to stop the leak, contain the oil, and kill the well ("Containing," n.d.).

BP used news media and social media to disseminate messages dealing with the state of the spill, their efforts to contain and stop the spill, and to provide information on how those affected could seek help and compensation from the trusts. While communicating its messages with publics through Twitter, BP used several different frames to address the oil spill. Because social media are relatively new and emerging communication tools, especially in terms of crisis communication, it is important to analyze and understand how they were used by the acting corporation held responsible for the worst oil spill in U.S. history (Blackburn & Gutman, 2010).

Therefore, this study aimed to analyze how BP used Twitter during the oil spill crisis. Specifically, this research explored what kinds of strategic frames were developed and used in the official BP Twitter account. Since Twitter appeared in 2006, public relations practitioners have adopted and used Twitter as a communication tool for their campaigns because it has the ability to keep a constant two-way flow of communication with publics.

Thus, it is crucial to conduct public relations research on social media (Twitter in particular) due to their widespread use and ability to reach millions of users with real-time information in a quick and relatively easy manner. Social media, especially in times of a crisis, are a platform for ongoing and transparent communication between the organization involved in the crisis and the millions of people connected to social media outlets (Borremans, 2010).

As such, the following research questions were posed in this study:

RQ1. Which frames were used by the official BP Twitter account in describing the Gulf of Mexico oil spill?

RQ2. What kinds of keywords were used by the official BP Twitter account?

RQ3. How did BP use frames and themes to address the oil spill crisis?

Literature review

The BP oil spill crisis

The oil spill disaster in the Gulf started when an explosion and fire on the Deepwater Horizon (DWH) oil rig injured 17 workers and left 11 missing and presumably dead. Within a matter of days, the oil rig sank in 5,000 ft. of water, and oil was thought to be leaking out of the well at 1,000 barrels of oil per day (Guardian Research, 2010). U.S. agencies and departments began to quickly understand the magnitude of the oil, gas, and pollutants and their high risk of damaging the Gulf environment. To intensify the situation, the Coast Guard estimated the oil spill to be five times larger than was initially believed.

After numerous failed attempts by BP and its engineers to cap, plug, or contain the leaking oil, experts and publics became skeptical of the amounts of oil actually leaking from the well (Guardian Research, 2010). Subsequently, the U.S. government launched a criminal investigation into the oil spill.

By mid-June 2010, scientists raised their estimate of the number of barrels spilling per day to 40,000 (Guardian Research, 2010). Gulf coast residents had become victims of the worst oil spill in U.S. history. The spill cost many their jobs, livelihoods, and sense of security. To assist displaced workers and victims, BP agreed to establish a $20 billion trust intended to compensate those affected by the spill. BP announced that in the three months since the well began to leak, the spill had cost the company over $3 billion.

In mid-July 2010, a more secure containment cap was successfully installed on the well (Guardian Research, 2010). The flow of oil had ceased for the first time since the DWH rig sank. However, this cessation was short lived. Engineers soon discovered methane gas was continuing to seep through.

In early August 2010, BP attempted to "kill" the well by dumping heavy drilling mud and cement into the well (Guardian Research, 2010). On August 4, 2010, it was revealed that the static kill had successfully sealed the well, and oil, gas, and pollutants could no longer seep out of the well. The next day, more cement was pumped into the well to seal it off permanently.

After the crisis, BP's reputation had diminished dramatically as the people felt they could no longer trust BP. *PR Week*'s April 2011 survey of more than 2,000 members of the UK public found that 93% said the oil spill had been damaging or very damaging to BP's reputation (Magee, 2011). The lack of trust, transparency, and responsibility forced BP's reputation downhill. In late August 2010, BP's reputation was damaged so badly that the company had to withdraw a bid to drill in the Arctic.

Twitter as a social medium

Twitter is a text-based micro-blogging social media site that allows users to create individual statements of 140 characters or less, thereby forcing content to be direct and to the point (Miller & Vega, 2010). Users can "follow" a person or organization, and they can be alerted via their mobile device when a new tweet is posted, creating nearly instantaneous information distribution (Miller & Vega, 2010). Twitter was first created in 2006 and entered the social conscience at the annual South by Southwest festival/conference in Austin, Texas, where the average daily tweets reached over 60,000 (Miller & Vega, 2010).

In June 2011, Twitter had 200 million users who wrote 155 million tweets per day (Twitter, 2011b). What sets Twitter apart from other social media are its user-disseminated messages, called tweets. Hence, by using Twitter, communities, businesses, and individuals have the ability build relationships by communicating quickly and efficiently with publics. Unlike other communication vehicles (e.g., TV, radio, newspaper, magazine), Twitter allows users to get more information in a shorter amount of time (Miller & Helft, 2010). Users are engaged and typically looking for a specific trending topic (Miller & Vega, 2010). Some corporations use Twitter as a way to maintain a real-time dialogue with customers and other publics (Miller & Vega, 2010). Organizations, from global corporations such as Disney and Nike to small local companies, use Twitter accounts to get information directly to their target audiences (Miller & Vega, 2010).

BP's official Twitter account, @BP_America, was created on August 12, 2008 (Gill, 2010). However BP did not really use the account until the oil spill crisis occurred. After the crisis happened, BP actively used the account to reach out to publics who were seeking information on the response, and to respond to attacks of incompetency and violation of social trust and values.

Studies in social media and crisis

An increasing number of communication studies focus on social media as tools for public relations campaigns, but few studies have examined social media (Twitter in particular) and crisis communication. Schultz, Utz, & Göritz (2011) conducted the first study that "experimentally addressed the role of social media and crisis responses in crisis communication" (p. 26) by analyzing their effects on crisis communication and public reactions.

Schultz et al. (2011) analyzed the effects of corporate response strategies, including apology, sympathy, and information, on crisis perceptions and secondary crisis communication via traditional (newspaper) and social (blogs and Twitter) media. The study was a random online experiment with 1,677 participants. Each respondent was given a crisis communication scenario in one of three experimental conditions. They then answered questions on their reaction to the message, the likelihood of forwarding the message (secondary crisis communication), and the organization's reputation.

The study found that the effect of reaction was strongest under the Twitter condition, and that this effect has a strong impact on behavioral intentions (Schultz et al., 2011). Twitter users were more likely to leave a comment, and "were more willing to share the message than both blog users and non-users of social media" (p. 25). Surprisingly, Twitter users indicated they would share a newspaper article before a blog post or tweet. Twitter users were also more likely than non-users to talk to their friends about the crisis.

It may be inferred that Twitter could aid in repairing an organization's reputation and preventing boycotts during crises because "crisis communication via Twitter led to less negative crisis reactions than blogs and newspaper articles" (Schultz et al., 2011, p. 25). Further, people who read tweets from the organization involved in the crisis are less likely to boycott or speak negatively about that particular organization. Consequently, the Schultz et al. study underlines the importance of Twitter to aid in the social deconstruction of crises, as well as the need to fill the gap in crisis communication literature, especially experimental studies of Twitter and other social media.

Several other studies have dealt with social media as a communication or public relations tool. Kelleher (2008) examined whether organizational contingencies, including technological orientation, led public relations practitioners to choose alternative communication vehicles, such as social media, to reach key publics. His study design was based on previous studies (e.g., Kelleher & Miller, 2006; Kent, 2008; Sweetser & Metzgar, 2007) that found blogging and social media are now among the factors practitioners must consider when making choices on how to relate to different publics in different scenarios. In his study, Kelleher (2008) found that the "publics perceive advantages for blogs as social media and as means for organizations to offer accommodations and to communicate with a conversational human voice, which positively influences relational outcomes and correlates with more favorable perceptions during crises" (p. 302).

In addition, Waters, Burnett, Lamm, and Lucas (2009) examined how nonprofit organizations use Facebook to engage their stakeholders and foster relationship growth. They found social networking sites could be "an effective way to reach stakeholder groups if organizations understand how their stakeholders use the sites" (p. 106). The nonprofits examined in this study did not take advantage of the interactive, two-way flow of social networking. Therefore, the authors suggested that before public relations practitioners use social media, they should understand how stakeholders use each medium so that they may meet "growing needs and expectations of stakeholders" (p. 106).

Moreover, Lariscy, Avery, Sweetser, and Howes (2009a) examined the use of social media content as agenda building for journalists. Two hundred business journalists were asked questions on demographics, uses, and perceived value of social media in telephone interviews. According to their study, journalists indicated social media such as YouTube and Twitter were quick resources for journalists to get story ideas and sources. Their study also revealed nearly 65% of journalists would use stories from public relations practitioners who maintain social media tools. The benefit of organizations using social media during crises is that their framed messages may provide journalists with story ideas, which could lead to more positive coverage (Lariscy et al., 2009a). Thus, their study suggested that public relations practitioners should begin "engaging social media in preparation for the day social media may contribute to agenda building" (p. 316).

There are also several studies of crisis communication that suggest practitioners should first focus on information dissemination early in a crisis. Coombs (2012) advised organizations to respond quickly to crises through information dissemination. In the early stage of a crisis, the information void created by the crisis takes place. Thus, when organizations face crises, they need to fill the information void by providing information (Coombs, 2012). If organizations fail to fill the information void in the early stage of crisis, "other sources will fill the void—often with rumor and speculation instead of facts" (Millner, Veil, & Sellnow, 2011, p. 74). In addition, if organizations fail to respond or are slow to respond to crises, they will lose credibility and "open the door for others to frame the crises" (Millner et al., 2011, p. 74). Therefore, when crises occur, organizations need to disseminate information related to their crises in the early stages to inform affected publics and news media.

Method

Sample

This study utilized content analysis to examine how BP used Twitter in the aftermath of the Gulf oil spill crisis. The researcher collected tweets from BP's official Twitter account (@BP_America) within 40 days (from April 20, 2010 to May 31, 2010) of the spill's inception because a crisis is most newsworthy when it first occurs. The researcher used the website www.backtweets.com to collect BP's tweets. At the time, this site was the only found resource capable of collecting past tweets from the @BP_America Twitter account. A total of 180 BP tweets were collected and analyzed.

Coding procedure

In a pre-test, coders first read through each of the 180 tweets. Coders then identified five main frames the BP Twitter account used in this initial period of the oil spill crisis. Thus, the following five frames identified through the pre-test were used to analyze the collected BP's tweets: social responsibility frame, attribution of responsibility frame, all that can be done frame, update frame, and information frame. The tweets were carefully read and coded using these five frames.

The "social responsibility" frame was assigned to tweets dealing with BP helping affected states, communities, residents, workers, and wildlife. Tweets pledging to enhance protection and safety plans also fell into this frame. Tweets dealing with the extreme challenges of fixing and stopping the leak, and/or BP's shift of responsibility to other causes and parties were classified in the "attribution of responsibility" frame.

The "all that can be done" frame highlighted what was being done in the short term, specifically those actions aimed at stopping the leak. The tweets in this frame emphasized how BP was trying to stop the well by using multiple BP resources and options to resolve the crisis. The "update" frame was used by BP to keep key publics informed about the crisis. Most tweets in this frame were linked to an official BP update. Tweets linking to photos, videos, and fact sheets fell under the "information" frame. This frame also included tweets announcing press briefings, upcoming events of public interest, where to find information, how to volunteer, where and how to make claims, and the like.

The coders re-read tweets to identify keywords of emphasis and gain a better understanding of the positioning and themes BP used in its Twitter account. Coders found 11 keywords (other than "oil" and "spill") from collected tweets: response, update, latest, effort, claim, information, operation, BP CEO, picture/photo, volunteer, and shoreline. Thereafter, collected tweets were coded using these 11 keywords.

In this study, two coders analyzed and coded all 180 BP tweets independently. Before coding began, an inter-coder reliability test was conducted in which two coders coded 50 randomly selected tweets. Results were compared using Holsti's method (Holsti, 1969). The inter-coder reliability was an acceptable 0.89.

Findings

Five Twitter frames (RQ1)

The BP Twitter account used five frames during the 40-day period after the oil spill crisis began. A total of 180 tweets were coded from BP's official Twitter account. As shown in Table 1, the most common frame found in BP's Twitter account was the "information" frame (68.33%; $n = 123$). The second most common frame found was the "update" frame (21.67%; $n = 39$). The third most common frame was the "social responsibility" frame (7.78%; $n = 14$). Fourth was a tie with the "attribution of responsibility" frame (1.11%; $n = 2$) and the "all that can be done" frame, both at 1.11% ($n = 2$).

Table 1. Frames used in BP tweets

FRAMES	NUMBERS	PERCENT
Information	123	68.33%
Update	39	21.67%
Social responsibility	14	7.78%
Attribution of responsibility	2	1.11%
All that can be done	2	1.11%
Total	180	100.00%

Eleven Twitter keywords (RQ2)

BP used 11 keywords while posting 180 tweets on its Twitter account during the 40-day period studied. Table 2 shows the frequency of keywords found. Among them, "response" ($n = 42$) was the most common, followed by "update" ($n = 21$); "latest" ($n = 19$); "effort" ($n = 18$); "claims" and "information" ($n = 16$ each); "operation", "BP CEO", and "picture/photo" ($n = 14$ each); and "volunteer" and "shoreline" ($n = 13$ each).

Functions of Twitter frames and keywords (RQ3)

BP used different frames to keep publics informed and up to date on their efforts to stop and contain the spill. Disseminating information about the spill, totaling 123 tweets in the "information" frame, appeared to be BP's priority in light of the media frenzy surrounding the crisis in its earliest days. Examples of tweets falling into this category included: "Hayward applauds President's statement on oil spill" (May 2); "BP has established a call center for receipt of oil spill related claims-the call in number is 1-800-440-0858" (May 3); "Video of BP CEO touring coffer dam construction" (May 3); "To report oiled shoreline or request volunteer information call the BP Community Support hotline (866)-448-5816" (May 6); "Check the latest NOAA trajectory maps" (May 9); "In pictures-Community work" (May 10); "Read the facts about identifying oil" (May 17).

Next in frequency were 39 BP "updates" to the public, including: "BP is attacking the oil spill on the seabed and on the coastline" (May 2); "Overnight ROVs stopped the flow from one of the 31 leaks on the damaged well and riser" (May 5); "The cofferdam (containment dome) is at the site and is getting ready to be offloaded into the water" (May 6); "Efforts continue: reduce oil flow via containment, further work using a 'top kill' option" (May 10); "The 'top hat' dome will be connected by drill pipe and riser lines to the Drillship Enterprise on the surface to collect and treat oil" (May 10); "ROVs are acting as the eyes and hands of the response team on the seafloor" (May 11); "Here's the latest BP response update, including info on the riser insertion tube tool and how it works" (May 17).

Examples of the 14 "social responsibility" tweets posted included: "BP steps up shoreline protection plans on US Gulf Coast" (April 30); "Safety is the #1 priority. We are going to figure out what hap-

Table 2. Keywords used in BP tweets

RANK	THEME	FREQUENCY	CLASSIFICATION
1	Response	42	Response to criticism
2	Update	21	Update on crisis; current efforts begun
3	Latest	19	BP's latest efforts
4	Effort	18	Clean up efforts
5	Claims	16	Information on claims
5	Information	16	Information on resources for spill (e.g., call center)
6	Operation	14	Current operations
6	BP CEO (Hayward)	14	Mentions from BP CEO
6	Picture/Photo	14	Links to pictures and photos
7	Volunteer	13	Call for volunteers
7	Shoreline	13	Cleaning the shoreline

Note: Table does not include the number of times oil (79) and spill (60) were used.

pened & that is going to help the industry get safer" (May 2); "$25m block grants to Louisiana, Alabama, Mississippi, Florida to help protect shorelines" (May 5); "BP has contracted local workers to remove booms from shore to sea, to prepare for the possible landfall" (May 9); "BP will do anything and everything we can to stop the leak, attack the spill off shore, and protect the shorelines of the Gulf Coast" (May 10); "Wildlife rehabilitation is a major part of our response effort" (May 15).

Tweets categorized in the less utilized frames of "attribution of responsibility" and "all that can be done" were: "Florida Keys tar balls not linked to Deepwater Horizon oil spill" (May 19); "BP continues to do everything possible to contain the oil spill and mitigate damage" (May 9); "Fishing vessels and local crews are working alongside BP and other agencies in the massive operation in the Gulf" (May 14).

Of the 11 keywords (except "oil" and "spill"), the most frequently used was "response," which appeared twice as often as the second most common keyword, "update." Use of these keywords (and others close in frequency) support BP's apparent efforts in the early days to focus on what the company was doing to respond to the crisis. Examples of "response" and "update" tweets include: "Daily update: The ongoing response to the Deepwater Horizon oil spill" (May 4); "For updates regarding ongoing activities, visit the BP Gulf of Mexico response website" (May 7); "BP will hold a technical briefing and update on the Gulf oil spill response at 12:00 p.m. CDT in Houston, TX" (May 10).

Conclusion

This study aimed to identify crisis frames and keywords used in BP's official Twitter account while the company responded to the April 2010 Gulf oil spill crisis. A content analysis of tweets revealed BP's use of five frames (information, update, social responsibility, attribution of responsibility, and all that can be done) and 11 keywords (response, update, latest, effort, claims, information, operation, BP CEO, picture/photo, volunteer, and shoreline).

Dissemination of information, including updates, appeared to be a higher priority for BP in the 40 days following the initial explosion, more so than touting its social responsibility, attributing responsibility, or defending its response actions. Specifically, BP's use of "information" and "update" frames 90% of the time revealed a focus on getting the latest information to the public in a timely manner, which aligns with literature discussing the importance of enacting such a strategy (e.g., Coombs, 2012; Millner, Veil, & Sellnow, 2011; Taylor, 2007).

Schultz et al. (2011) reported that crisis communication via social media (e.g., Twitter and blogs) resulted in fewer secondary crisis reactions because social media allow "immediate reactions to crisis situations" (p. 22). In line with this previous study, BP may have been able to reduce the number of secondary reactions by disseminating informational tweets, though such a claim is beyond the scope of this study.

Of the 11 keywords, "response" was used most frequently, indicating BP aimed to communicate and promote its efforts to control damage and, presumably, its worldwide reputation. Use of other keywords such as "update," "latest," "effort," and "claims," supports this assumption.

There is no doubt that Twitter is an effective medium for getting information out quickly. However, although BP was able to use Twitter to distribute basic information and updates about the spill, it was limited in its ability to provide a detailed explanation of its efforts because of Twitter's 140 character limit. To convey response details, BP included numerous website references in its tweets through which people could access detailed information about issues. Examples included: "For more info on claims see: http://bit.ly/cyisKB" (May 10); "Read more info on dispersants and spill response: http://bit.ly/bUrvVI#oilspill#bp" (May 10); "View BP's latest update on response effort. http://bit.ly/dAfcZk#oilspill#bp" (May 13).

Accordingly, Coombs and Holladay (2009) advised crisis managers that organizations should "deliver their messages in a variety of media as a means of reaching more people" (p. 5). BP used the Twitter site to keep the public informed, resulting in numerous tweets about "information" and "response" to the spill, as well as "updates" about clean-up efforts and claims information. Details on the BP website supplemented company tweets.

Although this study offers a replicable research design for Twitter analysis, it has several limitations. One is that its results may not be generalized to BP's Twitter strategy following the 40-day timeframe of interest herein. According to Coombs and Holladay (2004), there are three crisis response stages as time passes after a crisis. In the first stage, an organization acknowledges a crisis. In the second stage, an organization defines the crisis and places it in a larger context. In the final stage, an organization accepts responsibility but limits negativity and offers various forms of accommodation, including compensation, to counterbalance the crisis. This study also did not examine the perceptions and/or effects of these tweets on BP's followers, which would be essential in measuring the effectiveness of BP's Twitter strategy in the early days of the crisis. Interviews with BP communicators would have been tremendously insightful in understanding the company's Twitter goals at various points in time. Future studies are needed to establish how Twitter can be used over long periods of time to manage reputation before, during, and after crises.

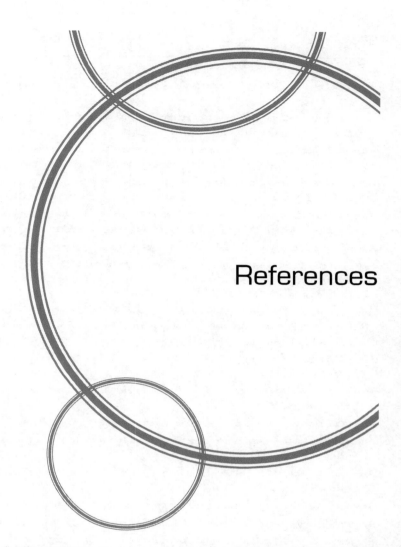

References

Aaker, J. L., & Smith, A. (2010). *The dragonfly effect: Quick, effective, and powerful ways to use social media to drive social change.* San Francisco, CA: Jossey-Bass.

Abaffy, L. (2011, February 28). Millennials bring new attitudes; their priorities are sustainability, high ethical standards and career fulfillment. They must be managed differently, or they jump ship. *Engineering News-Record.* Retrieved from Factiva database. (ENR00000020110307e72s00016)

About CIW (n.d.). Retrieved from http://ciw-online.org/about.html

Achenbach, J. (2011, March 17). Japan's "black swan": Scientists ponder the unparalleled dangers of unlikely disasters. Retrieved from http://www.washingtonpost.com/national/japans-black-swan-scientists-ponder-the-unparalleled-dangers-of-unlikely-disasters/2011/03/17/ABj2wTn_story.html

Adorno, T. W. (1973). *The jargon of authenticity.* Evanston, IL: Northwestern University Press.

Albrecht, T. L., & Adelman, M. B. (1987). *Communicating social support.* Newbury Park, CA: Sage.

Aldoory, L. (2001). Making health communications meaningful for women: Factors that influence involvement. *Journal of Public Relations Research, 13*(2), 163–185.

Aldoory, L., & Sha, B. (2007). The situational theory of publics: Practical applications, methodological challenges, and theoretical horizons. In E. L. Toth (Ed.), *The future of excellence in public relations and communication management* (pp. 339–355). Mahwah, NJ: Lawrence Erlbaum.

Alexa.com (2011, April 27). Top sites. Retrieved from http://www.alexa.com/topsites

Alexander, J. C. (1992). The promise of a cultural sociology: Technological discourse and the sacred and profane information machine. In R. Munch & N. J. Smelser (Eds.), *Theory of culture* (pp. 293–323). Berkeley: University of California Press.

Alexander, N. (2009). Brand authentication: Creating and maintaining brand auras. *European Journal of Marketing, 43*(3/4), 551–562.

Alfonso, G.-H., & Miguel, R.V. (2006). Trends in online media relations: Web-based corporate press rooms in leading international companies. *Public Relations Review, 32*(3), 267–275.

Allen, E. T. (2006, October). The role of attorneys in crisis communications. Retrieved from http://wordoflaw.com/newsletters/2006-10_The%20role%20of%20attorneys%20in%20crisis%20communications.pdf

American Cancer Society. (2011). Relay for Life of Second Life. Retrieved from http://main.acsevents.org/site/TR/RelayForLife/RFLFY11SL?sid=126717&type=fr_informational&pg=informational&fr_id=34810

American Red Cross. (n.d.). The case for integrating crisis response with social media. Retrieved from http://www.scribd.com/doc/35737608/White-Paper-The-Case-for-Integrating-Crisis-Response-With-Social-Media

American Red Cross. (2010a). Haiti earthquake one-month progress report. Retrieved from http://www.redcross.org/haiti

American Red Cross. (2010b, January 23). Telethon raises funds for Red Cross Haiti relief efforts. Retrieved from http://www.redcross.org/portal/site/en/menuitem.1a019a978f421296e81ec89e43181aa0/?vgnextoid=03ce9f4a01956210VgnVCM10000089f0870aRCRD

America's preparedness for disaster or emergency improves: Gains made in awareness & knowledge; seniors show major improvements. (2007, October 1). Retrieved from http://www.redcross.org/portal/site/en/menuitem.94aae335470e233f6cf911df43181aa0/vgnextoid=75d8c92887b6b110VgnVCM10000089f0870aRCRD

Ames, C. (2010). PR goes to the movies: The image of public relations improves from 1996 to 2008. *Public Relations Review, 36*(2), 164–170.

An, D. (2007). Advertising visuals in global brands' local websites: A six-country comparison. *International Journal of Advertising, 26*(3), 303–332.

Anderson, B. (1983). *Imagined communities: Reflections of the origin and spread of nationalism.* New York, NY: Verso.

Anderson, K. (2010, February 10). Labour launches social media campaign. Retrieved from http://www.guardian.co.uk/politics/blog/2010/feb/18/labour-twitter

Anderson, R., Reagan, J., Hill, S., & Sumner, J. (1989). Practitioner roles and the uses of new technology. *Public Relations Review, 15*(3), 53.

Argenti, P.A., & Druckenmiller, B. (2004). Reputation and the corporate brand. *Corporate Reputation Review, 7*(4), 368–74.

Arnett, J. S. (1999, September). From public enragement to engagement. *The School Administrator,* pp. 24–27.

Aronowitz, R. (2008). Framing disease: An underappreciated mechanism for the social patterning of health. *Social Science & Medicine, 67*(1), 1–9.

Aronson, J. (1994, Spring). A pragmatic view of thematic analysis. *The Qualitative Report, 2*(1). Retrieved from http://www.nova.edu/ssss/QR/BackIssues/QR2-1/aronson.html

Arora, H. (2009). A conceptual study of brand communities. *The Icfai University Journal of Brand Management, 6*(2), 7–21.

Arquilla, J., & Ronfeldt, D. (1999). *The emergence of noopolitik.* Santa Monica, CA: National Defense Research Institute-RAND.

Arthur, C. (2006, July 20). What is the 1% rule? Retrieved from http://www.guardian.co.uk/technology/2006/jul/20/guardianweeklytechnologysection2

Arts, B., Noortmann, M., & Reinalda, B. (2001). *Non-State actors in international relations.* Aldershot, England: Ashgate.

Association of Fundraising Professionals (2009, November 24). Most nonprofits experimenting with social media, but struggle to measure value for organization. Retrieved from http://www.afpnet.org/Audiences/NewsReleaseDetail.cfm?itemnumber=4257

Astroturfing. (2011). Retrieved from http://projects.washingtonpost.com/politicsglossary/Coined-Phrases/astroturfing/

Atwood, L., & Major, A. (1991). Applying situational communication theory to an international political problem: Two studies. *Journalism Quarterly, 68*(1/2), 200–210.

Auger, G.A. (2010). Using dialogic web site design to encourage effective grantor-grantee relationships. *Prism, 7*(2). Retrieved from http://www.prismjournal.org

Aula, P. (2011). Meshworked reputation: Publicists' views on the reputational impacts of online communication. *Public Relations Review, 37*(1), 28–36.

Avery, D., Tonidandel, S., & Phillips, M. (2008). Similarity on sports sidelines: How mentor-protege sex similarity affects mentoring. *Sex Roles, 58*(1/2), 72–80.

Avery, E., Lariscy, R., Amador, E., Ickowitz, T., Primm, C., & Taylor, A. (2010). Diffusion of social media among public relations practitioners in health departments across various community population sizes. *Journal of Public Relations Research*, 22(3), 336–358.

Avery, E., Lariscy, R., & Sweetser, K. D. (2010). Social media and shared - or divergent - uses? A coorientation analysis of public relations practitioners and journalists. *International Journal of Strategic Communication*, 4(3), 189–205.

Avery, E. J. (2010). Contextual and audience moderators of channel selection and message reception of public health information in routine and crisis situations. *Journal of Public Relations Research*, 22(4), 378–403. doi:10.1080/10627261003801404

Avidar, R. (2009). Social media, societal culture and Israeli public relations practice. *Public Relations Review*, 35(4), 437–439.

Ayish, M.I. (2005). Virtual public relations in the United Arab Emirates: A case study of 20 UAE organizations' use of the Internet. *Public Relations Review*, 31(3), 381–388.

Baca, J. (2011). Blake Griffin's dunk into new media. Unpublished undergraduate paper, Comm465: Entertainment PR, California State University, Fullerton.

Background on CTIA's semi-annual wireless industry survey. (n.d.). Retrieved from http://files.ctia.org/pdf/CTIA_Survey_Year_End_2010_Graphics.pdf

Baehr, C., & Alex-Brown, K. (2010). Assessing the value of corporate blogs: A social capital perspective. *IEEE Transactions on Professional Communication*, 53(4), 358–369.

Baehr, C., & Schaller, B. (2010). *Writing for the Internet: A guide to real communication in virtual space*. Santa Barbara, CA: Greenwood Press.

Bagin, R. (1998, May). A practical look at public engagement. (NSPRA) *Network*, pp. 7–8.

Bailey, A. A. (2004). Thiscompanysucks.com: The use of the Internet in negative consumer-to-consumer articulations. *Journal of Marketing Communications*, 10(3), 169–182. doi:10.1080/1352726042000186634

Bailey, S. (2010, January). Religious self-profiling: Identifying faith online proves challenging. *Christianity Today*, p. 14.

Bakardjieva, M. (2005). *Internet society: The Internet in everyday life*. London, England: Sage.

Baker, S., & Martinson, D. L. (2002). Out of the red-light district: Five principles for ethically proactive public relations. *Public Relations Quarterly*, 47(3), 15–19.

Barboza, D. (2007, February 5). Internet boom in China is built on virtual fun. Retrieved from http://www.nytimes.com/2007/02/05/world/asia/05virtual.html?oref=slogin

Bargh, J., & McKenna, K. (2004). The Internet and social life. *Annual Review of Psychology*, 55, 573–590.

Bargh, J., McKenna, K., & Fitzsimons, G. (2002). Can you see the real me? Activation and expression of the "true self" on the Internet. *Journal of Social Issues*, 58(1), 33–48.

Barnes, N. G. (2010a). The Fortune 500 and social media: A longitudinal study of blogging, Twitter, and Facebook usage by America's largest companies. Retrieved from http://www1.umassd.edu/cmr/studiesresearch/2010F500.pdf

Barnes, N. G. (2010b). The 2010 Inc. 500 update: Most blog, friend, and tweet, but some industries still shun social media. Retrieved from http://www1.umassd.edu/cmr/studiesresearch/2010inc500.pdf

Barnett, G. A., & Sung, E. (2005). Culture and the structure of the international hyperlink network. *Journal of Computer-Mediated Communication*, 11(1), 217–238.

Baron, G., & Philbin, J. (2009, March). Social media in crisis communication: Start with a drill. Retrieved from http://www.prsa.org/SearchResults/view/7909/105/Social_media_in_crisis_communication_Start_with_a

Barraket, J., & Henry-Waring, M. (2008). Getting it on(line): Sociological perspectives on e-dating. *Journal of Sociology*, 44(2), 149–165.

Barton, L. (2001). *Crisis in organizations II* (2nd ed.). Cincinnati, OH: College Divisions South-Western.

Barton, L. (2008). *Crisis leadership now*. New York, NY: McGraw Hill.

Bass, B. M. (1990). *Bass & Stogdill's handbook of leadership: Theory, research, and managerial applications* (3rd ed.). New York, NY: The Free Press.

Bauman, D. C. (2011). Evaluating ethical approach to crisis leadership: Insights from unintentional harm research. *Journal of Business Ethics*, 98(2), 281–295.

Bauman, Z. (2007). *Consuming life*. Cambridge, UK: Polity Press.

Bawin-Legros, B. (2004). Intimacy and the new sentimental order. *Current Sociology*, 52(2), 241–250.

Baxter, L. A., & Babbie, E. (2004). *The basics of communication research*. Belmont, CA: Wadsworth.

Beaubien, G. (2011, April 12). Crisis in Japan showcases vital role of social media during crises. Retrieved from http://prsa.org/SearchResults/view/9119/105/Crisis_in_Japan.html

Beckett, C. (2008). *SuperMedia: Saving journalism so it can save the world*. West Sussex, England: Blackwell Publishing.

Beckett, C. (2011, February 11). After Tunisia and Egypt: Towards a new typology of media and networked political change. Retrieved from http://www.charliebeckett.org/?p=4033

Beech, H. (2011, May 2). Katsunobu Sakurai: The mayor who shamed Japan. *Time, 177*(17), 63.

Bell, M. (2008). Toward a definition of virtual worlds. *Journal of Virtual Worlds Research, 1*(1), 2–5.

Benett, A., Gobhai, C., O'Reilly, A., & Welch, G. (2009). *Good for business: The rise of the conscious corporation*. New York, NY: Palgrave Macmillan.

Bennett, L. (2003). New media power: The Internet and global activism. In N. Couldry & J. Curran (Eds.), *Contesting media power: Alternative media in a networked world* (pp. 17–37). Lanham, MD: Rowman & Littlefield.

Bennett, L., & Manheim, J. (2001). The big spin: Strategic communication and the transformation of pluralist democracy. In L. Bennett & R. Entman (Eds.), *Mediated politics: Communication in the future of democracy* (pp. 279–298). Cambridge, England: Cambridge University Press.

Bennett, L., Pickard, V., Iozzi, D., Schroeder, C., Lagos, T., & Caswell, E. (2004). Managing the public sphere: Journalistic construction of the great globalization debate. *Journal of Communication, 54*(3), 437–455.

Benoit, W. L. (1995). *Accounts, excuses, and apologies: A theory of image restoration strategies*. Albany: State University of New York Press.

Benoit, W. L. (1997). Image repair discourse and crisis communication. *Public Relations Review, 23*(2), 177–186.

Benoit, W. L., & Henson, J. R. (2009). President Bush's image repair discourse on Hurricane Katrina. *Public Relations Review, 35*(1), 40–46.

Bergman, C. (2010, October 1). Local TV staffer loses job over tweet. Retrieved from http://www.lostremote.com/2010/10/01/local-tv-staffer-loses-job-over-tweet/

Beverland, M. (2005). Brand management and the challenge of authenticity. *Journal of Product & Brand Management, 14*(7), 460–461.

Beverland, M. B., & Farrelly, F. J. (2009). The quest for authenticity in consumption: Consumers' purposive choice of authentic cues to shape experience outcomes. *Journal of Consumer Research, 36*, 838–856.

Beverland, M. B., Lindgreen, A., & Vink, M. W. (2008, Spring). Projecting authenticity through advertising. *Journal of Advertising, 37*(1), 5–15.

Bhagat, P. S., Klein, A., & Sharma, V. (2009). The impact of new media on internet-based group consumer behavior. *Journal of Academy of Business and Economics, 9*(3), 83–94.

Bilton, N. (2011, April 21). Location apps generate privacy concerns, report says. Retrieved from http://bits.blogs.nytimes.com/2011/04/21/location-apps-generate-privacy-concerns-report-says/

Birmingham, L. (2011, April 1). Amid tweeted frustration, Japan may take control of TEPCO. Retrieved from http://www.time.com/time/world/article/0,8599,2062591,00.html#ixzz1LlSBwlbD

Bivins, T. (2011). *Public relations writing: The essentials of style and format* (7th ed.). New York, NY: McGraw-Hill.

BK Linden. (2011, May 6). Q1 2011 Linden dollar economy metrics up, users and usage unchanged. Retrieved from http://community.secondlife.com/t5/Featured-News/Q1-2011-Linden-Dollar-Economy-Metrics-Up-Users-and-Usage/ba-p/856693

Blackburn, B., & Gutman, M. (2010, May 27). BP oil spill called worst in U.S. history, as MMS official steps down. Retrieved from http://abcnews.go.com/WN/Media/bp-oil-leak-now-worst-history-surpassing-exxon/story?id=10759905

Blaney, J. R., & Benoit, W. L. (2001). *The Clinton scandals and the politics of image restoration*. Westport, CT: Praeger.

Blaney, J. R., Benoit, W. L., & Brazeal, L. M. (2002). Blowout! Firestone's image restoration campaign. *Public Relations Review, 28*(4), 379–392.

Blascovich, J., & Bailenson, J. (2011). *Infinite reality: Avatars, eternal life, new worlds, and the dawn of the virtual revolution*. New York, NY: HarperCollins.

Block, J. (2011). Slam-dunking a social media campaign: Blake Griffin All Star viral video. Unpublished graduate paper, Comm465: Entertainment PR, California State University, Fullerton.

Bloxham, E., & Nash, J. (2007, December). SOX: Why it happened and why it has been good for investors and companies— The child grows up! *Investor Relations Update*, pp. 14–15.

Boase, J., Horrigan, J. Wellman, B., & Rainie, L. (2006, January 25). The strength of Internet ties: The Internet and email aid users in maintaining their social networks and provide pathways to help when people face big decisions. Retrieved from http://www.pewinternet.org/~/media/Files/Reports/2006/PIP_Internet_ties.pdf

Boeder, P. (2005, September 5). Habermas' heritage: The future of the public sphere in the network society. *First Monday*, 10(9). Retrieved from http://firstmonday.org/htbin/cgiwrap/bin/ojs/index.php/fm/article/view/1280/1200

Bohman, J. (2004). Expanding dialogue: The Internet, the public sphere and prospects for transnational democracy. In N. Crossley & J. M. Roberts (Eds.), *After Habermas: New perspectives on the public sphere* (pp. 131–155). Oxford, England: Blackwell Publishing.

Boin, A., Hart, P., Stern, E., & Sundelius, B. (2005). *The politics of crisis management: Public leadership under pressure*. New York, NY: Cambridge University Press.

Bollen, K. A. (1989). *Structural equations with latent variables*. New York, NY: John Wiley & Sons.

Booz Allen Hamilton. (2009, July). Expert round table on social media and risk communication during times of crisis: Strategic challenges and opportunities. Retrieved from http://www.boozallen.com/media/file/Risk_Communications_Times_of_Crisis.pdf

Borins, S. (2009). From online candidate to online president. *International Journal of Public Administration*, 32(9), 753—758.

Borremans, P. (2010). Ready for anything. *Communication World*, 27(4), 31–33.

Bortree, D.S., & Seltzer, T. (2009). Dialogic strategies and outcomes: An analysis of environmental advocacy groups' Facebook profiles. *Public Relations Review*, 35(3), 317–319.

Boudreaux, C. (2009, December 16). Analysis of social media policies: Lessons and best practices. Retrieved from http://socialmediagovernance.com/blog/uncategorized/study-social-media-policy-analysis/

Boulos, M. K., & Wheeler, S. (2007). The emerging Web 2.0 social software: An enabling suite of sociable technologies in health and health care education. *Health Information and Libraries Journal*, pp. 2–23.

Bowen, S.A. (2010, Winter). An examination of applied ethics and stakeholder management on top corporate websites. *Public Relations Journal*, 4(1). Retrieved from http://www.prsa.org/Intelligence/PRJournal/

Boxer, S. (2004, November 10). Mudslinging weasels into online history. *The New York Times*, p. E1.

boyd, d. (2008a). Can social network sites enable political action? In A. Fine, M. L. Sifry, & A. Rasie (Eds.), *Rebooting America: Ideas for redesigning American democracy for the Internet age* (pp. 112–116). New York, NY: Personal Democracy Press.

boyd, d. (2008b). Digital handshakes in networked publics: Why Americans must interact, not broadcast. In B. Rigby (Ed.), *Mobilizing generation 2.0: A practical guide to using Web 2.0 technologies to recruit, organize, and engage youth* (pp. 91–94). San Francisco, CA: Jossey-Bass.

boyd, d. m., & Ellison, N. B. (2007). Social network sites: Definition, history, and scholarship. *Journal of Computer Mediated Communication*, 13(1), 210–230. Retrieved from http://jcmc.indiana.edu/vol13/issue1/boyd.ellison.html

Brady, D. (2011). Why your school's social media strategy is falling behind. Retrieved from http://www.fathomseo.com/resources/guides/social-media-white-paper.pdf

Branco, M. C., & Rodrigues, L. L. (2006). Communication of corporate social responsibility by Portuguese banks: A legitimacy theory perspective. *Corporate Communications: An International Journal*, 11(3), 232–248.

Branston, K., & Bush, L. (2010). The nature of online good networks and their impact on non-profit organisations and users. *Prism*, 7(2). Retrieved from http://www.prismjournal.org

Brashers, D. E. (2001). Communication and uncertainty management. *Journal of Communication*, 51, 477–497.

Brinton, H. (2010, June 21). Are social media changing religion? Retrieved from http://www.usatoday.com/news/opinion/forum/2010-06-21-column21_ST_N.htm

Briones, R.L., Kuch, B., Liu, B.F., & Jin, Y. (2011). Keeping up with the digital age: How the American Red Cross uses social media to build relationships. *Public Relations Review*, 37(1), 37–43.

Broom, G. M. (2009). *Cutlip & Center's effective public relations* (10th ed.). Upper Saddle River, NJ: Prentice Hall.

Brown, J., Broderick, A.J., & Lee, N. (2007). Word of mouth communication within online communities: Conceptualizing the online social network. *Journal of Interactive Marketing*, 21(3), 2–20.

Brown, K.C., & Roed, B. (2001). Delahaye Medilink's 2001 media reputation index results. *The Gauge, 16*(3), 1–2. Retrieved from http://www.thegauge.com/v16n3laydownlawprint.htm

Brown, R. (2010, July 16). EAS, IPAWS evolve to meet government mandates. Retrieved from http://broadcastengineering.com/infrastructure/eas-ipaws-evolve-meet-government-standards-20100718/

Brown, R.E. (2003). St. Paul as a public relations practitioner: A metatheoretical speculation on messianic communication and symmetry. *Public Relations Review, 29*(2), 229–241.

Bruneau, M., Chang, S. E., Eguchi, R. T., Lee, G. C., O'Rourke, T. D., Reinhorn, A. M., Shinozuka, M., Tierney, K., Wallace, W. A., & Winterfedlt, D. (2003). A framework to quantitatively assess and enhance the seismic resilience of communities. *Earthquake Spectra, 19*(4), 733–752.

Bruning, S. D., Dials, M., & Shirka, A. (2008). Using dialogue to build organization-public relationships, engage publics, and positively affect organizational outcomes. *Public Relations Review, 34*(1), 25–31.

Bruning, S. D., & Ledingham, J. (1999). Relationships between organizations and its publics: Development of a multi-dimensional organization-public relationship scale. *Public Relations Review, 25*(2), 157–170.

Brunner, B.R., & Boyer, L. (2008). Internet presence and historically black colleges and universities: Protecting their images on the World Wide Web? *Public Relations Review, 34*(1), 80–82.

Brus, B. (2011, March 17). Online oops: Social media blurs line between personal, professional lives. Retrieved from http://journalrecord.com/2011/03/17/online-oops-general-news/

Buber, M. (1965). *The knowledge of man: Selected essays.* New York, NY: Harper & Row Publishers, Inc.

Bucher, H. J. (2002). Crisis communication and the Internet: Risk and trust in a global media. *First Monday, 7*(4). Retrieved from http://131.193.153.231/www/issues/issue7_4/bucher/index.html

Buffery, H., & Marcer, E. (2011). *Historical dictionary of the Catalans.* Lanham, MD: The Scarecrow Press.

Buis, L. R., & Carpenter, S. (2009). Health and medical blog content and its relationships with blogger credentials and blog host. *Health Communication, 24*, 703–710.

Bulmer, D., & DiMauro, V. (2009). The new symbiosis of professional networks: Social media's impact on business and decision-making. Retrieved from http://sncr.org/wp-content/uploads/2010/02/NewSymbiosisReportExecSumm.pdf

Bumiller, E. (2002, July 31). Bush signs bill aimed at fraud in corporations. *The New York Times*, p. A1.

Burger, C. (1981). The edge of the communications revolution. *Public Relations Review, 7*(2), 3–12.

Burleson, B. R., Albrecht, T. L., Goldsmith, D. J., & Sarason, J. G. (1994). The communication of social support. In B. R. Burleson, T. L. Albrecht, & J. G. Sarason (Eds.), *Communication of social support: Messages, interactions, relationships, and community* (pp. xi–xxx). Thousand Oaks, CA: Sage.

Burns, K. S. (2008, October). The misuse of social media: Reactions to and important lessons from a blog fiasco. *Journal of New Communications Research, 3*(1), 41–54.

Burson and Marsteller. (2011). 2011 Fortune Global 100 Social Media Study. Retrieved from http://www.bursonmarsteller.com/Innovation_and_insights/blogs_and_podcasts/BM_Blog/Lists/Posts/Post.aspx?List75c7a224-05a3-4f25-9ce5-2a90a7c0c761&ID=254

Burton, M. S. (2005). Connecting the virtual dots: How the web can relieve our information glut and get us talking to each other. *Studies in Intelligence, 49*(3), 55–62.

Busch, L. (2011). To come to a correct understanding of Buddhism: A case study on spiritualizing technology, religious authority, and the boundaries of orthodoxy and identity in a Buddhist Web forum. *New Media & Society*, pp. 58–74.

Bush, M. (2010). Gulf Coast's $20B tourism biz asks BP to foot advertising bill. *Advertising Age, 81*(20), 1–40.

Byrne, B. M. (1994). *Structural equation modeling with EQS and EQS/Window: Basic concepts, applications, and programming.* Thousand Oaks, CA: Sage.

Cahill, B. (2009, Spring). Your attention please: The right way to integrate social media into your marketing plans. Retrieved from http://www.prsa.org/Intelligence/TheStrategist/Articles/download/6K020925/102/Your_Attention_Please_The_Right_Way_to_Integrate_S?

Cakim, I. (2007). Digital public relations: Online reputation management. In S. C. Duhé (Ed.), *New media and public relations* (pp. 135–144). New York, NY: Peter Lang.

Callison, C. (2003). Media relations and the Internet: How Fortune 500 company Web sites assist journalists in news gathering. *Public Relations Review, 29*(1), 29–41.

Calloway, L.J. (1991). Survival of the fastest: Information technology and corporate crises. *Public Relations Review, 17*(1), 85–92.

Campbell, H. (2010). *When religion meets new media.* New York, NY: Routledge.

Campbell Laidler, D. (2010, November). What's your social media strategy? *Black Enterprise, 41*(4), 74–76.

Canfield, S. (2011, March 8). In devastating complaint, Louisiana demands $1 million a day from BP & others. Retrieved from http://www.courthousenews.com/2011/03/08/34716.htm

Capriotti, P. (2007). Chemical risk communication through the Internet in Spain. *Public Relations Review, 33*(3), 326–329.

Capriotti, P. (2011). Communicating CSR through the Internet. In O. Ihlen, J.L. Bartlett, & S. May (Eds.), *The handbook of communication and CSR* (pp. 270–283). Thousand Oaks, CA: Wiley Blackwell.

Capriotti, P., & Moreno, A. (2007). Corporate citizenship and public relations: The importance and interactivity of social responsibility issues on corporate websites. *Public Relations Review, 33*(1), 84–91.

Carmines, E. G., & Zeller, R. A. (1979). *Reliability and validity assessment.* Thousand Oaks, CA: Sage.

Caroll, A. B. (1979). A three-dimensional conceptual model of corporate performance. *Academy of Management Review, 4*(4), 497–505.

Carroll, C.E., & McCombs, M. (2003). Agenda-setting effects of business news on the public's image and opinions about major corporations. *Corporate Reputation Review, 16,* 36–46.

Carter, D. (2008a). Living in virtual communities: An ethnography of human relations in cyberspace. *Information, Communication & Society, 8*(2), 148–167.

Carter, D. (2008b, November 20). Questions abound as emergency alert flops. Retrieved from http://www.eschoolnews.com/news/top-news/?i=56122

Carter, R. F. (1965). Communication and affective relation. *Journalism Quarterly, 42,* 203–212.

Carter, T. (2004). Recipe for growth. *ABA Journal, 90*(4), 85.

Castano, R., Sujan, M., Kacker, M., & Sujan, H. (2006). Managing consumer uncertainty in the adoption of new products: Temporal distance and mental simulation. *Journal of Marketing Research, 45,* 320–336.

Castells, M. (1996). *The rise of the network society.* Cambridge, England: Blackwell.

Castells, M. (2000a). *End of millennium* (2nd ed.) (The information age: Economy, society and culture, Vol. III). Malden, MA: Blackwell.

Castells, M. (2000b). *The rise of network society* (2nd ed.) (The information age: Economy, society and culture, Vol. I). Malden, MA: Blackwell.

Castells, M. (2001). *The Internet galaxy: Reflections on the Internet, business, and society.* Oxford, United Kingdom: Oxford University Press.

Castells, M. (2004a). Afterword: Why networks matter. In H. McCarthy, P. Miller, & P. Skidmore (Eds.), *Networked logic: Who governs in an interconnected world* (pp. 221–224). London, England: Demos.

Castells, M. (Ed.). (2004b). *The network society: A cross-cultural perspective.* North Hampton, MA: Edgar Elgar.

Castells, M. (2004c). *The power of identity* (2nd ed.) (The information age: Economy, society and culture, Vol. II). Malden, MA: Blackwell.

Castells, M. (2008a, January 26). Inventar naciones [Inventing nations]. *La Vanguardia,* p. 15.

Castells, M. (2008b). The new public sphere: Global civil society, communication networks, and global governance. *The Annals of the American Academy of Political and Social Science, 616,* 78–93.

Castells, M. (2009). *Communication power.* Oxford, United Kingdom: Oxford University Press.

Castronova, E. (2001). Virtual worlds: A first-hand account of market and society on the Cyberian frontier. *The Gruter Institute Working Papers on Law, Economics, and Evolutionary Biology, 2*(1). Retrieved from http://www.bepress.com/giwp/default/vol2/iss1/art1

Centers for Disease Control and Prevention (CDC). (2010a). CDC social media tools, guidelines & best practices. Retrieved from http://www.cdc.gov/SocialMedia/Tools/guidelines

Centers for Disease Control and Prevention (CDC). (2010b). The health communicator's social media toolkit. Retrieved from http://www.cdc.gov/healthcommunication/ToolsTemplates/SocialMediaToolkit_BM.pdf

charity: water. (2009). Twestival: Tweet, meet, give. Retrieved from http://www.charitywater.org/twestival/

Charlie Sheen's publicist quits. (2011, February 28). Retrieved from http://www.tmz.com/2011/02/28/charlie-sheen-two-and-a-half-men-chuck-lorre-argument-radio-talk-show-tirade-turd-thomas-jefferson-alcoholics-anonymous/print

Chen, J. (2011). H&R Block joins GM, Wrigley, Taco Bell in pulling advertisements from racy MTV show. Retrieved from http://www.nydailynews.com/entertainment/tv/2011/01/22/2011-01-22_hr_block_joins_gm_wrigley_taco_bell_in_pulling_advertisements_from_racy_mtv_show.html

Chen, N. (2003). From propaganda to public relations. *Asian Journal of Communication, 13*(2), 96–121.

Chen, Y.-N. K. (2007). A study of journalists' perception of candidates' websites and their relationships with the campaign organization in Taiwan's 2004 presidential election. *Public Relations Review, 33*(1), 103–105.

Cheney, G., Christensen, L., Zorn, T., & Ganesh, S. (2011). *Organizational communication in an age of globalization: Issues, reflections, practices* (2nd ed.). Prospect Heights, IL: Waveland Press.

Chernilo, D. (2007). *A social theory of the nation-state: The political forms of modernity beyond methodological nationalism*. London, England: Routledge.

Chiang, O. (2011, January 19). Twitter hits nearly 200M accounts, 110M tweets per day, focuses on global expansion. Retrieved from http://blogs.forbes.com/oliverchiang/2011/01/19/twitter-hits-nearly-200m-users-110m-tweets-per-day-focuses-on-global-expansion/

ChicagoHumanities.org. (2010). The virtual body: Coming of age in Second Life. Retrieved from http://www.youtube.com/watch?v=mmghOAYo1aI

Cho, S., & Hong, Y. (2009). Netizens' evaluations of corporate social responsibility: Content analysis of CSR news stories and online readers' comments. *Public Relations Review, 35*(2), 147–149.

Choi, Y., & Lin, Y.-H. (2009). Consumer responses to Mattel product recalls posted on online bulletin boards: Exploring two types of emotion. *Journal of Public Relations Research, 21*(2), 198–207.

Christ, P. (2005). Internet technologies and trends transforming public relations. *Journal of Website Promotion, 1*(4), 3–14.

Chun, A.C., Shulman, S., Sandoval, R., & Hovy, E. (2010). Government 2.0: Making connections between citizens, data and government. *Information Policy, 15*, 1–9.

Chung, W., Lee, T.D., & Humphrey, V.F. (2010). Academic institutions' electronic-recruitment efforts on academic diversity: A comparative analysis of websites of US, UK, and South Korean universities. *Prism, 7*(2). Retrieved from http://www.prismjournal.org

Church, G. (2008, July). To tweet or not to tweet: How and when to use Twitter in PR efforts. Retrieved from http://www.prsa.org/SearchResults/view/7446/105/To_tweet_or_not_to_tweet_How_and_when_to_use_Twitt

Cisco helps schools modernize campus safety. (2008, February). *Wireless News*, p. 1. Retrieved from http://search.proquest.com/docview/210169892?accountid=13158

Claiborne, T. (2011, May 11). New Google analytics: Overview reports overview. Retrieved from http://analytics.blogspot.com/2011/05/new-google-analytics-overview-reports.html?utm_source=feedburner&utm_medium=feed&utm_campaign=Feed%3A+blogspot%2FtRaA+%28Google+Analytics+Blog%29&utm_content=Google+Feedfetcher

Clapton, A. W., & Finch, B. (2010). Are college students "bowling alone?" Examining the contribution of team identification to the social capital of college students. *Journal of Sport Behavior, 33*(4), 377–402.

Cohen, N. (2011, May 23). Wikipedia. Retrieved from http://topics.nytimes.com/top/news/business/companies/wikipedia/index.html

Coleman, R., Lieber, P., Mendelson, A. L., & Kurpius, D. D. (2008). Public life and the Internet: If you build a better website, will citizens become engaged? *New Media and Society, 10*(2), 179–201.

Comeaux, E. (2010). Racial differences in faculty perceptions of collegiate student-athletes' academic and post-undergraduate achievements. *Sociology of Sport Journal, 27*, 390–412.

Commercial Mobile Alert System. (n.d.). Federal Communications Communication. Washington, D.C. Retrieved from http://www.fcc.gov/guides/commercial-mobile-alert-system

Compensating the people and communities affected. (n.d.). Retrieved from http://www.bp.com/sectiongenericarticle800.do?categoryId=9036584&contentId=7067605

Compete. (2011). Site profile. Retrieved from http://www.compete.com

Connolly-Ahern, C., & Broadway, S.C. (2007). The importance of appearing competent: An analysis of corporate impression management strategies on the World Wide Web. *Public Relations Review, 33*(3), 343–345.

Containing the leak. (n.d.). Retrieved from http://www.bp.com/sectiongenericarticle800.do?categoryId=9036583&contentId=7067603

Conway, L. (2008, November 1). Virgin Atlantic fired 13 staff for calling its flyers "chavs." Retrieved from http://www.independent.co.uk/news/uk/home-news/virgin-atlantic-sacks-13-staff-for-calling-its-flyers-chavs-982192.html

Coombs, W.T. (1992). The failure of the task force on food assistance: A case study of the role of legitimacy in issue management. *Journal of Public Relations Research, 4,* 101–122.

Coombs, W.T. (1998). The Internet as potential equalizer: New leverage for confronting social irresponsibility. *Public Relations Review, 24*(3), 289–303.

Coombs, W. T. (2002). Assessing online issue threats: Issue contagions and their effect on issue prioritisation. *Journal of Public Affairs, 2*(4), 215–229.

Coombs, W. T. (2007). Attribution theory as a guide for post-crisis communication research. *Public Relations Review, 33*(2), 135–139.

Coombs, W. T. (2008). Crisis communication and social media. Retrieved from http://www.instituteforpr.org./essential_knowledge/detail/crisis_communication_and_social_media

Coombs, W. T. (2010). Parameters for crisis communication. In W. T. Coombs & S.J. Holladay (Eds.), *Handbook of crisis communication* (pp. 17–53). Malden, MA: Blackwell Publishing.

Coombs, W. T. (2012). *Ongoing crisis communication: Planning, managing, and responding* (3rd ed.). Thousand Oaks, CA: Sage.

Coombs, W. T., & Holladay, S. J. (2004). Reasoned action in crisis communication: An attribution theory-based approach to crisis management. In D. P. Millar & R. L. Heath (Eds), *Responding to a crisis: A rhetorical approach to crisis communication* (pp. 95–115). Mahwah, NJ: Lawrence Erlbaum Associates.

Coombs, W. T., & Holladay, S. J. (2006). Halo or reputational capital: Reputation and crisis management. *Journal of Communication Management, 10*(2), 123–137.

Coombs, W.T., & Holladay, S.J. (2007a). Consumer empowerment through the Web: How Internet contagions can increase stakeholder power. In S. C. Duhé (Ed.), *New media and public relations* (pp. 173–187). New York, NY: Peter Lang.

Coombs, W. T., & Holladay, S.J. (2007b). The negative communication dynamic: Exploring the impact of stakeholder affect on behavioral intentions. *Journal of Communication Management, 11*(4), 300–312.

Coombs, W. T., & Holladay, S. J. (2009). Further explorations of post-crisis communication: Effects of media and response strategies on perceptions and intentions. *Public Relations Review, 35*(1), 1–6.

Coombs, W. T., & Holladay, S. J. (2010a). *Managing corporate social responsibility: A communication approach.* Malden, MA: Wiley-Blackwell.

Coombs, W.T., & Holladay, S.J. (2010b). *PR strategy and application: Managing influence.* Malden, MA: Wiley-Blackwell.

Cooper-Chen, A., & Tanaka, M. (2008). Public relations in Japan: The cultural roots of kouhou. *Journal of Public Relations Research, 20*(1), 94–114.

Corbin, J.M., & Strauss, A.C. (2008). *Basics of qualitative research* (3rd ed.). Thousand Oaks, CA: Sage.

Corcoran, A., Marsden, P., Zorbach, T., & Röthlingshöfe, B. (2006). Blog marketing. In J. Kirby & P. Marsden (Eds.), *Connected marketing: The viral, buzz, and wrd of mouth revolution* (pp. 148–158). Burlington, MA: Elsevier.

Corney, T., & du Plessis, K. (2010). Apprentices' mentoring relationships. *Youth Studies Australia, 29*(3), 18–26.

Corporate Intelligence Center. (2011). *From social media to social business.* Available at http://www.slideshare.net/CIC_China/from-social-media-to-social-business-topic-1

Corrigan, R. (2008). Back to the future: Digital decision making. *Information & Communications Technology Law, 17*(3), 199–220.

Cotten, S. R., & Gupta, S. S. (2004). Characteristics of online and offline health information seekers and factors that discriminate between them. *Social Science & Medicine, 59*(9), 1795–1806.

Cotugna, N., & Vickery, C. (1998). Reverse mentoring: A twist to teaching technology. *Journal of the American Dietetic Association, 98*(10), 1166.

Cova, B., & Cova, V. (2002). Tribal marketing: The tribalisation of society and its impact on the conduct of marketing. *European Journal of Marketing, 36*(5/6), 595–620.

Cowan, R. S. (1997). *A social history of American technology.* New York, NY: Oxford University Press.

Craig, J. (2007). Spokane County considering "reverse 911" calling system. Retrieved from http://search.proquest.com/docview/463807028?accountid=13158

Creswell, J. W. (2007). *Qualitative inquiry & research design: Choosing among five approaches.* Thousand Oaks, CA: Sage.

Crovitz, L. G. (2009, April 9). Information age: Wikipedia's old-fashioned revolution. *The Wall Street Journal*, p. A13.

Csikszentmihalyi, M. (1990). *Flow: The psychology of optimal experience.* New York, NY: Harper Perennial.

Curtin, P.A., & Gaither, T.K. (2004). International agenda-building in cyberspace: A study of Middle East government English-language websites. *Public Relations Review, 30*(1), 25–36.

Curtis, L., Edwards, C., Fraser, K.L., Gudelsky, S., Holmquist, J., Thornton, K., & Sweetser, K.D. (2010). Adoption of social media for public relations by nonprofit organizations. *Public Relations Review, 36*(1), 90–92.

Cutlip, S. M., Center, A. H., & Broom, G. M. (2004). *Effective public relations.* Upper Saddle River, NJ: Prentice Hall.

Czerniewicz, L., & Brown, C. (2009). A study of the relationship between institutional policy, organizational culture and e-learning use in four South African universities. *Computers & Education, 53*, 121–131.

Dando, N., & Swift, T. (2003). Transparency and assurance: Minding the credibility gap. *Journal of Business Ethics, 44*, 195–200.

Daniel, L. (2009). Church netiquette. *Christian Century*, pp. 26–28.

Danowski, J. A. (2008). Short-term and long-term effects of a public relations campaign on semantic networks of newspaper content: Priming or framing? *Public Relations Review, 34*(3), 288–290.

Dareini, A. A.(2009, May 25). Facebook block ahead of Iran vote hampers youth. Retrieved from http://www.thejakartapost.com/news/2009/05/25/facebook-block-ahead-iran-vote-hampers-youth.html

Darmon, K., Fitzpatrick, K., & Bronstein, C. (2008). Krafting the obesity message: A case study in framing and issues management. *Public Relations Review, 34*(4), 373–379.

Davies, G., Chun, R., da Silva, R. V., & Roper, S. (2003). *Corporate reputation and competitiveness.* New York, NY: Routledge.

Davis, D. (2011, February 16). Ten tips for church Facebook pages. *Baptist Trumpet*, pp. 1–2.

Day, S. (2003, June 2). Counteracting the Internet rumor. Retrieved from http://www.boycottwatch.org/misc/images/Starbucks1c.pdf

Deephouse, D. L. (2000). Media reputation as a strategic resource: An integration of mass communication and resource-based theories. *Journal of Management, 26*, 1091–1112.

Denzin, N., & Lincoln, Y. S. (2003). *Collecting and interpreting qualitative materials.* Thousand Oaks, CA: Sage.

Deutsche EuroShop (2011). The use of social media by European investment professionals. Retrieved from http://www.slideshare.net/DESAG/the-use-of-social-media-by-european-investment-professionals-7848326

Devaux, V. (2009, December). No networking allowed. *Training and Development, 63*(12), 20.

Dewdney, A., & Ride, P. (2006). *The new media handbook.* New York, NY: Routledge.

Dholakia, R.R., Zhao, M., Dholakia, N., & Fortin, D.R. (2000). Interactivity and revisits to websites: A theoretical framework. Available at http://ritim.cba.uri.edu/wp2001/wpdone3/Interactivity.PDF

Diana, A. (2010, June 28). Social media up 230% since 2007. Retrieved from http://www.informationweek.com/news/software/soa_webservices/225701600

DiFonzo, N. (2008). *The watercooler effect: A psychologist explores the extraordinary power of rumors.* New York, NY: Avery.

DiFonzo, N., & Bordia, P. (2000). How top PR professionals handle hearsay: Corporate rumors, their effects, and strategies to manage them. *Public Relations Review, 26*(2), 173–190.

Diga, M., & Kelleher, T. (2009). Social media use, perceptions of decision-making power, and public relations roles. *Public Relations Review 35*(4), 440–442.

Digital readiness report: Essential online public relations and marketing skills (2009). Retrieved from http://ipressroom.com/readinessreport

DiMaggio, A. R. (2008). *Mass media, mass propaganda: Examining American news in the "war on terror."* Lanham, MD: Lexington Books.

DiNardo, A.M. (2002). The Internet as a crisis management tool: A critique of banking sites during Y2K. *Public Relations Review, 28*(4), 367–378.

Dinich, H. (2009, July 9). How much is too much online? Retrieved from http://es.pn/NA2gR

Discovery (2011, March 13). In times of crisis, people turn to the Internet. Retrieved from http://news.discovery.com/tech/japan-tsunami-quake-internet-110313.html#mkcpgn=rssnws1

DiStaso, M. W., McCorkindale, T. M., & Wright, D. K. (in press). How public relations executives perceive and measure the impact of social media in their organizations. *Public Relations Review.*

DiStaso, M.W., & Messner, M. (2010). Forced transparency: Corporate image on Wikipedia and what it means for public relations. *Public Relations Journal, 4*(2). Retrieved from http://www.prsa.org/Intelligence/PRJournal/

DiStaso, M. W., Messner, M., & Stacks, D. W. (2007). The wiki factor: A study of reputation management. In S. C. Duhé (Ed.), *New media and public relations* (pp. 121–133). New York, NY: Peter Lang.

Dizikes, P. (2008, January 29). Tricky wiki: How public relations companies try to spin Wikipedia. Retrieved from http://www.infowars.com/tricky-wiki-how-public-relations-companies-try-to-spin-wikipedia/

Dodd, D.W. (2006). Have we learned the lesson of disaster preparedness? *College Planning & Management, 9*(4), 7.

Dolak, D. (2001). Building a strong brand: Brands and branding basics. Retrieved from http://www.davedolak.com/articles/dolak4.htm

Dowling, G. (1994). *Corporate reputations.* Melbourne, Australia: Longman Professional.

Downing, J. (2011). K-12 parents' attitudes about their school district's mass notification service. *Journal of School Public Relations, 32*(2), 90-121.

Dozier, D. M., Grunig, L. A., & Grunig, J. E. (1995). *Manager's guide to excellence in public relations and communication management.* Mahwah, NJ: Lawrence Erlbaum Associates.

Dozier, R. W. (2002). *Why we hate: Understanding, curbing, and eliminating hate in ourselves and our world.* Chicago, IL: Contemporary Books.

Drapeau, M., & Wells, L. (2009, April). Social software and national security: An initial net assessment. Retrieved from http://www.dtic.mil/cgi-bin/GetTRDoc?AD=ADA497525

Duhé, S. C. (2007). Public relations and complexity thinking in the age of transparency. In S. C. Duhé (Ed.), *New media and public relations* (pp. 57–75). New York, NY: Peter Lang.

Duke, S. (2002). Wired science: Use of World Wide Web and e-mail in science public relations. *Public Relations Review, 28*(3), 311-324.

Dutta-Bergman, M. J. (2003). Health communication on the web: The roles of web use motivation and information completeness. *Communication Monographs, 70*(3), 264–274.

Dutta-Bergman, M. J. (2004). Interpersonal communication after 9/11 via telephone and Internet: A theory of channel complementarity. *New Media & Society, 6*(5), 659–673.

Dutta-Bergman, M. J. (2006). Community participation and Internet use after September 11: Complementarity in channel consumption. *Journal of Computer-Mediated Communication, 11*(2), 469–484. Retrieved from http://jcmc.indiana.edu/vol11/issue2/dutta-bergman.html

Dworkin, D. (2011, March). *2010 FedEx/Ketchum social media benchmarking study highlights.* Paper presented at the 14th Annual International Public Relations Research Conference, Miami, FL.

A dynamic industry in a turbulent economy (2001, May 7). *PR News, 57*(18), 1.

Eastin, M. S. (2001). Credibility assessments of online health information: The effects of source expertise and knowledge of content. *Journal of Computer-Mediated Communication, 6*(4). doi: 10.1111/j.1083-6101.2001.tb00126.x

Edelman. (2011). 2011 Edelman Trust Barometer. Retrieved from http://www.edelman.com/trust/2011/uploads/Trust%20Executive%20Summary.PDF

Edgerton, R. B. (1985). *Rules, exceptions, and social order.* Berkeley: University of California Press.

Edmondson, E. (2010). CBS' sensational series *Survivor* successfully navigates new media waters. Unpublished undergraduate paper, Comm465: Entertainment PR, California State University, Fullerton.

Eid, M. (2007). Engendering the Arabic Internet: Modern challenges in the information society. In S. C. Duhé (Ed.), *New media and public relations* (pp. 247–268). New York, NY: Peter Lang.

Eid, M., & Ward, S. J. A. (2009). Editorial: Ethics, new media, and social networks. *Global Media Journal -- Canadian Edition, 2*(1), 1–4.

Eisenstadt, S. N. (1992). The order-maintaining and order-transforming dimensions of culture. In R. Munch & N. J. Smelser (Eds.), *Theory of culture* (pp. 64–88). Berkeley: University of California Press.

Eisenstein, C. (1994). *Meinungsbildung in der mediengesellschaft. Eine theoretische und empirische analyse zum Multi-Step Flow of Communication.* [Opinion making in media society: A theoretical and empirical analysis leading to a Multi-Step Flow of Communication.] Opladen, Germany: Westdeutscher Verlag.

Elliott, S. (2010, November 10). Nike harnesses "Girl Effect" again. Retrieved from http://www.nytimes.com/2010/11/11/giving/11VIDEO.html

Elliott, S. (2011, March 16). When the marketing reach of social media backfires. *The New York Times*, p. B3.

England, P. (2011, May). A digital revolution in Egypt and beyond. *New Internationalist*, 56.

Entman, R. (1993). Framing: Clarification of a fractured paradigm. *Journal of Communication*, 43(4), 51–58.

Entman, R. M. (2003). Cascading activism: Contesting the White House's frame after 9/11. *Political Communication*, 20, 415–432.

Esrock, S.L., & Leichty, G.B. (1998). Social responsibility and corporate Web pages: Self-presentation or agenda-setting? *Public Relations Review*, 24(3), 305–319.

Esrock, S.L., & Leichty, G.B. (2000). Organization of corporate Web pages: Publics and functions. *Public Relations Review*, 26(3), 327–344.

Evans, A., Twomey, J., & Talan, S. (2011, Winter). Twitter as a public relations tool. *Public Relations Journal*, 5(1). Retrieved from http://www.prsa.org/Intelligence/PRJournal/

Evans, D., Schmalz, J., Gainer, D., & Snider, J. (2010, November 15). Senior moment: Boomers turn 65. *USA Today*, p. 4A.

Eyrich, N., Padman, M., & Sweetser, K. (2008). PR practitioners' use of social media tools and communication technology. *Public Relations Review*, 34(4), 412–414.

Eysenbach, G. (2003). The impact of the Internet on cancer outcomes. *A Cancer Journal for Clinicians*, 53, 356–371. doi: 10.3322/canjclin.53.6.356

Face time. (2011). *Emmy*, 1, 9.

Facebook, Wikipedia execs brief Vatican on Web. (2009, November 12). Retrieved from http://www.usatoday.com/news/world/2009-11-12-vatican-web-briefing_N.htm#

Fairclough, N. (2003). *Analysing discourse: Textual research for social research.* London, England: Routledge.

Falsani, C. (2011, February 12). New apps help users on path to absolution. *The Advocate*, p. 3D.

Fans invited to help select hit song for Bon Jovi performance on the 52nd annual Grammy Awards. (2010, January 12). Retrieved from http://www.bonjovi.com/story/news-featured-backstage_jbj/fans_invited_to_help_select_hit_song_for_bon_jovi_performance_on_the_52nd_annual_grammy_awards

Farrow, H., & Yuan, Y. C. (2011). Building stronger ties with alumni through Facebook to increase volunteerism and charitable giving. *Journal of Computer-Mediated Communication*, 16(3), 445–464.

Fathi, S. (2008, October). From generating awareness to managing reputations: Why your company needs Twitter. Retrieved from http://www.affectstrategies.com/files/inthenews/PRSA_Tactics_Twitter_Article.pdf

Faulkner, X., & Cutwin, F. (2005, March). When fingers do the talking: A study of text messaging. *Interacting with Computers*, 17(2), 167–185.

Favoriso, J. (2008). *Sport publicity: A practical approach.* Oxford, UK: Butterworth-Heinemann.

FCC's public safety and Homeland Security Bureau sets timetable in motion for Commercial Mobile Service providers to develop a system that will deliver alerts to mobile devices. (2009, December 7). PS Docket No. 07–287. Federal Communications Commission, Washington, D.C. Retrieved from http://hraunfoss.fcc.gov/edocs_public/attachmatch/DA-09-2556A1.pdf

Fearn-Banks, K. (2002). *Crisis communications: A casebook approach* (2nd ed.). Mahwah, NJ: Lawrence Erlbaum Associates.

Federal Trade Commission. (2009, October 5). FTC publishes final guides governing endorsements, testimonials. Changes affect testimonial advertisements, bloggers, celebrity endorsements. Retrieved from http://www.ftc.gov/opa/2009/10/endortest.shtm

Feldner, S. B., & Meisenbach, R. J. (2007). Saving Disney: Activating publics through the Internet. In S. C. Duhé (Ed.), *New media and public relations* (pp. 189–201). New York, NY: Peter Lang Publishing.

Fenn, J., & Raskino, M. (2008). *Mastering the hype cycle: How to choose the right innovation at the right time.* Boston, MA: Harvard Business Press.

Fernando, A. (2010). Are you still not a Wikipedian? *Communication World*, 27(4), 8–9.

Fieseler, C., Fleck, M., & Meckel, M. (2010). Corporate social responsibility in the blogosphere. *Journal of Business Ethics, 91*(4), 599–614.

Fink, S. (1986). *Crisis management: Planning for the inevitable.* New York, NY: AMACOM.

Fisher, E. (2009). Fight of fance. *Sport Business Journal, 12*(7), 15-17.

Fisher, T. (2009). ROI in social media: A look at the arguments. *Journal of Database Marketing & Customer Strategy Management, 16*(3), 189–195.

Fitch, K. (2009). Making friends in the Wild West: Singaporean public relations practitioners' perceptions of working in social media. *Prism, 6*(2). Retrieved from http://www.prismjournal.org

Fitzpatrick, K. R. (1996). Public relations and the law: A survey of practitioners. *Public Relations Review, 22*(1), 1–8.

Fitzpatrick, K. R., & Rubin, M. S. (1995). Public relations vs. legal strategies in organizational crisis decisions. *Public Relations Review, 21*(1), 21–33.

Fleiss, J. L. (1971). Measuring nominal scale agreement among many raters. *Psychological Bulletin, 76*(5), 378–382.

Flint, J. (2011, April 2). Sheen defies damage control. *Los Angeles Times,* pp. D1, D6.

Foley, A. (2010) Using social media to your advantage. *Canadian Musician, 32*(3), 42–46.

Folkman, S. (1997). Positive psychological states and coping with severe stress. *Social Science & Medicine, 45*(8), 1207–1221.

Fombrun, C. J., & van Riel, C. B. M. (2004). *Fame & fortune: How successful companies build winning reputations.* New York, NY: Prentice Hall Financial Times.

Forrester Research. (2010). Consumer technographics data. Retrieved from http://www.forrester.com/empowered/tool_consumer.html

Fortson, C. (2003, February 14). Women's rights vital for developing world. Retrieved from http://www.yaledailynews.com/news/2003/feb/14/womens-rights-vital-for-developing-world/

Freberg, K., Graham, K., McGaughey, K., & Freberg, L.A. (2011). Who are the social media influencers? A study of public perceptions of personality. *Public Relations Review, 37*(1), 90–92.

Freedman, S. (2009, August 22). Among young Sikhs, expressions of faith mixing two worlds. *The New York Times,* p. 15.

Freeman, B., & Chapman, S. (2010). British American Tobacco on Facebook: Undermining Article 13 of the Global World Health Organization Framework Convention on Tobacco Control. *Tobacco Control, 19*(3), e1-e9. doi:10.1136/tc.2009.032847

Frenette, B., Woo, A., & Shaw, G. (2011, July 14). Circles, hangouts, sparks: A simple guide to getting started in Google Plus. Retrieved from http://www.vancouversun.com/news/Circles%20Hangouts%20Sparks%20simple%20guide%20getting%20started%20Google%20Plus/5096798/story.html?id=5096798

Friedman, W. (2011, March 11). "Super 8" film marketed on Twitter, Facebook. Retrieved from http://www.mediapost.com/publications/index.cfm?fa=Articles.showArticle&art_aid=146547

Friend, T. (2009, January 19). Letter from California. The cobra. Inside a movie marketer's playbook. *The New Yorker,* pp. 41–49.

Fritz, C. E., & Mathewson, J. H. (1957). Convergence behavior in disasters: A problem in social control. Retrieved from http://www.archive.org/stream/convergencebehavoofritrich/convergencebehavoofritrich_djvu.txt

Fugate, C. (2010, December 14). Welcome to the first-ever FEMA blog. Retrieved from http://blog.fema.gov/search?q=welcome+to+the+first-ever+fema+blog

Fullerton, L., & Ettema, J. (2009, May). *Striving for NPOV: Reconciling knowledge claims in Wikipedia.* Paper presented at the International Communication Association Conference, Chicago, IL.

Fussell-Sisco, H., & McCorkindale, T. (2011, March). *Communicating "Pink": A quantitative content analysis of the trustworthiness and communication strategies of breast cancer social media sites.* Paper presented at the International Public Relations Research Conference, Miami, FL.

Gainey, B.S. (2010, Fall). As the front-line responders, public sector organizations must be crisis ready. In S. Goldstein (Ed.), *PR News crisis management guidebook* (Vol. 4, pp. 112–114). Rockville, MD: PR News.

Galford, R., & Drapeau, A. S. (2003). The enemies of trust. *Harvard Business Review, 81*(2), 88–95.

Galloway, C. (2005). Cyber-PR and "dynamic touch." *Public Relations Review, 31*(4), 572–577.

Games for Change. (2011). 8th annual games for change festival: June 20–22, 2011. Retrieved from http://gamesforchange.org/festival2011/events/the-case-for-social-impact-games/

Garron, B. (2011). Fast friends. *Emmy, 1,* 9.

Gates, A. (2008, November 16). Motrin's email response to the onslaught of complaints over babywearing ad. Retrieved from http://crunchydomesticgoddess.com/2008/11/16/motrins-response-to-the-onslaught-of-complaints/

Generalitat de Catalunya. (2006). *Pla estratègic del Turisme a Catalunya* [Strategic Plan for tourism in Catalonia]. Barcelona, Spain: Deparatament de Comerç, Consum i Turisme.

Gerber, M., & Ferree, J. (2011, March). *New opportunities for public alerting.* Paper presented at the National Severe Weather Workshop, Norman, OK. Retrieved from http://www.norman.noaa.gov/nsww/wp-content/uploads/2011/03/Ferree_Gerber_NSWW2011.pdf

Giles, D., & Pitta, D. A. (2009). Internet currency. *Journal of Consumer Marketing, 26*(7), 529–530.

Giles, J. (2005). Internet encyclopedias go head to head. *Nature, 438,* 900–901.

Gill, K. (2010, May 25). Twitter account spoofs BP PR efforts. Retrieved from http://themoderatevoice.com/73827/twitter-account-spoofs-bp-pr-efforts/

Gillin, P. (2008). New media, new influencers and implications for the public relations profession. *Journal of New Communications Research, 2*(2). Available at http://iab.blogosfere.it/images/NewInfluencer.pdf

Gillmor, D. (2006). *We the media: Grassroots journalism by the people, for the people.* Sebastopol, CA: O'Reilly Media, Inc.

Gilmore, J. H., & Pine, J. B. (2007). *What consumers really want: Authenticity.* Boston, MA: Harvard Business School Press.

Gilpin, D. (2010a). Organizational image construction in a fragmented online media environment. *Journal of Public Relations Research, 22*(3), 265–287.

Gilpin, D. (2010b). Socially mediated authenticity. *Journal of Communication Management, 14*(3), 258–278.

Ginossar, T. (2008). Online participation: A content analysis of differences in utilization of two online cancer communities by men and women, patients and family members. *Health Communication, 23,* 1–12.

Gladwell, M. (2000). *The tipping point: How little things make a big difference.* New York, NY: Little, Brown and Company.

Glaser, B. G., & Strauss, A. L. (1967). The discovery of grounded theory: *Strategies for qualitative research.* Chicago, IL: Aldine Publishing Company.

Glenn, M.C., Gruber, W.H., & Rabin, K.H. (1982). Using computers in corporate public affairs. *Public Relations Review, 8*(3), 34–42.

Glionna, J. M., & Hall, K. (2011, April 29). Japanese drop their traditional politeness over nuclear crisis. Retrieved from http://articles.latimes.com/print/2011/apr/29/world/la-fg-japan-anger-20110429

Global Reporting Initiative. (2010). New GRI office helps companies showcase sustainability. Retrieved from http://www.globalreporting.org/NewsEventsPress/PressResources/2010/FocalPointUSA.htm

Goffman, E. (1959). *The presentation of self in everyday life.* Garden City, NY: Doubleday.

Gofton, K. (2000, November 30). PR firms keep on the takeover trail. *Marketing,* pp. 25–26.

Goldstein, E. (2011, April 25). Employee usage of social media—a toy or a tool? Retrieved from http://socialmediatoday.com/eric-goldstein/289381/employee-usage-social-media-toy-or-tool

Goldstein, P. (2009, October 13). The big picture: High marks: When CinemaScore grades a film, Hollywood is all ears. *Los Angeles Times,* pp. D1, D9.

Gomez, L.M., & Chalmeta, R. (2011). Corporate responsibility in U.S. corporate websites: A pilot study. *Public Relations Review, 37*(1), 93–95.

Gonzalez-Herrero, A., & Smith, S. (2008). Crisis communication management on the Web: How Internet-based technologies are changing the way public relations professionals handle business crises. *Journal of Contingencies and Crisis Management, 16*(3), 143–153.

Gordon, J., & Berhow, S. (2009). University websites and dialogic features for building relationships with potential students. *Public Relations Review, 35*(2), 150–152.

Gower, K. K. (2006). Truth and transparency. In K. Fitzpatrick & C. Bronstein (Eds.), *Ethics in public relations: Responsible advocacy* (pp. 89–106). Thousand Oaks, CA: Sage.

Grant, W. J., Moon, B., & Grant, J. B. (2010). Digital dialogue? Australian politicians' use of the social network tool Twitter. *Australian Journal of Political Science, 45*(4), 579–604.

Grayson, K., & Martinec, R. (2004). Consumer perceptions of iconicity and indexicality and their influence on assessments of authentic market offerings. *Journal of Consumer Research, 31,* 296–312.

Greengard, S. (2002). Moving forward with reverse mentoring. *Workforce, 81*(3), 15.

Greer, C.F., & Moreland, K.D. (2003). United Airlines' and American Airlines' online crisis communication following the September 11 terrorist attacks. *Public Relations Review, 29*(4), 427–441.

Greer, C.F., & Moreland, K.D. (2007). How Fortune 500 companies used the Web for philanthropic and crisis communication following Hurricane Katrina. *Public Relations Review, 33*(2), 214–216.

Gregory, A. (2004). Scope and structure of public relations: A technology driven view. *Public Relations Review, 30*(3), 245–254.

Griffin, R. J., Dunwoody, S., & Neuwirth, K. (1999). Proposed model of the relationship of risk information seeking and processing to the development of preventive behaviors. *Environmental Research, 80*, S230-S245.

Groves, I. (1998). *Mobile systems.* London, England: Chapman & Hall.

Grunig, J, E. (1989a). Publics, audiences and market segments: Segmentation principles for campaigns. In C. T. Salmon (Ed.), *Information campaigns: Balancing social values and social change* (pp. 199–223). Beverly Hills, CA: Sage.

Grunig, J.E. (1989b). Sierra Club study shows who become activists. *Public Relations Review, 15*(3), 3–24.

Grunig, J.E. (1989c). Symmetrical presuppositions as a framework for public relations theory. In C.H. Botan & V. Hazleton, Jr. (Eds.), *Public relations theory* (pp. 17–44). Hillsdale, NJ: Lawrence Erlbaum Associates, Inc.

Grunig, J.E. (1992). Communication, public relations, and effective organizations: An overview of the book. In J. E. Grunig (Ed.), *Excellence in public relations and communication management: Contributions to effective organizations* (pp. 18–19). Hillsdale, NJ: Lawrence Erlbaum Associates.

Grunig, J.E. (1993). Public relations and international affairs: Effects, ethics and responsibility. *Journal of International Affairs, 47*, 137–162.

Grunig, J. E. (1997). A situational theory of publics: Conceptual history, recent challenges and new research. In D. Moss, T. MacManus, & D. Vercic (Eds.), *Public relations research: An international perspective* (pp. 3–38). London, England: International Thomson Business Press.

Grunig, J.E. (2001). Two-way symmetrical public relations: Past, present, and future. In R.L. Heath (Ed.), *Handbook of public relations* (pp. 11–30). Thousand Oaks, CA: Sage Publications.

Grunig, J.E. (2006). Furnishing the edifice: Ongoing research on public relations as a strategic management function. *Journal of Public Relations Research, 18*(2), 151–176.

Grunig, J.E. (2009). Paradigms of global public relations in an age of digitalisation. *Prism, 6*(2). Retrieved from http://www.prismjournal.org

Grunig, J. E., & Grunig, L. A. (1992). Models of public relations and communication. In J. E. Grunig (Ed.), *Excellence in public relations and communication management* (pp. 285–326). Hillsdale, NJ: Lawrence Erlbaum Associates Inc.

Grunig, J. E., Grunig, L. A., & Dozier, D. M. (1996). Das situative model exzellenter public relations: Schlussfolgerungen aus einer internationalen studie [The contingency model of excellent public relations: Conclusions from an international study]. In G. Bentele, H. Steinmann & A. Zerfass (Eds.), *Dialogorientiene unternehmenskommunikation* [Corporate communications](pp. 199–228). Berlin, Germany: Vistas.

Grunig, J. E., & Huang, Y. H. (2000). From organizational effectiveness to relationship indicators: Antecedents of relationships, public relations strategies, and relationship outcomes. In J. A. Ledingham & S. D. Bruning (Eds.), *Public relations as relationship management: A relational approach to the study and practice of public relations* (pp. 23–54). Mahwah, NJ: Erlbaum.

Grunig, J. E., & Hunt, T. (1984). *Managing public relations.* Fort Worth, TX: Harcourt Brace Jovanovich.

Grunig, L. A. (1992). Strategic public relations constituencies on a global scale. *Public Relations Review, 18*, 127–136.

Grunig, L. A., Grunig, J. E., & Dozier, D. (2002). *Excellent public relations and effective organizations: A study of communication management in three countries.* Mahwah, NJ: Erlbaum.

Guardian Research. (2010, July 22). BP oil spill timeline. Retrieved from http://www.guardian.co.uk/environment/2010/jun/29/bp-oil-spill-timeline-deepwater-horizon

Gudykunst, W. B., & Ting-Toomey, S. (1988). *Culture and interpersonal communication.* Newbury Park, CA: Sage.

Guentheroth, H., Schoenert, U., & Rodtmann, E. (2007). Wikipedia: Wissen fuer alle [Knowledge for all]. *Stern, 60*(50), 30–44.

Guibernau, M. (1999). *Nations without states: Political communities in a global age.* Cambridge, MA: Polity Press.

Guibernau, M. (2010). Catalonia: Nationalism and intellectuals in nations without states. In M. Guibernau & J. Rex (Eds.), *The ethnicity reader: Nationalism, multiculturalism & migration* (2nd ed., pp. 138–153). Cambridge, MA: Polity Press.

Gustafsson, C. (2006). Brand trust and authenticity: The link between trust in brands and the consumer's role on the market. *European Advances in Consumer Research, 7,* 522–527.

Gustavsen, P.A., & Tilley, E. (2003). Public relations communication through corporate websites: Towards an understanding of the role of interactivity. *Prism, 1*(1). Retrieved from http://www.prismjournal.org

Habermas, J. (1981). *The theory of communicative action.* Boston, MA: Beacon Press.

Habermas, J. (1962/1989). *The structural transformation of the public sphere: An inquiry into a category of bourgeois society.* Cambridge, MA: MIT Press.

Hachigian, D., & Hallahan, K. (2003). Perceptions of public relations Web sites by computer industry journalists. *Public Relations Review, 29*(1), 43–62.

Haigh, D., & Knowles, J. (2004). How to define your brand and determine its value. *Marketing Management, 13*(3), 22–28.

Hall, E. T. (1989). *Beyond culture.* New York, NY: Anchor Books/ Doubleday.

Hall, E. T., & Hall, M. R. (1990). *Hidden differences.* New York, NY: Anchor Books/Doubleday.

Hall, R., & Jaugietis, Z. (2011). Developing peer mentoring through evaluation. *Innovative Higher Education, 36*(1), 41–52.

Hall, S. (1975). Introduction. In A. C. H. Smith (Ed.), *Paper voices: The popular press and social change, 1935–1965* (pp. 1–24). Totowa, NJ: Rowman and Littlefield.

Hallahan, K. (1999). Seven models of framing: Implications for public relations. *Journal of Public Relations Research, 11*(3), 205–242.

Hallahan, K. (2001a). The dynamics of issues activation and response: An issues process model. *Journal of Public Relations Research, 13*(1), 27–59.

Hallahan, K. (2001b). Improving public relations Web sites through usability research. *Public Relations Review, 27*(2), 223–239.

Hallahan, K. (2004). Protecting an organization's digital public relations assets. *Public Relations Review, 30*(3), 255–268.

Hamilton, P. K. (1992). Grunig's situational theory: A replication, application, and extension. *Journal of Public Relations Research, 4*(3), 123–149.

Han, G., & Zhang, A. (2009). Starbucks is forbidden in the Forbidden City: Blog, circuit of culture and informal public relations campaign in China. *Public Relations Review, 35*(4), 395–401.

Han, R. (2009, July 15). Pepsi, so-so social media campaign. Retrieved from http://www.littleredbook.cn/2009/07/15/pepsi-social-media-campaign/

Han, X. L. (2010, August 26). Kaixin SNS marketing model innovation and transformation. Retrieved from http://media.people.com.cn/GB/40628/199046/12550362.html

Harfoush, R. (2009). *Yes we did: An inside look at how social media built the Obama brand.* Barkley, CA: New Riders.

Harlow, W. F., Brantley, B. C., & Harlow, R. M. (2011). BP initial image repair strategies after the Deepwater Horizon spill. *Public Relations Review, 37*(1), 80–83.

Harrington, A. (2008). Digital diversity: Marketing to minority audiences online. *njbiz, 21*(20), 18.

Harris, A., Garramone, G., Pizante, G., & Komiya, M. (1985). Computers in constituent communications. *Public Relations Review, 11*(3), 34–39.

Harris, F., & de Chernatony, L. (2001). Corporate branding and corporate brand performance. *European Journal of Marketing, 35,* 441–57.

Hathi, S. (2007, April/May). Study reveals social media use. Retrieved from http://findarticles.com/p/articles/mi_hb5797/is_200704/ai_n32218933/

Hathi, S. (2009). Communicators remain unclear on business case for social media. *Strategic Communication Management, 14*(1), article 9. Retrieved from http://www.internalcommshub.com/open/news/smpollresults.shtml?mxmroi=26359396/24774879/false

Haymarket Media, Inc. (2010). *PR Week's press & public relations handbook.* New York, NY: Author.

Hays, B., & Swanson, D.J. (2011). *Success strategies and challenges for reverse mentoring in public relations.* Unpublished manuscript.

Hazleton. V., Harrison-Rexrode J., & Kennan, W.R. (2007). New technologies in the formation of personal and public relations: Social capital and media. In S. C. Duhé (Ed.), *New media and public relations.* (pp. 91–105). New York, NY: Peter Lang.

Heaps, D. (2011, April 14). IR with leading edge technology. Retrieved from http://www.slideshare.net/darrell_heaps/ir-with-leading-edge-technology-ir-websites-social-media

Heath, R.L. (1992). The wrangle in the marketplace: A rhetorical perspective on public relations. In E.L. Toth & R.L. Heath (Eds.), *Rhetorical and critical approaches to public relations* (pp. 17–36). Hillsdale, NJ: Lawrence Erlbaum.

Heath, R.L. (1998). New communication technologies: An issues management point of view. *Public Relations Review, 24*(3), 273–288.

Heath, R. L. (2001). A rhetorical enactment rationale for public relations. In R. L. Heath (Ed.), *Handbook of public relations* (pp. 31–59). Thousand Oaks, CA: Sage.

Heath, R.L. (2006). Best practices in crisis communication: Evolution of practice through research. *Journal of Applied Communication Research, 34*(3), 245–248.

Heath, R. L., & Douglas, W. (1991). Effects of involvement on reactions to sources of messages and to message clusters. In L. A. Grunig & J. E. Grunig (Eds.), *Public relations research annual* (Vol. 3, pp. 179–193). Hillsdale, NJ: Lawrence Erlbaum Associates, Inc.

Heath, R. L., & Nathan, K. (1990). Public relations' role in risk communication: Information, rhetoric and power. *Public Relations Quarterly, 35*(4), 15–22.

Heath, R. L., & Palenchar, M. J. (2009). *Strategic issues management: Organizations and public policy changes* (2nd ed.). Thousand Oaks, CA: Sage.

Heeter, C. (1992). Being there: The subjective experience of presence. *Presence, Teleoperators and Virtual Environments, 1*, 262–271.

Helm, S. (2007). One reputation or many? Comparing stakeholders' perceptions of corporate reputation. *Corporate Communications: An International Journal, 12*(3), 238–254.

Henderson, A. (2010). Authentic dialogue? The role of "friendship" in a social media recruitment campaign. *Journal of Communication Management, 14*(3), 237–257.

Henle, C., Kohut, G., & Booth, R. (2009). Designing electronic use policies to enhance employee perceptions of fairness and to reduce cyberloafing: An empirical test of justice theory. *Computers in Human Behavior, 25*(4), 902–910.

Heverin, T., & Zach, L. (2010). *Microblogging for crisis communication: Examination of Twitter use in response to a 2009 violent crisis in the Seattle-Tacoma, Washington area.* Paper presented at the annual meeting of the International Community on Information Systems for Crisis Response and Management, Seattle, WA.

Hickerson, C.A., & Thompson, S.R. (2009). Dialogue through wikis: A pilot exploration of dialogic public relations and wiki websites. *Prism, 6*(1). Retrieved from http://www.prismjournal.org

Hickey, K. (2011, February 7). Next-gen 911 app for Android includes video, audio and location with emergency calls. Retrieved from http://gcn.com/Articles/2011/02/07/U-of-Maryland-next-gen-911.aspx?p=1

Hiebert, R.E. (2005). Commentary: New technologies, public relations and democracy. *Public Relations Review, 31*(1), 1–9.

Hill, L.N., & White, C. (2000). Public relations practitioners' perception of the World Wide Web as a communications tool. *Public Relations Review, 26*(1), 31–51.

Hinson, M. D, & Wright, D. K. (2010, March). *Exploring the impact of social media on public relations practice.* Paper presented at the 13th Annual International Public Relations Research Conference, Miami, FL. Retrieved from http://www.instituteforpr.org/wp-content/uploads/IPRRC_13_Proceedings.pdf

Historic tornado outbreak April 27, 2011. (n.d.). Retrieved from http://www.srh.noaa.gov/bmx/?n=event_04272011

Hjavard, S. (2006). The mediatization of religion. *Journal of Adolescence, 31*(1), 125–146.

Hockerts, K., & Moir, L. (2004). Communicating corporate responsibility to investors: The changing role of the investor relations function. *Journal of Business Ethics, 52*(1), 85–98.

Holahan, C. (2007, May 21). Social networking for the faithful. Retrieved from www.businessweek.com/technology/content/may2007/tc20070521_126201.htm

Holling, C. S. (1973). Resilience and stability of ecological systems. *Annual Review of Ecology and Systematics, 4*, 1–23.

Holmes, P., & Sudhaman, A. (2011, January 31). Top 10 crises of 2010. Retrieved from http://www.holmesreport.com/featurestories-info/9821/Top-10-Crises-of-2010.aspx

Holson, L.M. (2011, April 17). When publicists say "Shh!" Retrieved from http://www.nytimes.com/2011/04/17/fashion/17PUBLICISTS.html

Holsti, O. (1969). *Content analysis for the social sciences and humanities.* Menlo Park, CA: Addison-Wesley.

Hon, L., & Grunig, J.E. (1999). *Guidelines for measuring relationships in public relations.* Gainesville, FL: The Institute for Public Relations.

Hong, S.Y., & Rim, H. (2010). The influence of customer use of corporate websites: Corporate social responsibility, trust, and word-of-mouth communication. *Public Relations Review, 36*(4), 389–391.

Hoover's Business Press. (2010). *Hoover's handbook of industry profiles*. Austin, TX: Author.

Horton, A. (2010, July 5). Spreading the Word with new media. *Jet*, p. 20.

Horyn, C. (2011, March 3). Exiting the way that he entered: John Galliano was always good at shocking people. Too good, as it turned out. *The New York Times*, pp. E1, E8.

Hovey, W.L. (2010, Spring). Examining the role of social media in organization-volunteer relationships. *Public Relations Journal, 4*(2). Retrieved from http://www.prsa.org/Intelligence/PRJournal/

How chemical industry rewrote history of banned pesticide. (1998). Retrieved from http://www.ewg.org/node/8005

Hu, Y. (2010, June 1). BBS sites on China's changing Web. Retrieved from http://cmp.hku.hk

Hughes, A. L., & Palen, L. (2009). Twitter adoption and use in mass convergence and emergency events. *International Journal of Emergency Management, 6*(3), 248–260.

Hughes, A. L., Palen, L., Sutton, J., Liu, S. B., & Vieweg, S. (2008, May). "Site-seeing" in a disaster: An examination of on-line social convergence. *Proceedings of the 5th International ISCRAM Conference*, Washington, D.C.

Hung, C. J. F. (2006). Toward the theory of relationship management in public relations: How to cultivate quality relationships? In E. L. Toth (Ed.), *The future of excellence in public relations and communication management* (pp. 443–476). Mahwah, NJ: Erlbaum.

Huston, L., & Sakkab, N. (2006, March). Connect and develop: Inside Procter and Gamble's new model for innovation. *Harvard Business Review, 84*(3), 58–66.

Hutchby, I. (2001). Technologies, texts and affordances. *Sociology, 35*(2), 441–456.

Hwang, S., & Cameron, G. T. (2008). Public's expectation about an organization's stance in crisis communication based on perceived leadership and perceived severity of threats. *Public Relations Review, 34*(1), 70–73.

Hyland, K. (1999). Academic attribution: Citation and the construction of disciplinary knowledge. *Applied Linguistics, 20*(3), 341–367.

Ifukor, P. (2010). "Elections" or "selections"? Blogging and twittering the Nigerian 2007 general elections. *Bulletin of Science, Technology & Society, 30*(6), 398–414.

Ingenhoff, D., & Koelling, A.M. (2009). The potential of Web sites as a relationship building tool for charitable fundraising NPOs. *Public Relations Review, 35*(1), 66–73.

International Telecommunication Union. (2010, October 19). ITU estimates two billion people online by end 2010. Retrieved from http://www.itu.int/net/pressoffice/press_releases/2010/39.aspx

Internet world stats. (2011). Retrieved from http://www.internetworldstats.com/stats.htm

Jacobson, L. (2000). Trying to reinvent a PR firm. *National Journal, 32*(42), 3261–3262.

Jacques, A. (2009, August 17). Domino's delivers during crisis: The company's step-by-step response after a vulgar video goes viral. Retrieved from http://www.prsa.org/Intelligence/TheStrategist/Articles/view/8226/102/Domino_s_delivers_during_crisis_The_company_s_step

Jahansoozi, J. (2006). Organization-stakeholder relationships: Exploring trust and transparency. *Journal of Management Development, 25*(10), 942–955.

James, C. (2011). Social media boot camp. Unpublished undergraduate paper, Comm465: Entertainment PR, California State University, Fullerton.

James, M. (2007). A review of the impact of new media on public relations: Challenges for terrain, practice and education. *Asia Pacific Public Relations Journal, 8*. Retrieved from http://www.pria.com.au/journal/categories?id=16

Jansen, B. J., Sobel, K., & Cook, G. (2011). Classifying ecommerce information sharing behavior by youths on social networking sites. *Journal of Information Science, 37*(2), 120–136.

Jeanne Clery Disclosure of Campus Security Policy and Campus Crime Statistics Act, 20 U.S.C. § 1092(f) (1990).

Jewett, R., & Dahlberg, L. (2009). The trouble with twittering: Integrating social media into mainstream news. *International Journal of Media and Cultural Politics, 5*, 233–246.

Jin, Y. (2010). Emotional leadership as a key dimension of public relations leadership: A national survey of public relations leaders. *Journal of Public Relations Research, 22*(2), 159–181.

Jin, Y., & Liu, B.F. (2010). The Blog-Mediated Crisis Communication Model: Recommendations for responding to influential external blogs. *Journal of Public Relations Research, 22*(4), 429–455.

Jin, Y., Liu, B. F., & Austin, L. L. (2011). *The effects of crisis attribution, information form, and source on publics' crisis responses: Examining the role of social media in effective crisis management.* Paper presented at the 14th International Public Relations Research Conference, Miami, FL.

Jo, S., & Kim, Y. (2003). The effect of Web characteristics on relationship building. *Journal of Public Relations Research, 15*(3), 199–223.

Johnson, M.A. (1997). Public relations and technology: Practitioner perspectives. *Journal of Public Relations Research, 9*(3), 213–236.

Johnson, T. J., & Kaye, B. K. (2010). Choosing is believing? How Web gratifications and reliance affect Internet credibility among politically interested users. *Atlantic Journal of Communication, 18*, 1–21.

Jones, B., Temperley, J., & Lima, A. (2009). Corporate reputation in the era of Web 2.0: The case of Primark. *Journal of Marketing Management, 25*(9–10), 927–939.

Jones, C. (2008, May 29). Text messages tested as solicitation tool. Retrieved from http://philanthropy.com/article/Text-Messages-Tested-as/61462/

Jones, D. (2011, April 14). Social media investor relations reaches tipping point. Retrieved from http://irwebreport.com/20110414/social-media-investor-relations-reaches-tipping-point

Jonsson, P. (2007, May 8). To raise an alarm, use cellphones? *The Christian Science Monitor*, p. 3.

Judd, L.R. (1995). An approach to ethics in the information age. *Public Relations Review, 21*(1), 35–44.

Kan, M. (2011, May 11). Chinese top microblog hits over 90 million active users. Retrieved from http://www.cio.com/article/682064/Chinese_Top_Microblog_Hits_Over_90_Million_Active_Users

Kang, D.S., & Mastin, T. (2008). How cultural difference affects international tourism public relations websites: A comparative analysis using Hofstede's cultural dimensions. *Public Relations Review, 34*(1), 54–56.

Kang, S., & Norton, H.E. (2004). Nonprofit organizations' use of the World Wide Web: Are they sufficiently fulfilling organizational goals? *Public Relations Review, 30*(3), 279–284.

Kang, S., & Norton, H.E. (2006). Colleges and universities' use of the World Wide Web: A public relations tool for the digital age. *Public Relations Review, 32*(4), 426–428.

Kanter, B., & Fine, A. (2010). *The networked nonprofit.* San Francisco, CA: Jossey-Bass.

Kaplan, A. M., & Haenlein, M. (2010). Users of the world, unite! The challenges and opportunities of social media. *Business Horizons, 53*(1), 59–68.

Katz, E., & Lazarsfeld, P. F. (1955). *Personal influence: The part played by people in the flow of mass communications.* Glencoe, IL: Free Press.

Kauffman, J. (2008). When sorry is not enough: Archbishop Cardinal Bernard Law's image restoration strategies in the statement on sexual abuse of minors by clergy. *Public Relations Review, 34*(3), 258–262.

Kaufman, A. (2011, January 20). Staying "connected": Helmer designs "kit" to keep auds talking. *Daily Variety*, pp. 4, 11.

Keane, B. (2009). Watching the slow death of traditional political TV, part 1. Retrieved from http://www.crikey.com.au/2009/08/31/watching-the-slow-death-of-traditional-political-tv-part-1/4

Keane, J. (1995). Structural transformations of the public sphere. *Communications Review, 1*(1), 1–22.

Keating, A. (2010, November 22). Losing my religion: Do digital self-portraits resemble our authentic selves? *America*, pp. 19–20.

Keating, M. (1996). *Nations against the state: The new politics of nationalism in Quebec, Catalonia and Scotland.* London, England: Macmillan.

Keating, M. (1997). Stateless nation-building: Quebec, Catalonia and Scotland in the changing state system. *Nations and Nationalism, 3*(4), 689–717.

Keating, M. (2001). Nations without state: Minority nationalism in the global era. In F. Requejo (Ed.), *Democracy and national pluralism* (pp. 40–58). London, England: Routledge.

Keen, J. (2010, January 16). Facebook, Twitter, 2-way "lifeline" for Haiti. Retrieved from http://www.usatoday.com/tech/webguide/internetlife/2010-01-13-haitisocial_N.htm

Kelleher, T. (2008). Organizational contingencies, organizational blogs and public relations practitioner stance toward publics. *Public Relations Review, 34*(3), 300–302.

Kelleher, T. (2009). Conversational voice, communicated commitment, and public relations outcomes in interactive online communication. *Journal of Communication, 59*(1), 172–188. doi:10.1111/j.1460–2466.2008.01410.x

Kelleher, T. (2010). Editor's note. *Journal of Public Relations Research, 22*(3), 239–240.

Kelleher, T., & Miller, B. M. (2006). Organizational blogs and the human voice: Relational strategies and relational outcomes. *Journal of Computer-Mediated Communication, 11*(2), 395–414.

Kelly, K. S. (1998). *Effective fund-raising management.* Mahwah, NJ: Lawrence Erlbaum.

Kendra, J. M., & Wachtendorf, T. (2003). Reconsidering convergence and converger: Legitimacy in response to the World Trade Center disaster. *Research in Social Problems and Public Policy, 11,* 97–122.

Kennan, W.R., Hazleton, V., Janoske, M., & Short, M. (2008, Spring). The influence of new communication technologies on undergraduate preferences for social capital formation, maintenance, and expenditure. *Public Relations Journal, 2*(2). Retrieved from http://www.prsa.org/Intelligence/PRJournal/

Kennedy, H. (2008). New media's potential for personalization. *Information, Communication & Society, 11*(3), 307–325.

Kent, M.L. (2008). Critical analysis of blogging in public relations. *Public Relations Review, 34*(1), 32–40.

Kent, M.L. (2010). Directions in social media for professionals and scholars. In R.L. Heath (Ed.), *The Sage handbook of public relations* (2nd ed., pp. 643–656). Thousand Oaks, CA: Sage.

Kent, M.L., & Taylor, M. (1998). Building dialogic relationships through the World Wide Web. *Public Relations Review, 24*(3), 321–334.

Kent, M. L., & Taylor, M. (2002). Toward a dialogic theory of public relations. *Public Relations Review, 28*(1), 21–37.

Kent, M.L., Taylor, M., & White, W.J. (2003). The relationship between Web site design and organizational responsiveness to stakeholders. *Public Relations Review, 29*(1), 63–77.

Kepcher, C. (2011, February 11). Partnering up with career coach, mentor or advisor can give you inside track to career success. Retrieved from http://www.nydailynews.com/money/2011/02/04/2011-02-04_partnering_up_for_success.html#ixzz1EKYets9E

Ki, E-J., & Hon, L. C. (2009). A measure of relationship cultivation strategies. *Journal of Public Relations Research, 21*(1), 1–24.

Kim, D., Nam, Y., & Kang, S. (2010). An analysis of corporate environmental responsibility on the global corporate Web sites and their dialogic principles. *Public Relations Review, 36*(3), 285–288.

Kim, E., Kwak, D., & Yun, M. (2010). Investigating the effects of peer association and parental influence on adolescent substance use: A study of adolescents in South Korea. *Journal of Criminal Justice,* pp. 17–24.

Kim, H., Coyle, J. R., & Gould, S. J. (2009). Collectivist and individualist influences on website design in South Korea and the U.S.: A cross-cultural content analysis. *Journal of Computer-Mediated Communication, 14*(3), 581–601.

Kim, H., Rao, A.R., & Lee, A.Y. (2009). It's time to vote: The effect of marching message orientation and temporal frame on political persuasion. *Journal of Consumer Research, 35*(6), 877–889.

Kim, J.-N., & Grunig, J. E. (2011). Problem solving and communicative action: A situational theory of problem solving. *Journal of Communication, 61,* 120–149. doi:10.1111/j.1460–2466.2010.01529.x

Kim, J.-N., Grunig, J. E., & Ni, L. (2010). Reconceptualizing the communicative action of publics: Acquisition, selection, and transmission of information in problematic situations. *International Journal of Strategic Communication, 4*(2), 126–154. doi: 10.1080/15531181003701913

Kim, J.-N., Lee, S., & Guild, J. (2009, May). Healthy communication: Mere communication effect on managing chronic health problems. Paper presented to the Health Communication Division, International Communication Association, Chicago, IL.

Kim, J.-N. & Ni, L. (2010). Seeing the forest through the trees: The behavioral, strategic management paradigm in public relations and its future. In R. H. Heath (Ed.), *The Sage handbook of public relations* (pp. 35–57). Thousand Oaks, CA: Sage.

Kim, J-N., Park, S.-C., Yoo, S., & Shen, H. (2010). Mapping health communication scholarship: Breadth, depth, and agenda of published scholarship in "Health Communication." *Health Communication, 25,* 487–503. doi: 10.1080/10410236.2010.507160

Kim, S., Park, J.-H., & Wertz, E.K. (2010). Expectation gaps between stakeholders and web-based corporate public relations efforts: Focusing on Fortune 500 corporate web sites. *Public Relations Review, 36*(3), 215–221.

Kim, Y. (2001). Searching for the organization-public relationship: A valid and reliable instrument. *Journalism and Mass Communication Quarterly, 78,* 799–815.

King, G. (2006). Image restoration: An examination of the response strategies used by Brown and Williamson after allegations of wrongdoing. *Public Relations Review, 32*(2), 131–136.

Kingston, J. (2011) *Contemporary Japan: History, politics and social change since the 1980s.* Malden, MA: Wiley-Blackwell.

Kiousis, S. (2002). Interactivity: A concept explication. *New Media & Society, 4*(3), 355–383.

Kiousis, S., & Dimitrova, D.V. (2006). Differential impact of Web site content: Exploring the influence of source (public relations versus news), modality, and participation on college students' perceptions. *Public Relations Review, 32*(2), 177–179.

Kiousis, S., & Stromback, J. (2011). Political public relations research in the future. In J. Stromback & S. Kiousis, (Eds.), *Political public relations: Principles and applications.* New York, NY: Routledge.

Kirat, M. (2007). Promoting online media relations: Public relations departments' use of Internet in the UAE. *Public Relations Review, 33*(2), 166–174.

Kirkpatrick, M. (2010, January 9). Facebook's Zuckerberg says the age of privacy is over. Retrieved from http://www.readwriteweb.com/archives/facebooks_zuckerberg_says_the_age_of_privacy_is_ov.php

Kitchen, P.J., & Panopoulos, A. (2010). Online public relations: The adoption process and innovation challenge, a Greek example. *Public Relations Review, 36*(3), 222–229.

Klein, P. (n.d.). Cell broadcast technology for emergency alert notifications. Retrieved from http://transition.fcc.gov/pshs/docs/advisory/cmsaac/pdf/CellCastComment070307.pdf

Kline, L. B. (1998). *Principles and practice of structural equation modeling.* New York, NY: Guilford.

Klout: The standard for influence. (2011). Retrieved from http://corp.klout.com/kscore

Kluth, A. (2007, November 15). The rediscovery of discretion. Retrieved from http://www.economist.com/node/10125781

Knight Commission on the Information Needs of Communities in a Democracy. (2009). Informing communities: Sustaining democracy in the digital age. Retrieved from http://www.knightcomm.org/wp-content/uploads/2010/02/Informing_Communities_Sustaining_Democracy_in_the_Digital_Age.pdf

Knoke, D. (1994). Networks of elite structure and decision making. In S. Wasserman & J. Galaskiewicz (Eds.), *Advances in social network analysis: Research in the social and behavioral sciences* (pp. 3–25). Thousand Oaks, CA: Sage.

Know your enemy: Why we contemplate catastrophe. (2006, Fall). *Harvard International Review,* p. 36.

Kotler, P. (1986). The prosumer movement: A new challenge for marketers. *Advances in Consumer Research, 13,* 510–513.

Kotler, P., & Lee, N. (2005). *Corporate social responsibility: Doing the most good for your company and your cause.* New York, NY: Wiley.

Kovic, I., Lulic, I., & Brumini, G. (2008). Examining the medical blogosphere: An online survey of medical bloggers. *Journal of Medical Internet Research, 10*(3). doi: 10.2196/jmir.1118

Kozinets, R. V., de Valck, K., Wojnicki, A., & Wilner, S. J. (2010). Networked narratives: Understanding word-of-mouth marketing in online communities. *Journal of Marketing, 74*(2), 71–89.

Kruckeberg, D., & Starck, K. (1988). *Public relations and community: A reconstructed theory.* New York, NY: Praeger.

Kruckeberg, D., & Starck, K. (2004). The role and ethics of community building for consumer products and services. *Journal of Promotion Management, 10*(1/2), 133–146.

Kruckeberg, D., & Tsetsura, K. (2009, November). *The role of public relations and technology in global society.* Keynote presentation at the International Anniversary Conference, 20th Anniversary, University of Bucharest, College of Journalism and Mass Communication Studies, Bucharest, Romania.

Kruckeberg, D., & Tsetsura, K. (2011, May). *Benefits of collaboration between academia and industry—the perfect public relations world.* Keynote presentation at the First International Conference of the (Russian) Association of Public Relations Educators, St. Petersburg and Moscow, Russia.

Kwak, H., Lee, C., Park, H., & Moon, S. (2010, April). *What is Twitter, a social network or a news media?* Paper presented at World Wide Web Conference 2010, Raleigh, NC.

Kweon, S. (2000). A framing analysis: How did three U.S. news magazines frame about mergers or acquisitions? *The International Journal on Media Management, 2*(3/4), 165–177.

Kymlicka, W. (2001). The new debate over minority rights. In F. Rquejo (Ed.), *Democracy and national pluralism* (pp. 15–39). London, England: Routledge.

Lai, E. (2010). Getting in step to improve the quality of in-service teacher learning through mentoring. *Professional Development in Education, 36*(3), 443–469.

Lamy, P. (2010, April 21). L'Oreal found its luxury voice in Chinese social media. Retrieved from http://adage.com/china/article/viewpoint/loreal-found-its-luxury-voice-in-chinese-social-media-says-philippe-lamy/143261/

Langheinrich, M., & Karjoth, G. (2010). Social networking and the risk to companies and institutions. *Information Security Technical Report, 5*, 51–56.

Langlois, G., & Elmer, G. (2009). Wikipedia leeches? The promotion of traffic through a collaborative Web format. *New Media & Society, 11*(5), 773–794.

Lariscy, R.W., Avery, E.J., Sweetser, K.D., & Howes, P. (2009a). An examination of the role of online social media in journalists' source mix. *Public Relations Review, 35*(3), 314–316.

Lariscy, R.W., Avery, E.J., Sweetser, K.D., & Howes, P. (2009b, Fall). Monitoring public opinion in cyberspace: How corporate public relations is facing the challenge. *Public Relations Journal, 3*(4). Retrieved from http://www.prsa.org/Intelligence/PRJournal/

Laskin, A.V. (2009). A descriptive account of the investor relations profession: A national study. *Journal of Business Communication, 46*(2), 208–233.

Laskin, A.V. (2010a). Investor relations. In R.L. Heath (Ed.), *The Sage handbook of public relations* (2nd ed., pp. 611–622). Thousand Oaks, CA: Sage.

Laskin, A.V. (2010b). *Managing investor relations: Strategies for effective communication.* New York, NY: Business Expert Press.

Laskin, A.V. (2011a, March). The Community Age: The end of Information Age and the future of journalism. Retrieved from http://investor-relations.blogspot.com/2011/03/network-age-end-of-information-age-and.html

Laskin, A.V. (2011b, April). What is wrong with Citigroup investor relations or fish where the fish are. Retrieved from http://investor-relations.blogspot.com/2011/04/what-is-wrong-with-citigroup-investor.html

Laskin, A.V. (2011c). How investor relations contributes to the corporate bottom line. *Journal of Public Relations Research, 23*(3), 302–324.

Latimer, D. (2008, May/June). Text messaging as emergency communication superstar? Nt so gr8. *Educause Review*, pp. 84–85. Retrieved from http://www.educause.edu/EDUCAUSE+Review/EDUCAUSEReviewMagazineVolume43/TextMessagingasEmergencyCommun/162894

Lauria, P., & Grove, L. (2010, August 3). How Newsweek blew it. Retrieved from http://www.thedailybeast.com/articles/2010/08/03/newsweek-losses-revealed.html

Lauzen, M. M. (1992). Public relations roles, intraorganizational power, and encroachment. *Journal of Public Relations Research, 4*(2), 61–80.

Lavidge, R., & Steiner, G.A. (1961). A model for predictive measurement of advertising effectiveness. *Journal of Marketing, 25*, 59–62.

Lavrusik, V. (2009, July 15). 10 ways universities share information using social media. Retrieved from http://mashable.com/2009/07/15/social-media-public-affairs/

Lazarsfeld, P. F., Berelson, B., & Gaudet, H. (1948). *The people's choice: How the voter makes up his mind in a presidential campaign.* New York, NY: Columbia University Press.

Lazarus, R. S. (1993). From psychological stress to the emotions: A history of changing outlooks. *Annual Reviews in Psychology, 44*, 1–21.

Leadership Directories, Inc. (2010). *Corporate yellow book.* New York, NY: Author.

League, C. E., Díaz, W., Philips, B., Bass, E. J., Kloesel, K., Gruntfest, E., & Gessner, A. (2010). Emergency manager decision-making and tornado warning communication. *Meteorological Applications, 17*, 163–172. doi:10.1002/met.201

Learmonth, M. (2008). How Twittering critics brought down Motrin mom campaign. Retrieved from http://adage.com/article/digital/twittering-critics-brought-motrin-mom-campaign/132622/

Ledingham, J. A. (2001). Government-community relationships: Extending the relational theory of public relations. *Public Relations Review, 27*(3), 285–295.

Ledingham, J. A. (2006). Relationship management: A general theory of public relations. In C. H. Botan & V. Hazleton (Eds.), *Public relations theory II* (pp. 465–483). Mahwah, NJ: Lawrence Erlbaum Associates.

Lee, J., Jares, S. M., & Heath, R. L. (1999). Decision-making encroachment and cooperative relationships between public relations and legal counselors in the management of organizational crisis. *Journal of Public Relations Research, 11*(3), 243–270.

Lee, J., & Lee, H. (2010). The computer-mediated communication network: Exploring the linkage between the online community and social capital. *New Media & Society, 12*(5), 711–727.

Lee, J., Woeste, J.H., & Heath, R.L. (2007). Getting ready for crises: Strategic excellence. *Public Relations Review, 33*(3), 334–336.

Lee, K. (2008). Making environmental communications meaningful to female adolescents. *Science Communication, 30*(2), 147–176.

Lee, K. M. (2004). Presence, explicated. *Communication Theory, 14*(1), 27–50.

Lee, M. L., Geistfeld, L. V., & Stoel, L. (2007). Cultural differences between Korean and American apparel web sites. *Journal of Fashion Marketing and Management, 11*(4), 511–528.

Leh, A. (2005). Lessons learned from service learning and reverse mentoring in faculty development: A case study in technology training. *Journal of Technology and Teacher Education, 13,* 25.

Lenckus, D. (2007). College tragedy highlights need to spot threats. *Business Insurance, 41*(17), 20–21. Retrieved from http://search.proquest.com/docview/233511685?accountid=13158

Lenhart, A. (2009, October). *The democratization of online social networks: A look at the change in demographics of social network users over time.* Paper presented at the Association of Internet Researchers 10th Annual Conference, Milwaukee, WI. Retrieved from http://www.pewinternet.org/Presentations/2009/41--The-Democratization-of-Online-Social-Networks.aspx

Lenhart, A., Purcell, K., Smith, A., & Zickuhr, K. (2010, February 3). Social media and young adults. Retrieved from http://pewinternet.org/Reports/2010/Social-Media-and-Young-Adults.aspx

Leonard, A. (2010). Faced with an angry mob, a functionary commits, then apologizes for, the ultimate P.R. sin—acting human. Retrieved from http://www.salon.com/news/social_media/index.html?story=/tech/htww/2010/03/19/nestle_s_brave_facebook_flop

Leonhardt, D. (2011, March 27). A better way to measure Twitter influence. *The New York Times Magazine,* p. 18.

Lerbinger, O. (1997). *The crisis manager: Facing risk and responsibility.* Mahwah, NJ: Lawrence Erlbaum.

Lerner, J. S., Gonzalez, R. M., Small, D. A., & Fischhoff, B. (2003). Effects of fear and anger on perceived risks of terrorism: A national field experiment. *Psychological Science, 14*(2), 144–150.

Lester-Roushanzamir, E. P., & Raman, L. (1999). The global village in Atlanta: A textual analysis of Olympic news coverage for children in the *Atlanta Journal-Constitution. Journalism & Mass Communication Quarterly, 76*(4), 699–712.

L'Etang, J. (2008). *Public relations: Concepts, practice and critique.* London, England: Sage.

Leung, L. (2009). User-generated content on the Internet: An examination of gratifications, civic engagement and psychological empowerment. *New Media & Society, 11*(8), 1327–1347.

Levenshus, A. (2010). Online relationship management in a presidential campaign: A case study of the Obama campaign's management of its Internet-integrated grassroots effort. *Journal of Public Relations Research, 22*(3), 313–335.

Levine, R., Locke, C., Searls, D., & Weinberger, D. (2000). *The cluetrain manifesto: The end of business as usual.* New York, NY: Perseus Books.

Lewis, B.K. (2010, Summer). Social media and strategic communication: Attitudes and perceptions among college students. *Public Relations Journal, 4*(3). Retrieved from http://www.prsa.org/Intelligence/PRJournal/

Lewis, R., & Kitchin, P. (2010). New communications: Media for sport. In M. Hopwood, P. Kitchin, & J. Skinner (Eds.), *Sport public relations and communication* (pp. 187–214). Oxford, UK: Elsevier.

Li, C. (2010). *Open leadership.* San Francisco, CA: Jossey-Bass.

Liao, Q.Y., Pan, Y.X., Zhou, M.X., & Ma, F. (2010, April). Chinese online communities: Balancing management control and individual autonomy. *Proceedings of the 28th International Conference on Human factors in computing systems (CHI '10,* pp. 2193–2202), New York, NY. doi: 10.1145/1753326.1753658

Liberman, N., Trope, Y., & Stephan, E. (2007). Psychological distance. In A. W. Kruglanski, & E. T. Higgins (Eds.), *Social psychology: Handbook of basic principles* (Vol. 2, pp. 353–383) New York, NY: Guilford Press.

Lievrouw, L. A. (2002). Determination and contingency in new media development: Diffusion of innovations and social shaping of technology perspectives. In L. A. Lievrouw & S. Livingstone (Eds.), *Handbook of new media* (pp. 183–199). London, England: Sage.

Lievrouw, L. A. (2010). Social media and the production of knowledge: A return to little science? *Social Epistemology, 24*(3), 219–237.

Linden Lab. (2008). Case study: How meeting in Second Life transformed IBM's technology elite into virtual world believers. Retrieved from http://secondlifegrid.net.s3.amazonaws.com/docs/Second_Life_Case_IBM_EN.pdf

Lindlof, T. R. (1995). *Qualitative communication research methods. Current communication: An advanced text series* (Vol. 3). Thousand Oaks, CA: Sage.

Littleton, C., & Weisman, J. (2011, March 8). A sobering move: Ax falls, lawsuit looms. *Daily Variety*, pp. 1, 11.

Liu, B. F. (2009). An analysis of U.S. government and media disaster frames. *Journal of Communication Management, 13*(3), 268–283.

Liu, B.F. (2010). Distinguishing how elite newspapers and A-list blogs cover crises: Insights for managing crises online. *Public Relations Review, 36*(1), 28–34.

Liu, B. F., Austin, L. L., & Jin, Y. (2011, May). *How audiences respond to crisis communication strategies: The interplay of information form and source*. Paper presented at the International Communication Association Conference, Boston, MA.

Liu, B. F., Jin, Y., & Austin, L. A. (in press). The tendency to tell: Understanding publics' communicative responses to crisis information form and source. *Journal of Public Relations Research*.

Liu, B. F., Jin, Y., Briones, R., & Kuch, B. (in press). Managing turbulence online: Evaluating the Blog-Mediated Crisis Communication model with the American Red Cross. *Journal of Public Relations Research*.

Liu, B. F., & Kim, S. (in press). How organizations framed the 2009 H1N1 pandemic via social and traditional media: Implications for U.S. health communicators. *Public Relations Review*.

Liu, B. F., & Levenshus, A. B. (2011, August). *Nearly a decade after September 11: Navigating current and future counterterrorism communication research*. Paper presented at the Association for Education in Journalism and Mass Communication Convention, St. Louis, MO.

Livingstone, S. (1999). New media, new audiences? *New Media & Society, 1*(1), 59–66.

Lo, J. L., Bishop, J., & Eloff, J. H. P. (2008). SMSSec: An end-to-end protocol for secure SMS. *Computers & Security, 27*, 154–167. doi:10.1016/j.cose.2008.05.003

Locke, J. (1689). *Two treatises of government* (Thomas Hollis, Ed.). London, England: A. Millar et al.

Loechner, J. (2011). Retailer followup on negative review pays off. Retrieved from http://www.mediapost.com/publications/?fa=Articles.showArticle&art_aid=146630

Loewenfeld, F. von. (2006). *Brand communities: Erfolgsfaktoren und ökonomische relevanz von markengemeinschaften*. [Brand communities: Factors of success and economic relevance.] Wiesbaden, Germany: Deutscher Universitäts-Verlag.

Longstaff, P.H., & Yang, S. (2008). Communication management and trust: Their role in building resilience to "surprises" such as natural disasters, pandemic flu, and terrorism. Retrieved from http://www.ecologyandsociety.org/vol13/iss1/art3/

Longwell, K. (2009, November 4). Pastor has come to appreciate "sign of the times." *Sideroads South Simcoe*, p. 384.

Luce, E. (2010, June 2). BP faces public relations disaster. Retrieved from http://www.ft.com/cms/s/0/42659bee-6e74-11df-ad16-00144feabdco.html

Lukoff, K. (2010, April 7). China's top four social networks: RenRen, Kaixin001, Qzone, and 51.com. Retrieved from http://venturebeat.com/2010/04/07/china%E2%80%99s-top-4-social-networks-renren-kaixin001-qzone-and-51-com

Lukoff, K. (2011, March 8). China's top 15 social networks. Retrieved from http://techrice.com/2011/03/08/chinas-top-15-social-networks/

Lum, C., Kennedy, L., & Sherley, A. (2006). Are counter-terrorism strategies effective? The results of the Campbell systematic review on counter-terrorism evaluation research. *Journal of Experimental Criminology, 2*(4), 489–516.

Lyncheski, J. (2010, October). Social media in the workplace. *Long-Term Living Magazine, 59*(10), 32–35.

M. A. (2010, June 10). Travelling with a little help from friends Retrieved from http://www.economist.com/blogs/gulliver/2010/06/online_travel

Macias, W., Hilyard, K., & Freimuth, V. (2009). Blog functions as risk and crisis communication during Hurricane Katrina. *Journal of Computer-Mediated Communication, 15*(1), 1–31. doi: 10.1111/j.1083–6101.2009.01490.x

Macias, W., Lewis, L. S., & Smith, T. L. (2005). Health-related message boards/chat rooms on the Web: Discussion content and implications for pharmaceutical sponsorships. *Journal of Health Communication, 10*(3), 209–223.

Macnamara, J. (2010). Public communication practices in the Web 2.0–3.0 mediascape: The case for PRevolution. *Prism, 7*(3). Retrieved from http://www.prismjournal.org

Madden, M. (2010, August 27). Older adults and social media. Retrieved from http://pewinternet.org/Reports/2010/Older-Adults-and-Social-Media.aspx

Madera, J. M., & Smith, D. B. (2009). The effects of leader negative emotions on evaluations of leadership in a crisis situation: The role of anger and sadness. *The Leadership Quarterly, 20*(2), 103–114.

Magee, K. (2011, April 27). Reputation survey: BP - Ghost of Mexico still haunts BP. Retrieved from http://www.prweek.com/uk/news/1067250/Reputation-Survey-BP---Ghost-Mexico-haunts-BP/

Majchrzak, A., Jarvenpaa, S., & Hollingshead, A. (2007). Coordinating expertise among emergent groups responding to disasters. *Organization Science, 18*(1), 147–161.

Mandl, T. (2009, June). Comparing Chinese and German blogs. *Proceedings of the 20th ACM Conference on Hypertext and Hypermedia (HT '09*, pp. 299–308), Torino, Italy. doi: 10.1145/1557914.1557964

Manpower Inc. (2010a). Social networks vs. management? Harness the power of social media. Retrieved from http://us.manpower.com/us/en/research/whitepapers/harness-the-power-of-social-media.jsp

Manpower Inc. (2010b, January). Employer perspectives on social networking: Global key findings. Retrieved from http://www.manpower.com.hk/pdf/Social_Networking_Global_Key_Findings.pdf

Manyena, S. B. (2006). The concept of resilience revisited. *Disasters, 30*(4), 433–450.

Marani, P. (2011). The contagious campaign: A case study on J.J. Abrams' viral marketing tactics for *Super 8*. Unpublished undergraduate paper, Comm465: Entertainment PR, California State University, Fullerton.

Marvin, C. (1988). *When old technologies were new: Thinking about electric communication in the late nineteenth century*. New York, NY: Oxford Press.

Marvin Austin breaks silence with Facebook post. (2010). Retrieved from http://bit.ly/1QfhfV

Marx, G. (2000). *Ten trends: Educating children for a profoundly different future*. Arlington, VA: Educational Research Service.

Maryott, M. (2010, Spring). Social networking in the workplace creates peril for employers. *In-House Litigator, 24*(3), 3–6.

Mastrodicasa, J. (2008, Winter). Technology use in campus crisis. *New Directions for Student Services, 124*, 37–53. doi: 10.1002/ss.294

Mattson, E., & Barnes, N. G. (2007). Blogging for the hearts of donors: Largest U.S. charities use social media. Retrieved from http://www.umassd.edu/cmr/studiesandresearch/bloggingfortheheartsofdonors/

Maynard, M., & Tian, Y. (2004). Between global and glocal: Content analysis of the Chinese Web sites of the 100 top global brands. *Public Relations Review, 30*(3), 285–291.

McAlexander, J. H., Schouten, J. W., & Koenig, H. F. (2002). Building brand community. *Journal of Marketing, 66*, 38–54.

McAllister, S.M, & Taylor, M. (2007). Community college Web sites as tools for fostering dialogue. *Public Relations Review, 33*(2), 230–232.

McAllister-Spooner, S.M. (2008, Winter). User perceptions of dialogic public relations tactics via the Internet. *Public Relations Journal, 2*(1). Retrieved from http://www.prsa.org/Intelligence/PRJournal/

McAllister-Spooner, S.M. (2009). Fulfilling the dialogic promise: A ten-year reflective survey on dialogic Internet principles. *Public Relations Review, 35*(3), 320–322.

McAllister-Spooner, S.M. (2010, Spring). Whose site is it anyway? Expectations of college Web sites. *Public Relations Journal, 4*(2). Retrieved from http://www.prsa.org/Intelligence/PRJournal/

McCaffrey, S. (2006, April 30). Political dirty-tricksters are using Wikipedia. Retrieved from http://www.post-gazette.com/pg/06120/685924-96.stm

McClurg, S. D. (2003). Social networks and political participation: The role of social interaction in expanding political participation. *Political Research Quarterly, 56*(4), 449–464.

McCombs, M. (2005). A look at agenda-setting: Past, present and future. *Journalism Studies, 6*(4), 543–557.

McCorkindale, T. (2009, March). Can you see the writing on my wall? A content analysis of the Fortune 50's Facebook social networking sites. *Proceedings of the 12th Annual International Public Relations Research Conference*, Miami, FL. Retrieved from http://www.instituteforpr.org/events/iprrc/proceedings/

McCorkindale, T. (2010a, Summer). Can you see the writing on my wall? A content analysis of the Fortune 50's Facebook social networking sites. *Public Relations Journal, 4*(3). Retrieved from http://www.prsa.org/Intelligence/PRJournal/

McCorkindale, T. (2010b, August). *Twitter me this, Twitter me that: A quantitative content analysis of the 40 Best Twitter Brands*. Paper presented at the Association for Education and Journalism Convention, Denver, CO.

McCurry, J. (2011, June 16). Japan's gangs move in on tsunami clean-up: Yakuza changes focus after helping disaster victims battle for contracts to remove debris and rebuild. *The Guardian* (London), p. 21.

McEntire, D. A., & Meyers, A. (2004). Preparing communities for disasters: Issues and processes for government readiness. *Disaster Prevention and Management, 13*(2), 140–152.

McGary, M. (2008). Get connected with ASHA Social Networks. *The Asha Leader, 13*(16), 4.

McGonigal, J. (2011). *Reality is broken: Why games make us better and how they can change the world.* New York, NY: Penguin Press.

McKee, A. (2003). *Textual analysis: A beginner's guide.* Thousand Oaks, CA: Sage.

McKeown, C.A., & Plowman, K.D. (1999). Reaching publics on the Web during the 1996 presidential campaign. *Journal of Public Relations Research, 11*(4), 321–347.

McLennan, A., & Howell, G.V.J. (2010). Social networks and the challenge for public relations. *Asia Pacific Public Relations Journal, 11*(1), 11–19.

McLuhan, M. (1964). *Understanding media: The extensions of man* (1st ed.). New York, NY: McGraw-Hill (reissued by MIT Press, 1994)

McLuhan, M. (1967). *The medium is the message.* New York, NY: Random House.

McMillan, S. J., Hoy, M. G., Kim, J., & McMahan, C. (2008). A multifaceted tool for a complex phenomenon: Coding web-based interactivity as technologies for interaction evolve. *Journal of Computer-Mediated Communication, 13*(4), 794–826.

McNulty, C. (2011, April 4). Performance review: "Torpedo" is a bomb, all right. *Los Angeles Times*, pp. D1, D5.

Media myths and realities. (2010, February 15). Retrieved from http://newsroom.ketchum.com/multimedia-center/podcasts/media-myths-and-realities

Mejias, U. A. (2010). The limits of networks as models for organizing the social. *New Media & Society, 12*(4), 603–617.

Melanson, M. (2010, March 15). Why Wikipedia should be trusted as a breaking news source. Retrieved from http://www.readwriteweb.com/archives/why_wikipedia_should_be_trusted_or_how_to_consume.php

Melanson, M. (2011, January 31). Google & Twitter team up to offer speech-to-tweet service for Egypt. Retrieved from http://www.readwriteweb.com/archives/google_twitter_team_up_to_offer_speech-to-tweet_se.php

Meng, X., Zerfos, P., Samanta, V., Wong, S. H. Y., & Lu, S. (2007). Analysis of the reliability of a nationwide short message service. Retrieved from http://www.cs.ucla.edu/wing/publication/papers/Meng.INFOCOM07.pdf

Messner, M., & South, J. C. (2011). Legitimizing Wikipedia: How U.S. national newspapers frame and use the online encyclopedia in their coverage. *Journalism Practice, 5*(2), 145–160.

Miami Hurricanes players ordered to shut down Twitter accounts. (2010). Retrieved from http://bit.ly/dryNXR

Michaelson, D., & Griffin, T. L. (2005). *A new model for media content analysis.* Gainesville, FL: Institute for Public Relations.

Mickey, T.J. (1998). Selling the Internet: A cultural studies approach to public relations. *Public Relations Review, 24*(3), 335–349.

Middlemiss, N. (2003). Authentic not cosmetic: CSR as brand enhancement. *Brand Management, 10*(4–5), 353–361.

Miles, M. B., & Huberman, A. M. (1994). *Qualitative data analysis* (2nd ed.) Thousand Oaks, CA: Sage.

Mileti, D. S. (1999). *Disasters by design: A reassessment of natural hazards in the United States.* Washington, DC: Joseph Henry Press.

Mill, J. S. (1867). *Utilitarianism.* London, England: Longmans, Green, Reader, and Dyer.

Miller, C. C., & Helft, M. (2010, September 14). Twitter revamps its Web site. Retrieved from http://www.nytimes.com/2010/09/15/technology/15twitter.html

Miller, C. C., & Vega, T. (2010, October 10). After building an audience, Twitter turns to ads. Retrieved from http://www.nytimes.com/2010/10/11/business/media/11twitter.html?ref=business

Miller, L. (2010, January 13). Social media sites rife with activity after Haiti quake. Retrieved from http://www.ragan.com/Main/Articles/38974.aspx

Miller, S. (2009). Phased retirement keeps Boomers in the workforce. *HR Magazine/2009 HR Trend Book*, pp. 61–62.

Millner, A. G., Veil, S. R., & Sellnow, T. L. (2011). Proxy communication in crisis response, *Public Relations Review, 37*(1), 74–76.

Minksy, M. (1980, June). *Omni*, pp. 44–52.

Miracle, G., Chang, K. Y., & Taylor, C. R. (1992). Culture and advertising executions: A comparison of selected characteristics of Korean and U.S. television commercials. *International Marketing Review, 9*(4), 5–17.

Mishra, K. E. (2006). Help or hype: Symbolic or behavioral communication during Hurricane Katrina. *Public Relations Review, 32*(4), 358–366.

Mitchell, R. K., Agle, R. A., & Wood, S. J. (1997). Toward a theory of stakeholder identification and salience: Defining the principle of who and what really counts. *Academy of Management Review, 22*(4), 853–886.

Mo, P. K., & Coulson, N. S. (2008). Exploring the communication of social support within virtual communities: A content analysis of messages posted to an online HIV/AIDS support group. *Cyberpsychology & Behavior, 11*(3), 371–374. doi:10.1089/cpb.2007.0118

Moayeri, M. (2010). Collecting online data with usability testing software. *Prism, 7*(3). Retrieved from http://www.prismjournal.org

Mobile Giving Foundation. (2009). For non-profit organizations: Power of mobile giving. Retrieved from http://mobilegiving.org/?page_id=26

Mobile networks bear blast strain. (2005, July 7). Retrieved from http://news.bbc.co.uk/2/hi/technology/4659737.stm

Molleda, J. C. (2010). Authenticity and the construct's dimensions in public relations and communication research. *Journal of Communication Management, 14*(3), 223–236.

Money, K., & Hillenbrand, C. (2006). Using reputation measurement to create value: An analysis and integration of existing measures. *Journal of General Management, 32*(1), 1–12.

Montgomery, M. (2001). Defining "authentic talk." *Discourse Studies, 3*, 397–405.

Moore, S., & Seymour, M. (2005). *Global technology and corporation crisis: Strategies, planning and communication in the information age.* New York, NY: Routledge.

Moreau-Defarges, P. (2008). *La gouvernance* [Governance]. Paris: Presses Universitaires de France.

Morgan, J. (2009, December 8). President's note. *IR Weekly*, pp. 1–2.

Mountz, A. (2011, February 23). PSUTXT test a success. Retrieved from http://live.psu.edu/story/51596

MTV's VMA Twitter tracker lets stars, fans control the conversation. (2010, September 12). Retrieved from http://www.mtv.com/news/articles/1647640/mtvs-vma-twitter-tracker-lets-stars-fans-control-conversation.jhtml

Mui, Y., & Whoriskey, P. (2010, December 31). Facebook passes Google as most popular site on the Internet, two measures show. Retrieved from http://www.washingtonpost.com/wp-dyn/content/article/2010/12/30/AR2010123004645.html

Muniz, A. M., Jr., & O'Quinn, T. C. (2001). Brand community. *The Journal of Consumer Research, 27*(4), 412–432.

Muniz, A. M., Jr., & O'Quinn, T. C. (2005). Marketing communications in a world of consumption and brand communities. In A. J. Kimmel (Ed.), *Marketing communication: New technologies, approaches and styles* (pp. 63–85). Oxford, England: Oxford University Press.

Muralidharan, S., Rasmussen, L., Patterson, D., & Shin, J.-H. (2011). Hope for Haiti: An analysis of Facebook and Twitter usage during the earthquake relief efforts. *Public Relations Review, 37*(2), 175–177.

Murdock, G. (2010). Shifting anxieties, altered media: Risk communication in networked times. *Catalan Journal of Communication & Cultural Studies, 2*(2), 159–176. doi:10.1386/cjcs.2.2.159_1

Murphree, V., Reber, B. H., & Blevens, F. (2009). Superhero, instructor, optimist: FEMA and the frames of disaster in hurricanes Katrina and Rita. *Journal of Public Relations Research, 21*(3), 273–294.

Murray, E., Lo, B., Pollack, L., Donelan, K., Catania, J., Lee, K., et al. (2003). The impact of health information on the Internet on healthcare and the physician-patient relationship: National U.S. survey among 1.050 U.S. physicians. *Journal of Medical Internet Research, 5*(3). doi:10.2196/jmir.5.3.e17

Nacos, B. L. (2012). *Terrorism and counterterrorism* (4th ed.). New York, NY: Longman.

Nakamura, J., & Csikszentmihalyi, M. (2002). The concept of flow. In C. R. Snyder & S. J. Lopez (Eds.), *Handbook of positive psychology* (pp. 89–105). Oxford, England: Oxford University Press.

Nathan, K., Heath, R. L., & Douglas, W. (1992). Tolerance for potential environmental health risks: The influence of knowledge, benefits, control, involvement, and uncertainty. *Journal of Public Relations Research, 4*, 235–258.

National Investor Relations Institute. (2003). About us. Retrieved from http://www.niri.org/about/mission.cfm

National Research Council. (1989). *Improving risk communication.* Washington, DC: National Academies Press.

National Research Council. (2011). *Public response to alerts and warnings on mobile devices: Summary of a workshop on current knowledge and research gaps.* Washington, DC: National Academies Press.

Naudé, A.M.E., Froneman, J.D., & Atwood, R.A. (2004). The use of the Internet by ten South African non-governmental organizations—a public relations perspective. *Public Relations Review, 30*(1), 87–94.

Naughton, J. (2008, January 27). Thanks, Gutenberg—but we're too pressed for time to read. Retrieved from http://www.guardian.co.uk/media/2008/jan/27/internet.pressandpublishing

Nededog, J. (2011). MTV responds to "Skins" controversy. Retrieved from http://tv.msn.com/mom-pop-culture/mtv-responds-to-skins-controversy/story/feature/?gt1=28103

Neef, D. (2003). *Managing corporate reputation and risk: Developing a strategic approach to corporate integrity using knowledge management.* Burlington, MA: Elsevier Science.

Neff, B. D., & Hansen-Horn, T. L. (2010). Public relations theory: Translating in a global economy. *Business research yearbook: Global business perspectives, 2010*(2), 621–625.

Nelson, K. (2011). BP says to have processed 54% of claims. Retrieved from http://www.sunherald.com/2011/03/17/2952567/bp-says-to-have-processed-54-of.html

Nelson, R.A., & Heath, R.L. (1984). Corporate public relations and new media technology. *Public Relations Review, 10*(3), 27–38.

New emergency system to alert NYC and DC residents by cell phone. (2011, May 10). Retrieved from http://www.cnn.com/2011/US/05/10/us.emergency.alert/index.html?iref=allsearch

New Media Consortium (NMC). (2008). 2008 survey of educators in Second Life: Summary of findings. Retrieved from http://www.nmc.org/pdf/2008-sl-survey-summary.pdf

Ng, E., Schweitzer, L., & Lyons, S. (2010). New generation, great expectations: A field study of the Millennial generation. *Journal of Business & Psychology, 25*(2), 281–292.

Ni, L., & Kim, J.-N. (2009). Classifying publics: Communication behaviors and problem-solving characteristics in controversial issues. *International Journal of Strategic Communication, 3*(4), 1–25.

Nike Foundation. (2008). Girls: An unexpected answer. Retrieved from http://www.nikefoundation.org/index.html

Nike Foundation improving the lives of adolescent girls. (2011). Retrieved from http://www.nikebiz.com/responsibility/nike_foundation/

Noam, E. M. (2001, September 24). Straining communications systems to the limit. *New York Times*, p. C4.

Norris, F. H., Stevens, S. P., Pfefferbaum, B., Wyche, K. F., & Pfefferbaum, R. L. (2008). Community resilience as a metaphor, theory, set of capacities, and strategy for disaster readiness. *American Journal of Community Psychology, 41*, 127–150.

#numbers. (2011, March 14). Retrieved from http://blog.twitter.com/2011/03/numbers.html

Nye, J. S., Jr. (2002). *The paradox of American power: Why the world's only superpower can't go it alone.* New York, NY: Oxford University Press.

Nye, J. S., Jr. (2004). *Soft power: The means to success in world politics.* New York, NY: Public Affairs Press.

Nyland, R., & Near, C. (2007, February). Jesus is my friend: Religiosity as a mediating factor in Internet social networking use. *Proceedings of the AEJMC Midwinter Conference*, Reno, NV. Available at http://www.mendeley.com/research/jesus-friend-religiosity-mediating-factor-internet-social-networking/

Oblinger, D. (2003, July/August). Boomers, Gen-Xers & Millennials. *EduCause Review*, pp. 37–45.

Oceana. (2010). The beacon. Retrieved from http://na.oceana.org/en/blog

O'Dell, J. (2011, January 12). Are we too obsessed with Facebook? Retrieved from http://mashable.com/2011/01/12/obsessed-with-facebook-infographic/

O'Hair, H., & Heath, R. (2005). Conceptualizing communication and terrorism. In H. D. O'Hair, R. L. Heath, & G. R. Ledlow (Eds.), *Community preparedness and response to terrorism* (Vol. 3, pp. 1–12). Westport, CT: Praeger.

Oil spill and World Cup among Yahoo's most searched terms. (2010, December 1). Retrieved from http://www.cnn.com/2010/TECH/web/12/01/yahoo.most.searched.terms/

Okabe, R. (1983). Cultural assumptions of East and West: Japan and the United States. In W. B. Gudykunst (Ed.), *Intercultural communication theory: Current perspectives* (pp. 21–44). Beverly Hills, CA: Sage.

Olivarez-Giles, N. (2011, March 3). Sheen a major draw on Twitter. *Los Angeles Times*, p. B3.

Oliveira, M. (2007). Creating a dream, changing reality: A Brazilian Web site as a public relations tool for social change. In S. C. Duhé (Ed.), *New media and public relations* (pp. 234–246). New York, NY: Peter Lang.

O'Neil, J. (2008). Linking public relations tactics to long-term success: An investigation of how communications contribute to trust, satisfaction, and commitment in a nonprofit organization. *Journal of Promotion Management*, pp. 263–274.

Oprah, Ashton Kutcher mark Twitter "turning point." (2009, April 17). Retrieved from http://articles.cnn.com/2009-04-17/tech/ashton.cnn.twitter.battle_1_twitter-co-founder-biz-stone-first-tweet-social-media?_s=PM:TECH

O'Reilly, T., & Battelle, J. (2009). Web squared: Web 2.0 five years on. Retrieved from http://www.web2summit.com/web2009/public/schedule/detail/10194

Ostrow, A. (2009, July 28). Number of social networking users has doubled since 2007. Retrieved from http://mashable.com/2009/07/28/social-networking-users-us/

Owyang, J. (2010a, April 15). Framework and matrix: The five ways companies organize for social business. Retrieved from http://www.web-strategist.com/blog/2010/04/15/framework-and-matrix-the-five-ways-companies-organize-for-social-business/

Owyang, J. (2010b, July 19). Greenpeace vs. brands: Social media attacks continue. Retrieved from http://www.forbes.com/2010/07/19/greenpeace-bp-nestle-twitter-facebook-forbes-cmo-network-jeremiah-owyang_3.html

Oyer, S. (2010, Summer). Effects of crisis type and interactive online media type on public trust during organizational crisis. *Public Relations Journal*, 4(3). Retrieved from http://www.prsa.org/Intelligence/PRJournal/

Packer, R., & Jordan, K. (2001). *Multimedia: From Wagner to virtual reality*. New York, NY: W.W. Norton & Company, Inc.

Paek, H-J, Hilyard, K., Freimuth, V., Barge, J. K., & Mindlin, M. (2010). Theory-based approaches to understanding public emergency preparedness: Implications for effective health and risk communication. *Journal of Health Communication*, 15(4), 428–444.

Paine, K. (2007). Introduction. In S. C. Duhé (Ed.), *New media and public relations*. (pp. xiii-xiv). New York, NY: Peter Lang.

Paine, K. (2011). *Measure what matters: Online tools for understanding customers, social media, engagement and key relationships*. Hoboken, NJ: John Wiley & Sons.

Palen, L., & Liu, S. B. (2007). Citizen communications in crisis: Anticipating a future of ICT-supported public participation. *CHI '07: Proceedings of the SIGCHI conference on human factors in computing systems*, 727–736. doi:10.1145/1240624.1240736

Palen, L., Vieweg, S., Liu, S. B., & Hughes, A. L. (2009). Crisis in a networked world: Features of computer-mediated communication in the April 16, 2007, Virginia Tech event. *Social Science Computer Review*, 27(4), 467–480. doi:10.1177/0894439309332302

Palenchar, M. J. (2005). Risk communication. In R. L. Heath (Ed.), *Encyclopedia of public relations* (pp. 752–755). Thousand Oaks, CA: Sage.

Palenchar, M. J. (2010). Historical trends of risk and crisis communication. In R. L. Heath & D. O'Hair (Eds.), *Handbook of risk and crisis communication* (pp. 31–52). New York, NY: Routledge.

Palenchar, M. J., & Heath, R. L. (2002). Another part of the risk communication model: Analysis of communication processes and message content. *Journal of Public Relations Research*, 14(2), 127–158.

Palenchar, M.J., & Heath, R.L. (2006). Responsible advocacy through strategic risk communication. In K. Fitzpatrick & C. Bronstein (Eds.), *Ethics in public relations: Responsible advocacy* (pp. 131–154). Thousand Oaks, CA: Sage.

Pan, P.-L., & Xu, J. (2009). Online strategic communication: A cross-cultural analysis of U.S. and Chinese corporate websites. *Public Relations Review*, 35(3), 251–253.

Pandya, G. (2011, June 13). Weekend box office (June 10–12, 2011). Retrieved from http://www.boxofficeguru.com/weekend.htm

Papacharissi, Z. (2002). The virtual sphere: The Internet as the public sphere. *New Media and Society*, 4(1), 9–27.

Parekh, R. (2007). Unilever youth teach vets the ABCs of digital. *Advertising Age*, 78(40), 3–50.

Park, H., & Lee, H. (2011, March). *The use of human voice as a relationship building strategy on social networking sites*. Paper presented at the International Public Relations Research Conference, Miami, FL.

Park, H., & Reber, B.H. (2008). Relationship building and the use of Web sites: How Fortune 500 corporations use their Web sites to build relationships. *Public Relations Review*, 34(4), 409–411.

Park, N., & Lee, K.M. (2007). Effects of online news forum on corporate reputation. *Public Relations Review*, 33(3), 346–348.

Parker, S. (2010, September 16). Lady Gaga's "Don't ask, don't tell" activism is getting results. Retrieved from http://www.politicsdaily.com/2010/09/16/lady-gagas-dont-ask-dont-tell-activism-is-getting-results/print/

Parks, M. R., & Roberts, L. D. (1998). "Making MOOsic": The development of personal relationships on-line and a comparison to their off-line counterparts. *Journal of Social and Personal Relationships*, 15, 517–537.

Paton, D. (2006). Disaster resilience: Building capacity to co-exist with natural hazards and their consequences. In D. Paton & D. M. Johnston (Eds.), *Disaster resilience: An integrated approach* (pp. 3–10). Springfield, IL: Charles C. Thomas Publisher, Ltd.

Paton, D., & Johnston, D. M. (2006). Identifying the characteristics of a disaster resilient society. In D. Paton & D. M. Johnston (Eds.), *Disaster resilience: An integrated approach* (pp. 11–18). Springfield, IL: Charles C. Thomas Publisher, Ltd.

Paton, D., McClure, J., & Bürgelt, P. T. (2006). Natural hazard resilience: The role of individual and household preparedness. In D. Paton & D. M. Johnston (Eds.), *Disaster resilience: An integrated approach* (pp. 105–124). Springfield, IL: Charles C. Thomas Publisher, Ltd.

Paton, D., Millar, M., & Johnston, D. (2001). Community resilience to volcanic hazard consequences. *Natural Hazards, 24*(2), 157–169.

Patton, M. (1990). *Qualitative evaluation and research methods* (2nd ed.). Newbury Park, CA: Sage Publications.

Pauchant, T.C., & Mitroff, I.I. (1992). *Transforming the crisis-prone organization*. San Francisco, CA: Jossey-Bass Publishers.

Paulden, P. (2006). Selling spin, spin, spin. *Institutional Investor, 40*(10), 12–13.

Pavlik, J. V. (2007). Mapping the consequences of technology on public relations. Gainesville, FL: Institute for Public Relations.

Pearson, R. (1989). Business ethics as communication ethics: Public relations practice and the idea of dialogue. In C. H. Botan & V. Hazleton Jr. (Eds.), *Public relations theory* (pp. 111–131). Hillsdale, NJ: Erlbaum.

Pegoraro, A. (2010). Look who's talking - athletes on Twitter: A case study. *International Journal of Sport Communication, 3*(4), 501–514.

Peroune, D. (2007). Tacit knowledge in the workplace: The facilitating role of peer relationships. *Journal of European Industrial Training, 31*, 244.

Perry, D. C., Taylor, M., & Doerfel, M. L. (2003). Internet-based communication in crisis management. *Management Communication Quarterly, 17*(2), 206–232.

Perry, R., & Greene, M. (1982). The role of ethnicity in the emergency decision-making process. *Sociological Inquiry, 52*, 306–334.

Perry, R., & Lindell, M. (1991). The effects of ethnicity on evacuation decision-making. *International Journal of Mass Emergencies and Disasters, 9*(1), 47–68. Retrieved from http://search.proquest.com/docview/61243816?accountid=13158

Personal Localized Alerting Network (PLAN). (n.d.). Retrieved from http://www.fcc.gov/encyclopedia/personal-localized-alerting-network-plan

Peters, B. (2009). And leads us not into taking the new is new: A bibliographic case for new media history. *New Media & Society, 11*(1&2), 13–30.

Petersen, B.K., & Martin, H.J. (1996). CEO perceptions on investor relations function: An exploratory study. *Journal of Public Relations Research, 8*(3), 173–209.

Peterson, P. G. (2002). Public diplomacy and the war on terrorism. *Foreign Affairs, 81*(5), 74–94.

Petroff, M. (2010). Social media policy guide for higher ed. Retrieved from http://doteduguru.com/id6144-social-media-policy-resource-guide-from-simtech10.html

Pettigrew, J.E., & Reber, B.H. (2010). The new dynamic in corporate media relations: How Fortune 500 companies are using virtual press rooms to engage the press. *Journal of Public Relations Research, 22*(4), 404–428.

Pew Internet & American Life Project. (2010a). Generations 2010. Retrieved from http://www.pewinternet.org/Reports/2010/Generations-2010.aspx

Pew Internet & American Life Project. (2010b). Web 2.0. Retrieved from http://www.pewinternet.org/topics/Web-20

Pew Research Center (2000). The online healthcare revolution: How the Web helps Americans take better care of themselves. Retrieved from http://www.pewinternet.org

Pew Research Center for the People & the Press. (2007, April 15). Public knowledge of current affairs little changed by news and information revolutions. Retrieved from http://people-press.org/2007/04/15/public-knowledge-of-current-affairs-little-changed-by-news-and-information-revolutions/

Pew Research Center for the People & the Press. (2011, January 4). Internet gains on television as public's main news source: More young people cite Internet than TV. Retrieved from http://pewresearch.org/pubs/1844/poll-main-source-national-international-news-internet-television-newspapers

Pfeffer, J., & Salancik, G. R. (1978). *The external control of organizations: A resource dependence perspective*. New York, NY: Harper and Row.

Pfeiffer, M., & Zinnbauer, M. (2010). Can old media enhance new media? How traditional advertising pays off for an online social network. *Journal of Advertising Research, 50*(1), 42–49. doi: 10.2501/S0021849910091166

Philanthropy News Digest. (2010, January 16). Social media, text messaging bring in record donations for Haiti relief efforts. Retrieved from http://foundationcenter.org/pnd/news/story.jhtml?id=281400038

Plank, B.A. (1983). The revolution in communication technology for public relations. *Public Relations Review, 9*(1), 3–10.

Pleil, T. (2007). *Online-PR im Web 2.0: Fallbeispiele aus wirtschaft und politik.* [Online-PR in Web 2.0: Case studies out of economy and politics.] Konstanz, Germany: UVK.

Polat, R. K. (2005). The Internet and political participation: Exploring the explanatory links. *European Journal of Communication, 24*(4), 435–459.

Porter, L.V., & Sallot, L.M. (2005). Web power: A survey of practitioners' World Wide Web use and their perceptions of its effects on their decision-making power. *Public Relations Review, 31*(1), 111–119.

Porter, L.V., Sweetser-Trammell, K.D., Chung, D., & Kim, E. (2007). Blog power: Examining the effects of practitioner blog use on power in public relations. *Public Relations Review, 33*(1), 92–95.

Porter, M.E., & Kramer, M.R. (2002, December). The competitive advantage of corporate philanthropy. *Harvard Business Review*, pp. 5–16.

Porter, M.E., & Kramer, M.R. (2011, January-February). Creating shared value. *Harvard Business Review*, pp. 62–77.

Posetti, J. (2010). The spill effect: Twitter hashtag upends Australian political journalism. Retrieved from http://www.pbs.org/mediashift/2010/03/the-spill-effect-twitter-hashtag-upends-australian-political-journalism061.html

Prairie, M., Garfield, T., & Herbst, N.L. (2010). *College and school law.* Chicago, IL: American Bar Association Publishing.

Press, A.L., & Williams, B.A. (2010). *The new media environment: An introduction.* Malden, MA: Wiley-Blackwell.

Procopio, C. H., & Procopio, S. T. (2007). Do you know what it means to miss New Orleans? Internet communication, geographic community, and social capital in crisis. *Journal of Applied Communication Research, 35*(1), 67–87.

PRSA Task Force. (1988). Public relations body of knowledge task force report. *Public Relations Review, 14*(1), 3–40.

Psacharopoulos, G., & Patrinos, H. A. (2002). Returns to investment in education: A further update. Policy Research Working Paper 2881. Washington, D.C.: World Bank. Available at http://siteresources.worldbank.org/EDUCATION/Resources/278200-1099079877269/547664-1099079934475/547667-1135281504040/Returns_Investment_Edu.pdf

Public Relations Society of America. (2000). Retrieved from http://www.prsa.org/AboutPRSA/Ethics/CodeEnglish/index.html

Public's state of the union: Look for work. (2011, January). Retrieved from http://pewresearch.org/databank/dailynumber/?NumberID=1186

Purcell, K. (2011, March 29). Information 2.0 and beyond: Where are we, where are we going? Retrieved from http://www.pewinternet.org/Presentations/2011/Mar/APLIC.aspx

Putnam, R. D. (1995). Tuning in, tuning out: The strange disappearance of social capital in America. *PS: Political Science and Politics, 28*(4), 664–683.

Putnam, R. D. (2000). *Bowling alone: The collapse and revival of American community.* New York, NY: Simon & Schuster.

Pyle, K. (2005). Youth are the present. *Telephony, 246,* 40.

Qualman, E. (2011). Chapter Four: Obama's success driven by social media. In E. Qualman (Ed.), *Socialnomics: How social media transforms the way we live and do business* (pp. 64–90). Hoboken, NJ: John Wiley & Sons.

Rae, T. (2011, February 16). Colleges consider updated pager technology for emergency alerts. Retrieved from http://chronicle.com/blogs/wiredcampus/colleges-consider-updated-pager-technology-for-emergency-alerts/29837

Rainey, J. (2011, March 2). On the media: Wallowing in Sheen's meltdown. *Los Angeles Times,* pp. D1, D8.

Rainie, L., & Purcell, K. (2011, March 1). How the public perceives community information systems. Retrieved from http://www.pewinternet.org/Reports/2011/08-Community-Information-Systems/1-Report.aspx

Rains, S. A. (2007). Perceptions of traditional information sources and use of the World Wide Web to seek health information: Findings from the health information national trends survey. *Journal of Health Communication, 12*(7), 667–680. doi: 10.1080/10810730701619992

Ramsey, S.A. (1993). Issues management and the use of technologies in public relations. *Public Relations Review, 19*(3), 261–275.

Randstad Holding. (2011, March). Social media use around the globe. Retrieved from http://www.randstad.com/press-room/research-reports

Ransford, M. (2010, June 15). Smart phones displace computers for more college students. Retrieved from http://www.bsu.edu/news/article/0,1370,7273-850-64351,00.html

Rasmussen, T. (2008). The Internet and differentiation in the political public sphere. *Nordicom Review, 29*(2), 73–83.

RAVENAlert™ (n.d.). RAVENAlert keychain. IntelliGuard Systems, Lewisville, TX. Retrieved from http://www.intelliguardsystems.com/pdfs/RavenAlert-Flash-drive_Product-Page.pdf

Rawlins, B. (2009). Give the emperor a mirror: Toward developing a stakeholder measurement of organizational transparency. *Journal of Public Relations Research, 21*(1), 71–99.

Raykov, T., & Marcoulides, G. A. (2000). *A first course in structural equation modeling* (2nd ed.). Mahwah, NJ: Erlbaum.

Raymond, D. (2003). Activism: Behind the banners. In S. John & S. Thomas (Eds.), *New activism and the corporate response* (pp. 207–225). New York, NY: Palgrave Macmillan.

Reber, B., & Berger, B. (2005). Framing analysis of activist rhetoric: How the Sierra Club succeeds or fails at creating salient messages. *Public Relations Review, 31*(2), 185–195.

Reber, B.H., Cropp, F., & Cameron, G. T. (2001). Mythic battles: Examining the lawyer-public relations counselor dynamic. *Journal of Public Relations Research, 13*(3), 187–218.

Reber, B.H., Gower, K.K., & Robinson, J.A. (2006). The Internet and litigation public relations. *Journal of Public Relations Research, 18*(1), 23–44.

Reber, B.H., & Kim, J.K. (2006). How activist groups use websites in media relations: Evaluating online press rooms. *Journal of Public Relations Research, 18*(4), 313–333.

Reeves, B., & Read, J. (2009). *Total engagement.* Boston, MA: Harvard Business Press.

Reeves, J. (2011). 5 students dead: Tornado devastates University of Alabama. Retrieved from http://www.huffingtonpost.com/2011/05/02/5-students-dead-tornado-d_n_856181.html

Reeves, P. (2011, March 24). Employers: Embrace social media—it's here to stay. *Legal Week, 13*(11), 17.

Reisinger, D. (2010, December 9). Youku video site makes strong NYSE debut. Retrieved from http://news.cnet.com/8301-13506_3-20025149-17.html

Rennie, E. (2007). Community media in the prosumer era. Retrieved from http://www.cbonline.org.au/3cmedia/3c_issue3/BarryERennie.pdf

Research on Web 2.0. (2011). Retrieved from http://www.pewinternet.org/topics/Web-20.aspx

Residents shun cell phone reverse 911 (2011, April 10). Retrieved from http://www.kulr8.com/news/local/Residents-Shun-Cell-Phone-Reverse-911-119568124.html

Rhodes, J., Liang, B., & Spencer, R. (2009). First do no harm: Ethical principles for youth mentoring relationships. *Professional Psychology: Research & Practice, 40*(5), 452–458.

Rice, J. (2009). *The church of Facebook: How the hyperconnected are redefining community.* Colorado Springs, CO: David C. Cook Distribution.

Richins, M. L. (1983). Negative word-of-mouth by dissatisfied consumers: A pilot study. *Journal of Marketing, 47*(1), 68–78.

Richtel, M. (2010, December 20). E-mail gets an instant makeover. Retrieved from http://www.nytimes.com/2010/12/21/technology/21email.html

Roach, D. (2011, January 27). *Poll: Churches are fans of Facebook, social media.* Retrieved from http://www.crosswalk.com/news/religion-today/poll-churches-are-fans-of-facebook-social-media-11644767.html

Robards, B. (2010). Randoms in my bedroom: Negotiating privacy and unsolicited contact on social network sites. *Prism, 7*(3). Retrieved from http://www.prismjournal.org

Roberts, A. (2008, October 8). VT emergency notification test successful. Retrieved from http://www.wsls.com/sls/news/local/new_river_valley/article/vt_emergency_notification_test_successful/18999/

Roberts, K. (2010, July 23). Save a girl, save the world. Retrieved from http://www.oprah.com/world/PSI-Interviews-Jennifer-Buffett-and-Maria-Eitel

Roberts, L. D., Smith, L. M., & Pollock, C. (1997, July). "u r a lot bolder on the net": The social use of text-based virtual environments by shy individuals. Paper presented at the International Conference on Shyness and Self-Consciousness, Cardiff, Wales.

Rogers, E.M. (1995). *The history of communication study: A biographical approach.* New York, NY: The Free Press.

Rousseau, J. J. (1762/1957). *The social contract.* New York, NY: Hafner.

Rowley, T. J. (1997). Moving beyond dyadic ties: A network theory of stakeholder influence. *Academy of Management Review, 22*(4), 887–910.

Ruane, L. (2010, December 17). Pastors are flocking to Facebook, Twitter. *USA Today,* p. 3A.

Ruggiero, T. E. (2000). Uses and gratifications theory in the 21st century. *Mass Communication & Society, 3*(1), 3–37.

Ruppenthal, A. (2010). College coaches finding ways to monitor athletes' social networking activity. Retrieved from http://bit.ly/bBqW3t

Ryan, C. (1991). *Prime time activism: Media strategies for grassroots organizing.* Boston, MA: South End Press.

Ryan, M. (2003). Public relations and the Web: Organizational problems, gender, and institution type. *Public Relations Review, 29*(3), 335–349.

Rybalko, S., & Seltzer, T. (2010). Dialogic communication in 140 characters or less: How Fortune 500 companies engage stakeholders using Twitter. *Public Relations Review, 36*(4), 336–341.

Ryzik, M. (2011, March 3). It's back to the studio, as Oscar season ends. *The New York Times,* pp. C1, C6.

Safko, L., & Brake, D. K. (2009). *The social media bible: Tactics, tools and strategies for business success.* Hoboken, NJ: John Wiley & Sons.

Sallot, L.M., Porter, L.V., & Acosta-Alzuru, C. (2004). Practitioners' Web use and perceptions of their own roles and power: A qualitative study. *Public Relations Review, 30*(3), 269–278.

Sandel, M. J. (1996). *Democracy's discontent: America in search of a public philosophy.* Cambridge, MA: Harvard University Press.

Sanders, K. (2008, January 28). UF looking to improve text message alerts. Retrieved from http://www.alligator.org/articles/2008/01/24/news/uf_administration/text.txt

Sanderson, J. (2010). Framing Tiger's troubles: Comparing traditional and social media. *International Journal of Sport Communication, 3,* 438–453.

Sassi, S. (2001). The transformation of the public sphere. In B. Axford & R. Huggins (Eds.), *New media and politics* (pp. 89–108). London, England: Sage.

Sawhney, M., Verona, G., & Prandelli, E. (2005). Collaborating to create: The Internet as a platform for customer engagement in product innovation. *Journal of Interactive Marketing, 19*(4), 4–17.

Scaffidi, C., Myers, B., & Shaw, M. (2007). Trial by water: Hurricane Katrina person locator Web sites. In S. Wisband (Ed.), *Leadership at a distance: Research in technologically-supported work* (pp. 209–222). Mahwah, NJ: Lawrence Erlbaum.

Scammell, M. (1999). Political marketing: Lessons for political science. *Journal of Political Studies, 47*(4), 719–739.

Scannell, P. (2001). Authenticity and experience. *Discourse Studies, 3,* 405–411.

Scheufele, D. A., & Tewksbury, D. (2007). Framing, agenda setting and priming: The evolution of three media effects models. *Journal of Communication, 57*(1), 9–20.

Schmid, A. P., & Jongman, A. J. (1988). *Political terrorism: A new guide to actors, authors, concepts, data bases, theories, & literature* (2nd ed.). New Brunswick, N.J: Transaction Publishers.

Schmitz, T. (2011, Winter). The limitations of Reverse 911 calling for modern municipalities. *Disaster Recovery Journal, 24,* 41–42.

Schoenberger-Orgad, M., & McKie, D. (2005). Sustaining edges: CSR, postmodern play, and SMEs. *Public Relations Review, 31*(4), 578–583.

Schöller, A.C.B. (2010). Brand communication in social media: Strategische nutzung von brand communities als kommunikationsinstrument durch unternehmen. [Brand communication in social media: Strategic use of brand communities as communication tool by enterprises]. Retrieved from http://epub.ub.uni-muenchen.de/11929/

Schultz, F., Utz, S., & Göritz, A. (2011). Is the medium the message? Perceptions of and reactions to crisis communication via Twitter, blogs and traditional media. *Public Relations Review, 37*(1), 20–27.

Scoble, R., & Israel, S. (2006). *Naked conversations: How blogs are changing the way businesses talk with customers.* Hoboken, NJ: John Wiley & Sons.

Scolari, C. A. (2009). Mapping conversations about new media: The theoretical field of digital communication. *New Media & Society, 11*(6), 943–964.

Scott, A.O. (2011, April 4). Critic's notebook: Belligerent and boozy, and that's just the audience. *The New York Times,* pp. C1, C5.

Scott, W. (1955). Reliability of content analysis: The case of nominal scale coding. *Public Opinion Quarterly, 17,* 321–325.

Searson, E.M., & Johnson, M.A. (2010). Transparency laws and interactive public relations: An analysis of Latin American government Web sites. *Public Relations Review, 36*(2), 120–126.

Securities and Exchange Commission. (2000, August 21). Final rule: Selective disclosure and insider trading. Retrieved from http://www.sec.gov/rules/final/33-7881.htm

Securities and Exchange Commission. (2008, August 7). Commission guidance on the use of company Web sites. Retrieved from http://www.sec.gov/rules/interp/2008/34-58288.pdf

Securities and Exchange Commission. (2011). The investor's advocate: How the SEC protects investors, maintains market integrity, and facilitates capital formation. Retrieved from http://www.sec.gov/about/whatwedo.shtml

Sellnow, T. (2010). BP's crisis communication: Finding redemption through renewal. *Communication Currents, 5*(4), 1–2.

Seltzer, T., & Mitrook, M.A. (2007). The dialogic potential of weblogs in relationship building. *Public Relations Review, 33*(2), 227–229.

Sennett, R. (1986). *The fall of public man.* London, England: Faber and Faber.

Seo, H., Kim, J.Y., & Yang, S.-U. (2009). Global activism and new media: A study of transnational NGOs' online public relations. *Public Relations Review, 35*(2), 123–126.

Settlement reached in case involving discharge for Facebook comments. (2011, February 8). Retrieved from http://www.nlrb.gov/news/settlement-reached-case-involving-discharge-facebook-comments

Sha, B.-L. (2004). Noether's theorem: The science of symmetry and the law of conservation. *Journal of Public Relations Research, 16*(4), 391–416.

Sha, B.-L. (2006). Cultural identity in the segmentation of publics: An emerging theory of intercultural public relations. *Journal of Public Relations Research, 18*(1), 45–65.

Sha, B.-L. (2007). Dimensions of public relations: Moving beyond traditional public relations models. In S. C. Duhé (Ed.), *New media and public relations* (pp. 3–25). New York, NY: Peter Lang.

Shankman, P. (2010). *Customer service: New rules for a social media world.* Indianapolis, IN: Que.

Shannon, C. (1949). *The mathematical theory of communication.* Urbana: The University of Illinois Press.

Sharma, A., & Vascellaro, J. E. (2008, November 21). Companies eye location-services market—Google, Nokia join start-ups to create tools to track phone users' whereabouts, offer location information. *Wall Street Journal,* p. B4.

Sheehan, Kim B. (2001). E-mail survey response rates: A review. *Journal of Computer-Mediated Communication, 6*(2). Retrieved from http://www.ascusc.org/jcmc/vol6/issue2/sheehan.htm

Shklovski, I., Palen, L., & Sutton, J. (2008, November). Finding community through information and communication technology during disaster events. Retrieved from http://spot.colorado.edu/~suttonj/cscw460-shklovski.pdf

Shoemaker, P.J., Tankard, J.W., & Lasorsa, D. (2004). *How to build social science theories.* Thousand Oaks, CA: Sage Publications.

Shumate, M., & O'Connor, A. (2010). The symbiotic sustainability model: Conceptualizing NGO-corporate alliance communication. *Journal of Communication, 60,* 577–609.

Siegel, R. (2007, February 9). How Wikipedia breaks news, and adjusts to it. Retrieved from http://www.npr.org/templates/story/story.php?storyId=7320255

Silverstone, R. (1999). What's new about new media? *New Media & Society, 1*(1), 10–12.

Simon, H. A. (1997). *Administrative behavior: A study of decision-making processes in administrative organizations* (4th ed.). New York, NY: The Free Press.

Simon, M. J. (1969). *Public relations law.* New York, NY: Meredith Corporation.

Singer, M. (2011, January 25). Study: Half of all Protestant churches are on Facebook. Retrieved from http://www.allfacebook.com/study-13-of-protestant-churches-are-on-facebook-2011-01

The size of the World Wide Web. (2011, June 27). Retrieved from http://www.worldwidewebsize.com/

Skoler, M. (2009). Why the news media became irrelevant—and how social media can help. Retrieved from http://www.nieman.harvard.edu/reportsitem.aspx?id=101897

Slater, D. (2001). Political discourse and the politics of need: Discourses on the good life in cyberspace. In L. Bennett & R. Entman (Eds.), *Mediated politics: Communication in the future of democracy* (pp. 117–140). Cambridge, England: Cambridge University Press.

Smigel, E. O. (1956). Public attitudes toward stealing as related to the size of the victim organization. *American Sociological Review, 21*(3), 320–327.

Smith, B.G. (2010a). The evolution of the blogger: Blogger considerations of public relations-sponsored content in the blogosphere. *Public Relations Review, 36*(2), 175–177.

Smith, B.G. (2010b). Socially distributing public relations: Twitter, Haiti, and interactivity in social media. *Public Relations Review, 36*(4), 329–335.

Smith, T. (2011, January 12). Annual report 2011: Welcome to social entertainment. Retrieved from http://globalwebindex.net/thinking/annual-report-2011-welcome-to-social-entertainment/

Smithsonian Institution. (2009). Smithsonian Latino Center inaugurates the Smithsonian Latino Virtual Museum in Second Life. Retrieved from http://newsdesk.si.edu/releases/smithsonian-latino-center-inaugurates-smithsonian-latino-virtual-museum-second-life

SMS over SS7. (2003, December). National Communications Systems. Technical Information Bulletin 03–02. Arlington, VA. Retrieved from http://www.ncs.gov/library/tech_bulletins/2003/tib_03-2.pdf

Snyder, P., Hall, M., Robertson, J., Jasinski, T., & Miller, J. S. (2006). Ethical rationality: A strategic approach to organizational crisis. *Journal of Business Ethics, 63*(4), 371–383.

Social Media Business Council. (2009). Disclosure best practices toolkit. Retrieved from http://www.socialmedia.org/disclosure/

Social networking sites overtake email in popularity. (2009, March 10). Retrieved from http://www.ctv.ca/servlet/ArticleNews/story/CTVNews/20090310/online_trends_090310/20090310?hub=TopStories

Sociable professors (2010, May 4). Retrieved from http://www.prweb.com/releases/2010/05/prweb3960844.htm

Sohn, D. (2009). Disentangling the effects of social network density on electronic-word-of-mouth (eWOM) intention. *Journal of Computer-Mediated Communication, 14*(2), 352–367. doi:10.1111/j.1083-6101.2009.0144.x

Solis, B. (2011). *Engage, revised and updated: The complete guide for brands and businesses to build, cultivate, and measure success in the New Web.* Hoboken, NJ: Wiley.

Solis, B., & Breakenridge, D. (2009). *Putting the public back in public relations: How social media is reinventing the aging business of PR.* Upper Saddle River, NJ: FT Press.

Solomon, M. (2001, January 29). Coaching the boss. *Computerworld,* p. 41.

Sommerfeldt, E. (2011). Activist e-mail action alerts and identification: Rhetorical relationship building strategies in collective action. *Public Relations Review, 37*(1), 87–89.

SOS from mayor of Minami Soma City, next to the crippled Fukushima nuclear power plant, Japan. (2011, March 26). Retrieved from http://www.youtube.com/user/p4minamisoma

Sparks, C. (2001). The Internet and the global public sphere. In L. Bennett & R. Entman (Eds.), *Mediated politics: Communication in the future of democracy* (pp. 75–98). Cambridge, England: Cambridge University Press.

Sparks, L., Kreps, G. L., Botan, C., & Rowan, K. (2005). Responding to terrorism: Translating communication research into practice. *Communication Research Reports, 22*(1), 1–5.

Springston, J.K., & Weaver-Lariscy, R. (2007). Health crises and media relations: Relationship management-by-fire. *Health Marketing Quarterly, 24*(3), 81–96.

Stafford, L., & Canary, D. J. (1991). Maintenance strategies and romantic relationship type, gender, and relational characteristics. *Journal of Social and Personal Relationships, 8,* 217–242.

Star, J. (2010). Chad Ochocinco's Twitter fine and the dumbest sports fines ever. Retrieved from http://bit.ly/aEqMaL

Stassen, W. (2010). Your news in 140 characters: Exploring the role of social media in journalism. *Global Media Journal— African edition, 4*(1), 1–17.

Statistics. (2011). Retrieved from http://www.facebook.com/press/info.php?statistics

Steinbach, P. (2009). Twitter rivals. *Athletic Business, 33*(10), 60–62.

Steinbach, P. (2010). Facebook value. *Athletic Business, 34*(8), 58–60.

Steinfield, C., Ellison, N. B., & Lampe, C. (2008). Social capital, self-esteem, and use of online social network sites: A longitudinal analysis. *Journal of Applied Developmental Psychology, 29,* 434–445.

Stelter, B. (2010a, April 7). Web Luddite no more: O'Brien hits Internet. *The New York Times,* p. C1.

Stelter, B. (2010b, August 28). With backstage webcast, Emmys enter new terrain. *The New York Times,* p. C7.

Stelter, B. (2011, January 21). Taco Bell pulls ads for MTV's "Skins." Retrieved from http://mediadecoder.blogs.nytimes.com/2011/01/21/taco-bell-pulls-ads-from-mtvs-skins/

Stelzner, M. (2011, April 7). Social media marketing industry report. Retrieved from http://www.socialmediaexaminer.com/social-media-marketing-industry-report-2011/

Stephen, P. (2010, March 29). Millennials do faith and politics their way. *USA Today*, p. 9A.

Stephens, K.K., & Malone, P.C. (2009). If the organizations won't give us information…:The use of multiple new media for crisis technical translation and dialogue. *Journal of Public Relations Research*, 21(2), 229–239.

Stepick, A., & Stepick, C. D. (1990). People in the shadows: Survey research among Haitians in Miami. *Human Organization*, 49(1), 64–77.

Stern, P.C. (2000). Toward a coherent theory of environmentally significant behavior. *Journal of Social Issues*, 56(3), 407–424.

Stevens, M. (2011). MSU's Stansbury bans Twitter for his players. Retrieved from http://bit.ly/gwkMpO

Stewart, A. (2011, April 27). New hitch for trailers. *Daily Variety*, pp. 1, 18.

Steyn, P., Salehi-Sangari, E., Pitt, L., Parent, M., & Berthon, P. (2010). The Social Media Release as a public relations tool: Intentions to use among B2B bloggers. *Public Relations Review*, 36(1), 87–89.

Stöber, R. (2004). What media evolution is—a theoretical approach to the history of new media. *European Journal of Communication*, 19(4), 483–505.

Strobbe, I., & Jacobs, G. (2005). E-releases: A view from linguistic pragmatics. *Public Relations Review*, 31(2), 289–291.

Stross, R. (2009, May 2). Encyclopedic knowledge, then vs. now. Retrieved from http://www.nytimes.com/2009/05/03/business/03digi.html

Subcommittee on Disaster Reduction (2005). *Grand challenges for disaster reduction*. Washington, DC: National Science and Technology Council.

Sullivan, E. (2011, April 7). US to use Facebook, Twitter to issue terror alerts: New advisory system designated to be easier to understand and more specific. Retrieved from http://www.msnbc.msn.com/id/42474206

Sun, E. (2011, May 11). Now Jesus Facebook page bumps Bieber and Lady Gaga. Retrieved from http://www.christiantoday.com/article/now.jesus.facebook.page.bumps.bieber.and.lady.gaga/27971.htms

Sunstein, C. (2001). The daily we: Is the Internet really a blessing for democracy? Retrieved from http://bostonreview.net/BR26.3/sunstein.php

Sunstein, C. (2006). *Infotopia: How many minds produce knowledge*. New York, NY: Oxford University Press.

Sutter, J. D. (2009, April 15). Ashton Kutcher challenges CNN to Twitter popularity contest. Retrieved from http://articles.cnn.com/2009-04-15/tech/ashton.cnn.twitter.battle_1_cnn-twitter-account-followers?_s=PM:TECH

Sutton, J. N. (2009). Social media monitoring and the Democratic National Convention: News tasks and emergent processes. *Journal of Homeland Security and Emergency Management*, 6(1), article 67.

Sutton, J. N. (2010, May). Twittering Tennessee: Distributed networks and collaboration following a technological disaster. *Proceedings of the 7th International ISCRAM Conference*, Seattle, WA.

Swallow, E. (2010, August 16). The future of public relations and social media. Retrieved from http://mashable.com/2010/08/16/pr-social-media-future/

Sweeney, S. (2010, May 25). Signs of life in the public relations job market. Retrieved from http://www.prsa.org/intelligence/thestrategist/articles/view/8643/1013/signs_of_life_in_the_public_relations_job_market?utm_campaign=PRSASearch&utm_source=PRSAWebsite&utm_medium=SSearch&utm_term=pr%20executive%20layoffs

Sweet success for Kit Kat campaign: You asked, Nestlé has answered. (2010, May 17). Retrieved from http://www.greenpeace.org/international/en/news/features/Sweet-success-for-Kit-Kat-campaign/

Sweetser, K.D. (2007). Blog bias: Reports, inferences, and judgments of credentialed bloggers at the 2004 nominating conventions. *Public Relations Review*, 33(4), 426–428.

Sweetser, K.D. (2010). A losing strategy: The impact of nondisclosure in social media on relationships. *Journal of Public Relations Research*, 22(3), 288–312.

Sweetser, K.D., & Metzgar, E. (2007). Communicating during crisis: Use of blogs as a relationship management tool. *Public Relations Review*, 33(3), 340–342.

Tankard, J. W. (2001). The empirical approach to the study of media framing. In S. D. Reese, O. H. Gandy, & A. E. Grant (Eds.), *Framing public life: Perspectives on media and our understanding of the social world* (pp. 95–106). Mahwah, NJ: Lawrence Erlbaum.

Tapscott, D. & Ticoll, D. (2003). *The naked corporation: How the age of transparency will revolutionize business*. New York, NY: Simon & Schuster.

Taylor, M. (2000a). Media relations in Bosnia: A role for public relations in building civil society. *Public Relations Review, 26,* 1–14.

Taylor, M. (2000b). Toward a public relations approach to nation building. *Journal of Public Relations Research, 12*(2), 179–210.

Taylor, M. (2007, November). *How the third party effect influences how organizations respond to a product recall crisis.* Paper presented at the National Communication Association 93rd Annual Convention, Chicago, IL.

Taylor, M., & Kent, M. L. (2006). Public relations theory and practice in nation building. In C. H. Botan & V. Hazleton (Eds.), *Public relations theory II* (pp. 341–359), Mahwah, NJ: Lawrence Erlbaum.

Taylor, M., & Kent, M. L. (2007). Taxonomy of mediated crisis responses. *Public Relations Review, 33*(2), 140–146.

Taylor, M., & Kent, M.L. (2010). Anticipatory socialization in the use of social media in public relations: A content analysis of PRSA's *Public Relations Tactics. Public Relations Review, 36*(3), 207–214.

Taylor, M., Kent, M.L., & White, W.J. (2001). How activist organizations are using the Internet to build relationships. *Public Relations Review, 27*(3), 263–284.

Taylor, M., & Perry, D.C. (2005). Diffusion of traditional and new media tactics in crisis communication. *Public Relations Review, 31*(2), 209–217.

Taylor, M.V., & Stephenson, P. (2009). Influenza A (H1N1) virus (swine influenza): A webliography. *Journal of Consumer Health on the Internet, 13,* 374–385.

Terilli, S.A., & Arnorsdottir, L.I. (2008, Fall). The CEO as celebrity blogger: Is there a ghost or ghostwriter in the machine? *Public Relations Journal, 2*(4). Retrieved from http://www.prsa.org/Intelligence/PRJournal/

Terilli, S.A., Driscoll, P.D., & Stacks, D.W. (2008, Spring). Business blogging in the fog of law: Traditional agency liability principles and less-than-traditional Section 230 immunity in the context of blogs about businesses. *Public Relations Journal, 2*(2). Retrieved from http://www.prsa.org/Intelligence/PRJournal/

Terilli, S.A., Stacks, D.W., & Driscoll, P.D. (2010, Winter). Getting even or getting skewered: Piercing the digital veil of anonymous Internet speech as a corporate public relations tactic. *Public Relations Journal, 4*(1). Retrieved from http://www.prsa.org/Intelligence/PRJournal/

Terlien, M. T. J., & Graham-Cumming, J. (1997). Hit and misses: A year watching the web. *Computer Networks and ISDN Systems, 29*(8), 1357–1365.

Thackeray, R., Neiger, B., Hanson, C., & McKenzie, J. (2008). Enhancing promotional strategies within social marketing programs: Use of Web 2.0 social media. *Health Promotion Practice, 9*(4), 338–343.

Thomas, M., Chandran, S., & Trope, Y. (2006). *The effects of information type and temporal distance on purchase intentions.* Unpublished manuscript.

Thomsen, S.R. (1995). Using online databases in corporate issues management. *Public Relations Review, 21*(2), 103–122.

Tierney, P. (2000). Internet-based evaluation of tourism web site effectiveness: Methodological issues and survey results. *Journal of Travel Research, 39*(2), 212–219.

Tilson, D. A., & Venkateswaran, A. (2004). Toward a peaceable kingdom: Public relations and religious diversity in the U.S. *Public Relations Quarterly, 49*(2), 37–44.

Timm, G. (2011, April 14). NAB Convention report—Part 1: Proposed new FCC EAS rules coming "very soon." Retrieved from http://www.awareforum.org/2011/04/nab-convention-report-%E2%80%93-part-1-proposed-new-fcc-eas-rules-coming-very-soon/

Tinker, T.L., Dumlao, M., & McLaughlin, G. (2009, Summer). Effective social media strategies during times of crisis: Learning from the CDC, HSS, FEMA, the American Red Cross and NPR. *The Public Relations Strategist,* pp. 25–26.

Tirkkonen, P., & Luoma-aho, V. (2011). Online authority communication during an epidemic: A Finnish example. *Public Relations Review, 37*(2), 172–174.

Tobin, G. A. (1999). Sustainability and community resilience: The holy grail of hazards planning? *Environmental Hazards, 1,* 13–25.

Tobin, G. A., & Whiteford, L. M. (2002). Community resilience and volcano hazard: The eruption of Tungurahua and evacuation of the *Faldas* in Ecuador. *Disasters, 26*(1), 28–48.

Tolson, A. (2001). "Being yourself": The pursuit of authentic celebrity. *Discourse Studies, 3*(4), 443–457.

Tomasello, T. K., Lee, Y., & Baer, A. P. (2010). "New media" research publication trends and outlets in communication. *New Media & Society, 12*(4), 531–548.

Tomer, J. F., & Sadler, T. R. (2007). Why we need a commitment approach to environmental policy. *Environmental Economics, 62,* 627–636.

Top 15 most popular social networking sites. (2011, July). Retrieved from http://www.ebizmba.com/articles/social-networking-websites

Top social media sites. (2010, May). Retrieved from http://www.prelovac.com/vladimir/top-list-of-social-media-sites

Trammell, K.D. (2006). Blog offensive: An exploratory analysis of attacks published on campaign blog posts from a political public relations perspective. *Public Relations Review*, 32(4), 402–406.

Trance, P. (2011, March 31). Meteo 360—iPhone Augmented Reality weather app. Retrieved from http://dailymobile.se/2010/03/31/meteo-360-iphone-augmented-reality-weather-app/

Traynor, P. (2008, September). Characterizing the limitations of third-party EAS over cellular text messaging services. Retrieved from http://www.4gamericas.org/documents/Characterizing_the_Limitations_of_3rd_Party_EAS-Traynor_Sept08.pdf

Traynor, P. (2010). Characterizing the security implications of third-party emergency alert systems over cellular text messaging services. In S. Jajodia and J. Zhour (Eds.), Security and privacy in communication networks: Lecture notes of the Institute for Computer Science, Social Informatics and Telecommunications Engineering (pp. 125–143). Berlin, Germany: Springer.

Treadwell, D., & Treadwell, J. B. (2000). *Public relations writing: Principles in practice*. Boston, MA: Allyn and Bacon.

Trivedi, C., & Stokols, D. (2011). Social enterprises and corporate enterprises: Fundamental differences and defining features. *Journal of Entrepreneurship*, 20, 1–32.

Tuman, J. S. (2010). *Communicating terror: The rhetorical dimensions of terrorism* (2nd ed.). Thousand Oaks, CA: Sage.

Tumasjan, A., Sprenger, T.O., Sandner, P. G., & Welpe, I. M. (2009). Predicting elections with Twitter: What 140 characters reveal about political sentiment (pp. 178–185). *Proceedings of the Fourth International AAAI Conference on Weblogs and Social Media*, Washington, DC. Available at http://www.aaai.org/ocs/index.php/ICWSM/ICWSM10/paper/viewFile/1441/1852

Tuna, C. (2009, April 27). Corporate blogs and "tweets" must keep SEC in mind. *The Wall Street Journal*, p. B4.

Turk, J. V. (2006, November). Public relations education for the 21st century: The professional bond. Retrieved from http://www.commpred.org/_uploads/report2-full.pdf

Twitter. (2011a). About Twitter. Retrieved from http://twitter.com/about#about

Twitter. (2011b). What is Twitter? Retrieved from http://business.twitter.com/basics/what-is-twitter

Tyma, A. W., Sellnow, D. D., & Sellnow, T. (2010, June). *Social media use in response to the Virginia Tech crisis: Moving from chaos to understanding*. Paper presented at the annual meeting of the International Communication Association, Suntec City, Singapore. Retrieved from http://www.allacademic.com/meta/p403670_index.html

Ulmer, R. R., Sellnow, T. L., & Seeger, M. W. (2011). *Effective crisis communication: Moving from crisis to opportunity* (2nd ed.). Thousand Oaks, CA: Sage.

Universal McCann (2008, September). When did we start trusting strangers? Retrieved from http://www.slideshare.net/mickstravellin/universal-mccanns-when-did-we-start-trusting-strangers-presentation

Update on deforestation and palm oil. (2010). Retrieved from http://www.nestle.com/Media/Statements/Pages/Update-on-deforestation-and-palm-oil.aspx

U. S. Census Bureau. (2000). Your gateway to Census 2000. Retrieved from http://www.census.gov/main/www/cen2000.html

U.S. Department of Health and Human Services (HHS). (2010). HHS general guidance for utilization of new and/or social media. Retrieved from http://www.hhs.gov/web/policies/webstandards/socialmedia.html

U.S. ditching terror color warning system. (2011, January 27). Retrieved from http://homelandsecuritynewswire.com/us-ditching-terror-color-warning-system

Utz, S. (2000). Social information processing in MUDs: The development of friendships in virtual worlds. *Journal of Online Behavior*, 1(1). Retrieved from http://www.behavior.net/JOB/v1n1/utz.html

Valentini, C. (2010, March). Handling social media with care. *Communication Director Magazine*, pp. 64–67.

Valkenburg, P., & Peter, J. (2008). Adolescents' identity experiments on the Internet: Consequences for social competence and self-concept unity. *Communication Research*, 35(2), 208–231.

Van de Donk, W., Loader, B. D., Nixon, P. G., & Rucht, D. (2004). *Cyberprotest: New media, citizens and social movements*. New York, NY: Routledge.

Van Dijk, J. (2006). *The network society: Social aspects of new media* (2nd ed.). London, England: Sage.

Van Hoosear, T. (2011, January 22). Never cry news. Retrieved from http://itsfreshground.com/2011/01/nevercrynews/

Van Leeuwen, T. (2001). What is authenticity? *Discourse Studies, 3,* 392–397.

Van Manen, M. (1990). *Researching lived experience: Human science for an action sensitive pedagogy* (2nd ed.) Albany, NY: State University of New York Press.

Van Woerkum, C. M. J., & Aarts, M. N. C. (2009). Visual transparency: Looking behind thick walls. *Public Relations Review, 35*(4), 434–436.

Vargas, L. (2011, February 17). Giving your privacy policy a second look. Retrieved from http://www.radian6.com/blog/2011/02/giving-your-privacy-policy-a-second-look/

Verčič, D., Grunig, J. E., & Grunig, L. A. (1996). Global and specific principles of public relations: Evidence from Slovenia. In H. Culbertson & N. Chen (Eds.), *International public relations: A comparative analysis* (pp. 31–65). Mahwah, NJ: Erlbaum.

Vieweg, S., Palen, L., Liu, S. B., Hughes, A. L., & Sutton, J. (2008, May). Collective intelligence in disaster: An examination of the phenomenon in the aftermath of the 2007 Virginia Tech shooting. *Proceedings of the 5th International ISCRAM Conference,* Washington, DC.

Vorvoreanu, M. (2006). Online organization-public relationships: An experience-centered approach. *Public Relations Review, 32*(4), 395–401.

Vujnovic, M., & Kruckeberg, D. (2010a). The local, national, and global challenges of public relations: A call for an anthropological approach to practicing public relations. In R. L. Heath (Ed.), *Handbook of public relations* (2nd ed., pp. 671–678). Thousand Oaks, CA: Sage Publications, Inc.

Vujnovic, M., & Kruckeberg, D. (2010b). Managing global public relations in the new media environment. In M. Deuze (Ed.), *Managing media work* (pp. 217–223). London, England: Sage.

Waddock, S. (2008). The development of corporate responsibility/corporate citizenship. *Organization Management Journal, 5,* 29–39.

Wakefield, R.I. (2008). Theory of international public relations, the Internet, and activism: A personal reflection. *Journal of Public Relations Research 20*(1), 138–157.

Wakefield, R., & Walton, S. (2010). The translucency corollary: Why full transparency is not always the most ethical approach. Retrieved from http://www.instituteforpr.org/wp-content/uploads/TranslucencyCorollary_BYU.pdf

Walther, J.B. (1996). Computer-mediated communication: Impersonal, interpersonal and hyperpersonal interaction. *Communication Research, 23*(1), 3–43.

Wang, J., & Chaudri, V. (2009). Corporate social responsibility engagement and communication by Chinese companies. *Public Relations Review, 35*(3), 247–250.

Ward, J. C., & Ostrom, A. L. (2006). Complaining to the masses: The role of protest framing in customer-created complaint Web sites. *The Journal of Consumer Research, 33*(2), 220–230.

Wardrip-Fruin, N., & Montfort, N. (2003). *The new media reader.* Cambridge, MA: MIT Press.

Warner, D., & Procaccino, J. D. (2007). Women seeking health information: Distinguishing the Web user. *Journal of Health Communication, 12*(8), 787–814. doi: 10.1080/10810730701672090

Warning, Alert and Response Network ("WARN") Act, Title VI of the Security and Accountability for Every Port Act of 2006, Pub. L. No. 109–347, 120 Stat. 1884 (2006).

Wasserman, S., & Galaskiewicz, J. (1994). *Advances in social network analysis: Research in the social and behavioral sciences.* Thousand Oaks, CA: Sage.

Waters, R. D. (2011a). Increasing fundraising efficiency through evaluation: Applying communication theory to the nonprofit organization-donor relationship. *Nonprofit & Voluntary Sector Quarterly, 40*(3), 458–475.

Waters, R.D. (2011b). Redefining stewardship: Examining how Fortune 100 organizations use stewardship with virtual stakeholders. *Public Relations Review, 37*(2), 129–136.

Waters, R.D., Burnett, E., Lamm, A., & Lucas, J. (2009). Engaging stakeholders through social networking: How nonprofit organizations are using Facebook. *Public Relations Review, 35*(2), 102–106.

Waters, R.D., Tindall, N.T.J., & Morton, T.S. (2010). Media catching and the journalist-public relations practitioner relationship: How social media are changing the practice of media relations. *Journal of Public Relations Research, 22*(3), 241–264.

Watson, T. (2006). Reputation and ethical behaviour in a crisis: Predicting survival. *Journal of Communication Management, 11,* 317–384.

Watson, T. (2011). The evolution of evaluation—public relations' erratic path to the measurement of effectiveness. Retrieved from http://www.bournemouth.ac.uk/lectures/professor/tom-watson.html

Watts, D. J. (2007). Challenging the influentials hypothesis. Retrieved from http://research.yahoo.com/pub/2112

Weber, L. (2007). *Marketing to the social web: How digital customer communities build your business*. Hoboken, NJ: John Wiley & Sons.

Weimann, G. (2009). The psychology of mass-mediated terrorism. *American Behavioral Scientist, 52*, 69–86.

Weiner, H. (1948). *Cybernetics: Or control and communication in the animal and the machine*. Cambridge, MA: MIT Press.

Weisæth, L., Knudsen, Ø., Jr., & Tønnessen, A. (2002). Technological disasters, crisis management and leadership stress. *Journal of Hazardous Materials, 93*(1), 33–45.

Werther, W.B., & Chandler, D. (2011). *Strategic corporate responsibility: Stakeholders in a global environment*. Los Angeles, CA: Sage.

Westhues, M., & Einwiller, S. (2006). Corporate foundations: Their role for corporate social responsibility. *Corporate Reputation Review, 9*, 144–153.

Whaling, H. (2011). The PR pro's guide to Twitter. Retrieved from http://mashable.com/author/heather-whaling/

What technology and methods can your company use to issue crisis alerts? (2007, October 1). Retrieved from http://business.highbeam.com/6079/article-1G1-169825149/technology-methods-can-your-co-use-issue-crisis-alerts

Wheatley, M. (2008). An era of powerful possibility. *The Nonprofit Quarterly*, pp. 44–46.

White, C., & Raman, N. (1999). The World Wide Web as a public relations medium: The use of research, planning, and evaluation in Web site development. *Public Relations Review, 25*(4), 405–419.

White, C., Vane, A., & Stafford, G. (2010). Internal communication, information satisfaction, and sense of community: The effect of personal influence. *Journal of Public Relations Research, 22*(1), 65–84.

Whitney, H. (1999, January 11). Out of the box. *Brandweek*, pp. 14–15.

Wight, M. (1994). *International theory: The three traditions*. Leicester, England: Leicester University Press.

Wigley, S., & Fontenot, M. (2010). Crisis managers losing control of the message: A pilot study of the Virginia Tech shooting. *Public Relations Review, 36*(2), 187–189. doi:10.1016/j.pubrev.2010.01.003

Wilcox, D. L. (2005). *Public relations writing and media techniques*. Boston, MA: Allyn and Bacon.

Wilcox, D.L., & Cameron, G.T. (2006). *Public relations: Strategies and tactics*. Boston, MA: Allyn and Bacon.

Will, E.M., & Callison, C. (2006). Web presence of universities: Is higher education sending the right message online? *Public Relations Review, 32*(2), 180–183.

Wilson, L. J. (1994). Excellent companies and coalition-building among the Fortune 500: A value- and relationship-based theory. *Public Relations Review, 20*(4), 333–343.

Wilson, M. (2011, April 25). Texas Instruments deputizes employees for social media duty. Retrieved from http://www.ragan.com/Main/Articles/42872.aspx

Wines, M. (2010, November 17). China's censors misfire in abuse-of-power case. Retrieved from http://www.nytimes.com/2010/11/18/world/asia/18li.html

Wise, N. (2011, January 11). 11 most useful augmented reality apps for iPhone and Android. Retrieved from http://onebiginternet.com/2011/01/11-most-useful-augmented-reality-apps-for-iphone-and-android/

Witmer, D.(2000). *Spinning the Web: A handbook for public relations on the Internet*. New York, NY: Longman.

Wolfram Alpha. (2011). Retrieved from http://www.wolframalpha.com/input/?i=Tianya.cn

Woloshin, M. A. (2009, July). Compromising position: Looking for senior-level jobs after a layoff. Retrieved from http://www.prsa.org/Intelligence/Tactics/Articles/view/6C-070914/101/Compromising_Position_Looking_for_senior_level_job?utm_campaign=PRSASearch&utm_source=PRSAWebsite&utm_medium=SSearch&utm_term=pr%20executive%20layoffs

Woo, C.W., An, S.-K., & Cho, S.H. (2008). Sports PR in message boards on Major League Baseball websites. *Public Relations Review, 34*(2), 169–175.

Woodly, D. (2008). New competencies in democratic communication? Blogs, agenda setting and political participation. *Public Choice, 134*(1/2), 109–123.

Word of Mouth Marketing Association. (2009). Code of ethics and standards of conduct for the Word of Mouth Marketing Association. Retrieved from http://womma.org/ethics/ethicscode.pdf

Worley, D. A. (2007). Relationship building in the Internet Age: How organizations use Web sites to communicate ethics, image, and social responsibility. In S. C. Duhe (Ed.), *New media and public relations* (pp. 145–158). New York, NY: Peter Lang.

Wright, J. (2005). Coaching mid-life, Baby Boomer women in the workplace. *Work, 25*(2), 179–183.

Wright, D.K., & Hinson, M.D. (2008a, March). *An analysis of the increasing impact of social and other new media on public relations practice.* Paper presented at 12th International Public Relations Research Conference, Miami, FL.

Wright, D.K., & Hinson, M.D. (2008b, Spring). How blogs and social media are changing public relations and the way it is practiced. *Public Relations Journal, 2*(2). Retrieved from http://www.prsa.org/Intelligence/PRJournal/

Wright, D.K., & Hinson, M.D. (2009a, Spring). An updated look at the impact of social media on public relations practice. *Public Relations Journal, 3*(2). Retrieved from http://www.prsa.org/Intelligence/PRJournal/

Wright, D.K., & Hinson, M.D. (2009b, Summer). Examining how public relations practitioners actually are using social media. *Public Relations Journal, 3*(3). Retrieved from http://www.prsa.org/Intelligence/PRJournal/

Wright, D.K., & Hinson, M.D. (2010a, Spring). An analysis of new communications media use in public relations: Results of a five-year trend study. *Public Relations Journal, 4*(2). Retrieved from http://www.prsa.org/Intelligence/PRJournal/

Wright, D.K., & Hinson, M.D. (2010b, Summer). How new communications media are being used in public relations: A longitudinal analysis. *Public Relations Journal, 4*(3). Retrieved from http://www.prsa.org/Intelligence/PRJournal/

Wyatt, E. (2011, May 9). Emergency alert system expected for cellphones. Retrieved from http://www.nytimes.com/2011/05/10/us/10safety.html

Xifra, J., & Grau, F. (2010). Nanoblogging PR: The discourse on public relations in Twitter. *Public Relations Review, 36*(2), 171–174.

Xifra, J., & Huertas, A. (2008). Blogging PR: An exploratory analysis of public relations weblogs. *Public Relations Review, 34*(3), 269–275.

Yang, A., & Taylor, M. (2010). Relationship-building by Chinese ENGOs' websites: Education, not activation. *Public Relations Review, 36*(4), 342–351.

Yang, S.-U., & Kang, M. (2009). Measuring blog engagement: Testing a four-dimensional scale. *Public Relations Review, 35*(3), 323–324.

Yang, S.-U., Kang, M., & Johnson, P. (2010). Effects of narratives, openness to dialogic-communication and credibility on engagement in crisis communication through organizational blogs. *Communication Research, 37*(4), 473–497. doi:10.1177/0093650210362682

Yang, S.-U., & Lim, J.S. (2009). The effects of Blog-Mediated Public Relations (BMPR) on relational trust. *Journal of Public Relations Research, 21*(3), 341–359.

Yee, N. (2006). The demographics, motivations, and derived experiences of users of massively multi-user online graphical environments. *Presence, 15*(3), 309–329.

Yee, N., Bailenson, J., Urbanke, M., Change, F, & Merget, D. (2007). The unbearable likeness of being digital: The persistence of nonverbal social norms in online virtual environments. *Cyberpsychology & Behavior, 10*(1), 115–121.

Young, C.L., Flowers, A.A., & Ren, N. (2011). Technology and crisis communication: Emerging themes from a pilot study of U.S. public relations practitioners. *Prism, 8*(1). Retrieved from http://www.prismjournal.org

Young, J. P. (2010, December 9). 18 percent of college students who go online use Twitter. Retrieved from http://chronicle.com/blogs/wiredcampus/18-percent-of-college-students-use-twitter/28642

Young, K. (2010a, January/February). Crafting an organizational model to combat employee Internet abuse. *Information Management, 44*(1), 34–38.

Young, K. (2010b). Policies and procedures to manage employee Internet abuse. *Computers in Human Behavior, 26*, 1467–1471.

Yu, R. (2011, April 26). Airlines turn to Twitter to ease passengers' frustrations. Retrieved from http://www.usatoday.com/money/industries/travel/2011-04-25-airlines-twitter.htm

Zakon, R.H. (2010). Hobbes' Internet timeline 10.1. Retrieved from http://www.zakon.org/robert/internet/timeline/

Zhang, J. (2010). To play or not to play: An exploratory content analysis of branded entertainment in Facebook. *American Journal of Business, 25*(1), 53–64.

Zhao, C., & Jiang, G. L. (2011, May). Cultural differences on visual self-presentation through social networking site profile images. *Proceedings of the 29th CHI Conference on Human Factors in Computing Systems*, Vancouver, BC, Canada. Retrieved from http://research.microsoft.com/en-us/um/beijing/groups/hci/pubs/1774_chi2011_chenzhao.pdf

Zhao, D., & Rosson, M.B. (2009, May). How and why people twitter: The role that micro-blogging plays in informal communication at work. *Proceedings of the ACM 2009 International Conference on Supporting Group Work*, Sanibel Island, FL. Available at http://portal.acm.org/citation.cfm?id=1531710&CFID=28061948&CFTOKEN=98115592

Zhong, B., Hardin, M., & Sun, T. (2011). Less effortful thinking leads to more social networking? The associations between the use of social network sites and personality traits. *Computers in Human Behavior, 27*(3), 1265–1271.

Zhong, W. (2008, June 6). Quake opens responsibility fault line. Retrieved from http://www.atimes.com/atimes/China_Business/JF06Cb01.html

Zickuhr, K, & Rainie, L. (2011, January 13). Wikipedia, past and present. Retrieved from http://www.pewinternet.org/Reports/2011/Wikipedia.aspx

Zimmerman, D. P. (1987). Effects of computer conferencing on the language use of emotionally disturbed adolescents. *Behavior Research Methods, Instruments & Computers, 19*, 224–230.

Zoch, L.M., Collins, E.L., & Sisco, H.F. (2008, Fall). From communication to action: The use of core framing tasks in public relations messages on activist organizations' Web sites. *Public Relations Journal, 2*(4). Retrieved from http://www.prsa.org/Intelligence/PRJournal/

Zoch, L.M., Collins, E.L., Sisco, H.F., & Supa, D.H. (2008). Empowering the activist: Using framing devices on activist organizations' Web sites. *Public Relations Review, 34*(4), 351–358.

Zoch, L. M., & Molleda, J. C. (2006). Building a theoretical model of media relations using framing, information subsidies, and agenda-building. In C. H. Botan & V. Hazleton (Eds.), *Public relations theory II* (pp. 279–309). Mahwah, NJ: Lawrence Erlbaum Associates.

Zuckerberg, R. (2010, Fall/Winter). Accidental activists: Using Facebook to drive change. *Journal of International Affairs, 64*(1), 177–180.

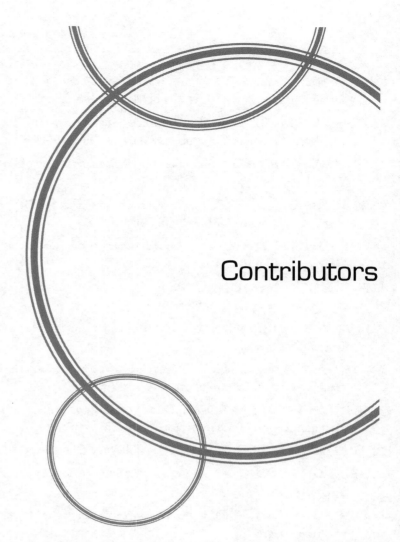

Contributors

CAROL AMES, Ph.D., teaches graduate and undergraduate courses in the Entertainment and Tourism and the Public Relations concentrations of the Department of Communications at California State University, Fullerton.

DEREK ANTOINE, M.A., is currently pursuing Ph.D. studies in Communication at Carleton University, Canada.

LUCINDA L. AUSTIN, Ph.D., is an assistant professor in the School of Communications at Elon University.

GEORGE BARTLETT is studying sociology and anthropology at the University of Queensland.

JENNIFER L. BARTLETT, Ph.D., is an academic in the School of Advertising, Marketing and Public Relations at the Queensland University of Technology, Australia.

KATI TUSINSKI BERG, Ph.D., is an assistant professor at Marquette University, where she teaches courses in public relations and corporate communication.

DENISE SEVICK BORTREE, Ph.D., is an assistant professor in the College of Communications at Penn State University and a Senior Research Fellow at the Arthur W. Page Center.

DAVID L. BRINKER, JR., is a graduate student and research assistant in the J. William and Mary Diederich College of Communication at Marquette University, Milwaukee, Wisconsin.

ROWENA L. BRIONES, M.A., is a doctoral student and instructor of record in the Department of Communication at the University of Maryland studying public relations and health communication.

RONALD LEE CARR, MFA, is an associate professor and director of the Tokyo branch of Temple University School of Communications and Theater.

MARK A. CASTEEL, Ph.D., is an associate professor of Psychology at Penn State University's York Campus.

JINBONG CHOI, Ph.D., is an assistant professor in the Department of Media & Communication at the Sungkonghoe University in South Korea.

WONJUN CHUNG, Ph.D., is an assistant professor in the Department of Communication at University of Louisiana at Lafayette.

W. TIMOTHY COOMBS, Ph.D., is a professor in the Nicholson School of Communication at the University of Central Florida.

DAVID CRIDER, M.A., is a doctoral student in the Mass Media and Communication program at Temple University.

DONNA Z. DAVIS, Ph.D., is currently a visiting assistant professor at the University of Oregon, teaching strategic communication.

MARCIA W. DISTASO, Ph.D., is an assistant professor of public relations in the College of Communications at Pennsylvania State University.

XUE DOU is a Ph.D. candidate in the College of Communications at Pennsylvania State University.

JOE R. DOWNING, Ph.D., is an assistant professor of Communication at Penn State University's York campus.

MAHMOUD EID, Ph.D., is an associate professor at the Department of Communication, University of Ottawa, Canada.

DARADIREK "GEE" EKACHAI, Ph.D., is associate professor and chair of the Department of Strategic Communication at Marquette University, Milwaukee, Wisconsin.

SARAH BONEWITS FELDNER, Ph.D., is an associate professor in the Communication Studies department at Marquette University, Milwaukee, Wisconsin.

ROMY FRÖHLICH, Ph.D., is full professor of communication science and media research at Ludwig-Maximilians-University Munich, Germany.

BARBARA S. GAINEY, Ph.D., is associate professor of communication, public relations, at Kennesaw State University, Georgia.

DEDRIA GIVENS-CARROLL, Ph.D., is an assistant professor of communication at the University of Louisiana at Lafayette, where she teaches public relations courses.

BETSY A. HAYS, APR, is an associate professor in the Department of Mass Communication and Journalism at California State University, Fresno.

IRENE CAROLINA HERRERA is a Ph.D. candidate at Nihon University, Tokyo, and an assistant professor at Temple University Japan.

SHERRY J. HOLLADAY, Ph.D., is a professor at the Nicholson School of Communication, University of Central Florida.

MELISSA JANOSKE, M.S., is currently pursuing her Ph.D. in Communication at the University of Maryland with emphases in public relations and crisis communication.

YAN JIN, Ph.D., is an assistant professor in the School of Mass Communications at the Virginia Commonwealth University.

TERRI L. JOHNSON, APR, ABC, ABD, is an associate professor at Eastern Illinois University.

EUNSEONG KIM, Ph.D., is an assistant professor in the journalism department at Eastern Illinois University.

JEONG-NAM KIM, Ph.D., is an assistant professor in the Department of Communication at Purdue University.

DEAN KRUCKEBERG, Ph.D., APR, Fellow PRSA, is executive director of the Center for Global Public Relations and a professor in the Department of Communication Studies at the University of North Carolina at Charlotte.

ALEXANDER V. LASKIN, Ph.D., is an assistant professor of public relations at Quinnipiac University.

LAUREN LAWSON is a recent graduate of the Masters in Strategic Communication program at North Carolina State University.

BROOKE FISHER LIU, Ph.D., is an assistant professor in the Department of Communication at the University of Maryland and an affiliated faculty member for the National Consortium for the Study of Terrorism and Responses to Terrorism (START).

YI LUO, Ph.D., is an assistant professor at the Department of Communication Studies at Montclair State University in New Jersey.

DALE MACKEY, M.S., is a program coordinator at the Office of Professional Development at North Carolina State University.

TINA MCCORKINDALE, Ph.D., is an assistant professor in communication specializing in public relations at Appalachian State University in Boone, NC.

LINJUAN RITA MEN is a doctoral student in public relations at the University of Miami's School of Communication.

SARAH MERRITT is a doctoral student in Communication at American University focusing on online visual media and foreign policy.

MARCUS MESSNER, Ph.D., is an assistant professor of mass communications at Virginia Commonwealth University.

DONNALYN POMPPER, Ph.D., is an associate professor in the Department of Strategic Communication at Temple University's School of Communications & Theater and director of the MS program in Communication Management.

CORNELIUS B. PRATT, Ph.D., APR, is a professor and head of the public relations concentration in the Department of Strategic Communication at Temple University, United States and also teaches at Temple University Japan.

CLARISSA SCHÖLLER, M.S., is international coordinator in the Department of Communication Science and Media Research at Ludwig-Maximilians-University (LMU) in Munich, Germany.

DON W. STACKS, Ph.D., is professor and director of the Public Relations Program in the School of Communication at the University of Miami.

DOUG J. SWANSON, Ed.D., is an associate professor in the Department of Communications at California State University, Fullerton.

NATALIE T.S. TINDALL, Ph.D., is an assistant professor in the Department of Communication at Georgia State University.

WAN-HSIU SUNNY TSAI, Ph.D., is an assistant professor in Advertising at the University of Miami's School of Communication.

CHIARA VALENTINI, Ph.D., is associate professor in Public Relations and Public Communication at Aarhus University, Business and Social Sciences, Denmark.

KELLY VIBBER is a doctoral student at Purdue University.

RICHARD D. WATERS, Ph.D., is an assistant professor in the School of Management at the University of San Francisco.

BROOKE WEBERLING, Ph.D., is an assistant professor in the School of Journalism and Mass Communications at the University of South Carolina where she teaches courses in public relations.

CHANG WAN WOO, Ph.D., is assistant professor of Public Relations in Division of Communication at University of Wisconsin, Stevens Point.

JORDI XIFRA, Ph.D., is professor of public relations at the Department of Communication of Pompeu Fabra University, Barcelona, Spain.

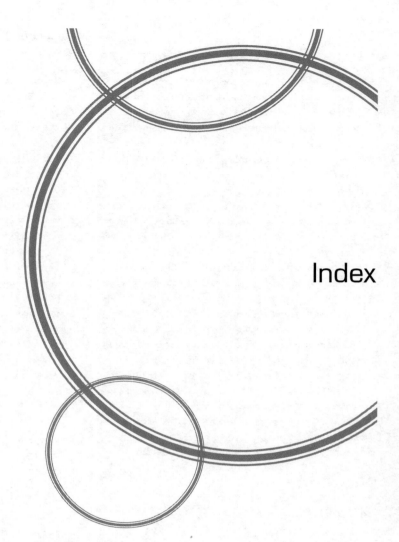

Index